MONOGRAPHS ON
PHYSICAL BIOCHEMISTRY

GENERAL EDITORS

W. HARRINGTON   A. R. PEACOCKE

NUCLEAR MAGNETIC RESONANCE (N.M.R.) IN BIOCHEMISTRY

APPLICATIONS TO ENZYME SYSTEMS

RAYMOND A. DWEK

Lecturer in Inorganic Chemistry
Christ Church and
Demonstrator in the Department of Biochemistry
University of Oxford

CLARENDON PRESS · OXFORD

1973

Oxford University Press, Ely House, London W. 1

GLASGOW NEW YORK TORONTO MELBOURNE WELLINGTON
CAPE TOWN IBADAN NAIROBI DAR ES SALAAM LUSAKA ADDIS ABABA
DELHI BOMBAY CALCUTTA MADRAS KARACHI LAHORE DACCA
KUALA LUMPUR SINGAPORE HONG KONG TOKYO

ISBN 0 19 854614 9

PRINTED IN NORTHERN IRELAND AT THE UNIVERSITIES PRESS, BELFAST

TO

SANDRA, JULIET, AND ROBERT

FOREWORD

MAGNETIC resonance spectra often contain a great deal of detailed information about molecular structures, molecular interactions, and about molecular motion in a sample; this young branch of spectroscopy has already established itself as of great power and versatility. Nuclear resonance is, however, an inherently insensitive technique, and the richness of the information contained in many spectra often limits the size and complexity of the molecules that can usefully be studied.

In recent years, advances in instrument design, and particularly the use of Fourier-transform spectroscopy, have greatly improved the sensitivity of spectrometers, so that resonances can now readily be observed in aqueous solutions at concentrations of the order of 1 mM. The development of superconducting magnets of high homogeneity makes possible the use of ever increasing magnetic fields; as the field is raised the separation of chemically-shifted resonances is increased so that molecules of greater molecular weight and complexity can usefully be studied.

These advances and many others now under development, make it possible to bring the power of molecular resonance to bear on problems of biochemical interest; its ability to give direct and detailed information of great diversity on small volumes of dilute aqueous solutions is probably unique. But the interpretation of the measurements is by no means straightforward, and without a thorough understanding of the physical processes involved, one can all too easily be misled.

Many biochemists will wish to make use of magnetic resonance, and some may find the literature on the subject, largely written for physicists and physical chemists, somewhat daunting; the standard textbooks written for organic chemists are not sufficient for the biochemist who will have to use every trick of the trade to get the information he wants. Dr Dwek, therefore, describes the more spohisticated ways in which magnetic resonance can be used to study macromolecular systems and has liberally illustrated every stage in the argument with specific examples. The interested biochemist can find in this book a real flavour of the exciting possibilities; if his appetite is whetted, a more detailed study of the book will be of tremendous assistance in preparing him to use the method himself.

R. E. RICHARDS

Merton College
Oxford.

PREFACE

MAGNETIC resonance can give molecular information on the structural level that is in principle, comparable to that obtainable from X-ray crystallography, but the technique has the additional advantage that it can be used in solution.

This book deals essentially with the application of magnetic resonance to some enzyme systems. The questions that I have tried to discuss are of a general type: how far can one go in using n.m.r. spectroscopy to obtain the complete three-dimensional structure in solution? Can the functional groups in catalysis be identified? What is the nature of the transition state and the role of any metal ion, co-factors, etc.? Additionally, n.m.r. spectroscopy affords the possibility of obtaining information on allosteric enzymes, such as the stereochemical relationships between the various sites, the number of conformational states, the magnitude (in Å) of structural changes in the different states and the distance over which these changes are transmitted. Clearly in the future one will want to study larger systems involving many enzymes and this too is possible by n.m.r. spectroscopy.

I have attempted here to determine the success of n.m.r. spectroscopy up to the present, to define some of the problems encountered, and to suggest how these may be overcome in the future. Some readers may be unfamiliar with some of the different aspects of n.m.r. spectroscopy and in many instances I have quoted experimental details in tables and figures so that the reader may obtain some idea of how large the observed effects are and what the actual experiments involve.

In chapter 2, I have presented only a skeleton outline of the basic principles of relaxation, chemical shifts spin–spin coupling and chemical exchange effects. This is because there are a number of excellent monographs available which deal with the more theoretical aspects of these. I have assumed that readers may already be familiar with some of the basic principles and applications of magnetic resonance spectroscopy. My presentation is therefore more empirical than it would otherwise be—serving, perhaps, to spotlight the main points in each topic and to act as a useful dictionary of some of the terms and jargon which are met in the more detailed mathematical texts. Nevertheless, I think that the theory given should be sufficient, on its own, to enable the reader, unfamiliar with magnetic resonance, to follow the applications which are described in this book. After all my aim in writing this book, is not to appeal solely to the pure n.m.r. spectroscopist but rather to suggest to the interested biochemists the type of experiments that could be done to help solve their particular problems.

The use of paramagnetic ions features prominently in this book. At present, however, there is no text available which deals adequately with the application of paramagnetic ions in n.m.r. spectroscopy from the point of view of the biochemist or biophysicist. I have therefore, attempted to provide a section which may serve as a 'handbook' on this subject. Readers unfamiliar with the applications of paramagnetic ions are advised, in the first reading, to concentrate mainly on the examples of their use.

In writing this book I have not sought to be totally comprehensive or to acknowledge all the people who, historically, have made a contribution to the application of n.m.r. spectroscopy. Rather I have drawn freely on what I believe are representative examples. Often, I have explained apparently obvious points—but I have been guided by the type of questions that I am constantly being asked by research workers in the various fields. I hope that this book may stimulate interested students into reading the original papers in the literature in a somewhat critical manner, and indeed even writing to me about any serious omissions or errors in this book.

I have deliberately avoided detailed discussion of small molecules partly because of space and also because some excellent reviews are available. The methods that I have mentioned, however, are applicable to most systems. For similar reasons, therefore, I have omitted discussion of the applications of magnetic resonance to membrane systems.

This is perhaps an appropriate place to record my debt to Dr. R. E. Richards F.R.S., who introduced me to the exciting possibilities of the use of n.m.r. spectroscopy. Drs G. K. Radda and R. J. P. Williams, F.R.S. have played a major role in stimulating my interests in biochemical problems and any credit from this book must also reflect on them. The mistakes however are my own!

Many of my colleagues in Oxford have contributed enormously to my understanding of magnetic resonance and have provided valuable criticism of the manuscript; among these I should like to acknowledge the help of Drs A. Bennick, J. S. Cohen, R. Freeman, S. J. Ferguson, G. Navon and A. V. Xavier, but particularly Drs Iain Campbell and Simon van Heyningen who made constructive suggestions at every stage of this book. I am grateful for the technical assistance of Mr. David Kozlow who so meticulously drew all the diagrams and to Mrs H. Holloway who, patiently and carefully, typed the original manuscript.

RAYMOND DWEK

Oxford.
November 10th, 1972

ACKNOWLEDGEMENTS

I should like to express my thanks to all those who have given me access to their work before publication. These include the research group of Dr. R. J. P. Williams, F.R.S. and in particular Drs C. M. Dobson, D. R. Martin and A. V. Xavier; the research groups of Dr. R. E. Richards, F.R.S. and Dr. G. K. Radda and in particular Dr. N. A. Thomas and Dr. R. W. Barker. Professor S. Forsén and Dr. T. Bull sent me preprints of their work which is described in Chapter 13. Dr. J. S. Cohen and his colleagues kept me supplied with up to the minute information on the story of ribonuclease. Drs J. S. Leigh Jr. and G. H. Reed from Professor Mildred Cohn's laboratory, gave me various diagrams and a computer program which appear in Chapters 11 and 12. Professor H. M. McConnell sent me a preprint of his lysozyme spin labelling work which is presented in Chapter 12.

In addition I should like to thank all those who granted permission to reproduce diagrams: details are included in the underlines and references.

RAYMOND DWEK

CONTENTS

1

INTRODUCTION

1.1. Introduction to n.m.r.

'MAGNETIC resonance studies can give structural and dynamic information'. This statement is well founded for small molecules but, in spite of increasingly sophisticated and expensive instrumentation, limited success has been achieved in applying this technique to macromolecules. Increasingly, however, methods are being devised which get round some of the inherent problems of magnetic resonance. In this Chapter we introduce some of the terms we shall meet later in the book and outline how magnetic resonance can be applied to macromolecular systems. We shall define some of the problems encountered and indicate the methods that may eventually make the initial statement true for large molecules, as well as small ones. The methods that are emphasized in this book are: the use of paramagnetic ions; chemical modifications of the macromolecule; and the use of spin labels.

It is convenient to consider the macromolecular system as consisting of several components: the macromolecule itself; the smaller molecules or ligands which interact with it; the solvent, water; and any cations and anions present in the system.

It is possible to study any of these components using n.m.r. spectroscopy. However, there are often severe limitations either in the observation of the resonances or the interpretation of the data. The requirements for a meaningful experiment may be defined as follows: there must be sufficient material and/or *sensitivity* to observe the resonances; individual atoms or chemical groups must be *resolved;* the observed resonances must be *assigned* to known chemical groups in the system; and the data must be *interpretable*, in terms of structure or motion of the system.

A completely successful experiment must therefore be designed so that all four criteria are fulfilled. Let us consider each in turn.

1.2. Sensitivity

Table 1.1 lists the relative sensitivities (at a constant magnetic field) of various biologically important *nuclei*. The table allows for the natural abundance of the elements. In certain cases, e.g. ^{13}C (which has a natural abundance of ca. 1 %) these figures can be improved by selective enrichment.

The sensitivity of the *instruments* varies of course, but with new signal-averaging techniques a concentration of hydrogen nuclei of ca. 1 mM (in less than 0·5 ml of solution) can be observed in ca. 20 min with quite a good signal-to-noise ratio. The sensitivity also increases with increasing magnetic field.

TABLE 1.1

Nucleus	Spin quantum number	Natural abundance %	Relative† sensitivity at constant field
^1H	$\frac{1}{2}$	100	100
^2H	1	$1\cdot5\times10^{-2}$	0·00015
^{13}C	$\frac{1}{2}$	1·1	0·018
^{14}N	1	99·6	0·1
^{17}O	$\frac{5}{2}$	$3\cdot7\times10^{-2}$	0·001
^{19}F	$\frac{1}{2}$	100	83·3
^{23}Na	$\frac{3}{2}$	100	9·25
^{25}Mg	$\frac{5}{2}$	10	0·028
^{31}P	$\frac{1}{2}$	100	6·63
^{35}Cl	$\frac{3}{2}$	75·5	0·35
^{39}K	$\frac{3}{2}$	93	0·047

† The product of (natural abundance) × (sensitivity).

1.3. Resolution

Once a spectrum has been obtained there is the problem of resolution of the individual components. The position of each resonance depends on the nature of the nucleus and its chemical (or electronic) environment. For nuclei of the same isotope the dependence by resonance positions on chemical environment leads to the concept of a *chemical shift* (Chapter 2). The chemical shift increases with increasing magnetic field and thus the resolution increases with increasing magnetic field. Clearly however the total range of chemical shifts is important. For example, ^1H chemical shifts occur over a very small range (ca. 12 p.p.m.) while fluorine-19 and carbon-13 chemical shifts occur over a far larger range (ca. 500 and 350 p.p.m. respectively). Judged on this criterion the resolution is better with ^{13}C and ^{19}F than it is for ^1H. However the ubiquity and high sensitivity of protons† make them an attractive starting point for any study. In relatively small molecules it may be possible to identify every chemically-distinct proton but as the size of the molecule increases this task becomes increasingly difficult.

Sometimes an intrinsic 'perturbation' of the system leads to some spectral simplification. For example, in favourable cases the presence of aromatic groups can lead to large chemical shifts (called *ring current shifts*) of certain groups (Chapter 3). Also the presence of a paramagnetic ion—such as in the haeme proteins—may result in large shifts (called *hyperfine shifts*) of selected resonances (Chapter 3). When these effects are not operative, it is necessary to use external perturbations, e.g. the addition of paramagnetic ions (Chapters 4 and 9). Alternatively techniques which involve chemical modification

† Although 'proton' has a distinct meaning, it is so widely used in n.m.r. to represent the hydrogen nucleus that we shall continue this usage here.

of particular groups, e.g. selective deuteriation can lead to simplification of the proton spectrum and thereby additional resolution (Chapter 5). Most observations of ^1H n.m.r. spectra of proteins are carried out in D_2O in order to suppress the very strong H_2O resonance. Exchangeable NH or OH hydrogens are not observed under these conditions. However the NH resonances often occur below the main spectral envelope and can be resolved even in the presence of H_2O (Chapter 5).

The resolution of individual resonances also depends on their line-widths. It is sufficient to state here that as the size of the molecule increases the slower motions associated with it result in broad resonance lines. The line-width $\Delta\nu$ (in Hz) is related to the spin–spin relaxation rate, $1/T_2$ ($1/T_2 = \pi\,\Delta\nu$) which is discussed below.

1.4. Assignment

Once a particular resonance has been detected and resolved in the spectrum there is the problem of assignment. Initially the chemical shift may give some information as to its nature and environment. For instance the histidine C(2) hydrogens occur to very low field of most other resonances and are often resolved (Chapter 5). If there are two or more histidines in different environments the C(2) hydrogens of each may still be resolved but the identification of each to a particular histidine in the molecule will not necessarily be obvious. A variety of perturbations, both intrinsic and extrinsic (such as changes in pH or temperature, chemical modifications, addition of ligands, double irradiation, changes in solvent from H_2O and D_2O, etc.), are then usually necessary to assign the resonances. The difficulties in resolution and assignment are only a major problem when spectrum of the *macromolecule* itself is observed. These problems are not usually severe when observing the spectra of other components of the system.

1.5. Interpretation of data

When a resonance has been resolved and assigned, it may be possible to derive from it the desired structural and dynamic information. In this section we will discuss the advantages and disadvantages of observing the various components of the system. First, however, we define the observable parameters in an n.m.r. spectrum.

The chemical shift. (δ) This has already been mentioned in Section 1.3. A resonance may shift (with respect to some standard) as a result of changes in temperature or pH, but such shifts are not usually interpretable in terms of structure or motion. However, the chemical shift may be dramatically influenced by the proximity of a paramagnetic centre and this shift may be related to the distance between the observed nucleus and the paramagnetic centre thus yielding structural information (Chapter 4).

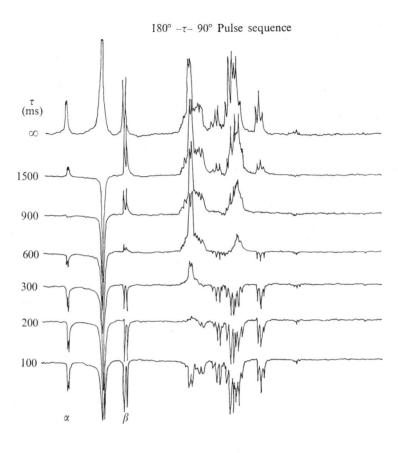

FIG. 1.1. Inversion and recovery of 270 MHz ¹H n.m.r. spectrum of a mixture of anomers of glucose-6-phosphate. The peaks labelled α and β represent the resonances of the anomeric protons. The β resonance has a shorter recovery rate (spin–lattice relaxation rate) than that of the α resonance. This arises because the C-(1) hydrogen of the β-anomer is close to both the C-(3) and C-(5) hydrogen whereas the C-(1) hydrogen of the α-anomer is only close to the C-(2) hydrogen. The hyperfine (spin–spin) splitting of the anomeric resonances arise from interactions with the C-(2) hydrogen; the different values reflect the different orientations of the anomeric hydrogens.

T_1 and T_2. In addition to being characterized by a chemical shift a particular resonance has two relaxation rates associated with it (Chapter 2). The *spin–spin relaxation rate* $(1/T_2)$ is related to the line-width of the resonance and its origin is similar to the 'uncertainty broadening' observed in other forms of spectroscopy. The *spin–lattice relaxation rate* $(1/T_1)$ characterizes the rate at which the system returns to its normal state after it has been perturbed. For example, the spectrum in Fig. 1.1 may be inverted by the application of suitable irradiation. When the radiation is turned off, the resonances return to their original state at their spin–lattice relaxation rates which depend on the time scale of the magnitude of dipolar interactions; thus the different relaxation rates of individual nuclei may make it possible to estimate that one particular nucleus or group is undergoing slower motions (i.e. is more rigid) than another. Also since the magnitude of the dipolar interaction depends on the distance between the dipoles (the nuclei), structural information may also be determined. However, because of the multitude of hydrogens which can contribute to dipolar interactions, information about the structure of, for example, an enzyme–substrate (ES) complex would be extremely difficult, if not impossible, without a detailed knowledge of the structure. The presence of a paramagnetic centre may overcome this difficulty. The large magnetic dipole of such a centre results in a dramatic increase in the relaxation rates. The effect depends on the sixth power of the distance between the centre and the observed nucleus and is thus very sensitive to that distance (Chapter 9).

Figure 1.1 also shows the presence of fine structure in the spectrum which arises from interaction of neighbouring nuclei possessing magnetic moments. This phenomenon is termed *spin–spin interaction*. The magnitude of the spin–spin interaction between two nuclei depends on the nature of the chemical bonds between them and their relative geometries. In *some* simple cases the magnitude of the interaction can be semi-empirically related to the geometries of the nuclei and by this means conformational analysis can be performed (Chapter 2). Such methods are dependent on the resolution of these small couplings.

1.5.1. *The resonances of the macromolecule*

As we have stated above the problems of resolution and assignment are often very severe when observations are carried out in the macromolecule. Useful observation of the hydrogen resonances has usually been limited to molecular weights of 20 000 or less. Representative examples of studies to date include haeme proteins, lysozyme, and ribonuclease (Chapter 5).

Once a resonance has been assigned, the usual range of experiments can be carried out, i.e. the effects of temperature, pH, ligands, etc. can be determined. However, a detailed interpretation of the results of such experiments usually relies heavily on the crystal structure being known. One

outstanding example of the combination of n.m.r. spectroscopy and crystallographic data is the study of the C(2) hydrogens of the four histidines in
ribonuclease which has led to the proposal of a mechanism of action
(Chapter 6).

The use of chemical modification or isotopic enrichment is often helpful
in the problems associated with resolution, assignment, and interpretation.
For example, selective deuteriation experiments of certain nuclei could lead
to simplification of the ^{1}H spectrum. Specific enrichment by ^{13}C may improve
the sensitivity and resolution of a ^{13}C spectrum (Chapter 7). Chemical
modifications for example by substitution of groups by CF_3 enables the
observation of ^{19}F resonances which may provide structural information
(Chapter 8). Alternatively specific groups such as site specific *spin labels*
(Chapter 12) can be covalently attached to strategic points. With any modification (not so much with isotopic substitution) there is always the danger that
the biological activity may be altered and this must be checked wherever
possible.

We have already noted that the presence of a paramagnetic ion in the
haeme proteins can result in selective *shifts* of certain resonances. Certain
paramagnetic ions can also cause dramatic changes in line-widths (or
relaxation rates). It is convenient to classify these two types of probes as
shift probes and *relaxation* probes (Chapter 9). The advantage of using a
paramagnetic probe is that the changes which result from its addition can be
interpreted relatively easily irrespective of the size of the macromolecule.
This is particularly useful when combined with techniques like difference
spectroscopy.

In cases where the resonances of the macromolecule can be observed, the
changes in relaxation rates of the resonance can give information on their
relative motions with respect to the paramagnetic probe and also to their
distance from it. Changes in chemical shifts induced by the probe can also
give structural information and/or the unpaired electron density at the
particular nucleus whose resonance is being examined. The concentration of
paramagnetic ions used in such experiments is generally an order of magnitude
less than that of the macromolecule because the effects are so large.

1.5.2. *The resonances of the ligand*

The problems of studying interactions between ligands and macromolecules
can also be approached by studying the ligand resonances themselves
(Chapter 6). In general, the concentration of macromolecules obtainable for
n.m.r. experiments is small. If the ligand binds to the macromolecule, then
the problems of sensitivity and resolution already encountered with macromolecules are applicable to the ligand resonances. However, there is a way
of increasing the resolution of the ligand resonances on the macromolecule.
The pure ligand resonances are usually quite narrow as it is a small molecule.

If we increase the concentration of ligand in the system such that there is a large excess over that of the macromolecule then we essentially observe only the free ligand. If rapid chemical exchange of the ligand between its two environments (i.e. free and bound) now exists, the observed parameters will be a weighted average of those in the two environments. Thus, if the effects in the bound site are reasonably large the observation of the bulk (free) resonance allows information concerning the bound site to be obtained. The practical limits for useful observation in such cases is ca. 1 mM per hydrogen resonance.

The analysis of the relaxation rates of the different ligand nuclei can give information on the mobility, and thus the mode of binding, of the ligand (Chapter 5). (For example, if it resides in a hydrophobic cleft.) Under certain circumstances the life-time of the ligand in the complex can also be obtained. Changes in chemical shifts as a result of changes in pH or temperature or the addition of ligands are more difficult to interpret unambiguously in the absence of a knowledge of the structure of the macromolecule. Nevertheless, whatever the reason, changes in relaxation rates or chemical shifts as a result of the addition of ligands can be used to monitor binding.

The effects of paramagnetic ion probes on the macromolecule can influence the ligand relaxation rates or chemical shifts when the ligand binds to the macromolecule. Further, because the effects caused by the paramagnetic ions are so intense, it is also possible to work at a lower macromolecular concentration, than in the diamagnetic case and still observe measurable effects. It is again possible to obtain the distance of the individual ligand nuclei from the paramagnetic ion, if conditions of fast exchange apply (Chapter 10). The concentration of ligand nuclei generally used in these studies is ca. 5 mM and the concentration of the paramagnetic ions is ca. 10 μM or even less (for relaxation probes) and ca. 1 mM for shift probes. If fast-exchange conditions do not apply, then information on the life-time of the metal–macromolecule–ligand complex may be obtained, i.e. kinetic information.

1.5.3. *The resonance of solvent water*

The concentration of hydrogen nuclei in water is 110 M. Thus there is no difficulty with sensitivity, resolution or assignment. The only problem is that of interpretation. If there is fast chemical exchange of the water molecules between the bulk and those bound to the macromolecule, we have again a means of studying the bound water. Because of the large excess of 'free' water, the effects of the bound site will be largely masked and the observed effects will be small. Additionally, if we note that water can exist in many different environments on the macromolecule, it becomes extremely difficult to interpret any changes quantitatively.

Much more information can be obtained when a paramagnetic ion is present (Chapters 10 and 11). Typical concentrations of metal ions are ca. 100 μM while the macromolecule concentrations required are of the same order. A larger concentration of probes may have to be used here because the water concentration is far greater than that of the ligands. The analysis of the relaxation rates can give information on the distance from the metal ion to the water molecules in the first hydration sphere, and the number and the lifetime of these water molecules. Surprisingly the relaxation rates of the water hydrogens in *some* enzyme systems seem to be dependent upon the conformational state of the enzyme. The determination of these conformational parameters in the presence of different ligands has led to the proposal of mechanisms of enzyme action in a number of cases. So not only can different conformational states be detected but also a quantitative estimate of any changes can be made by observation of the ligand resonances.

The possibility of binding studies in the presence of a paramagnetic ion may also be considered. As with any spectroscopic method in which a change in a spectral parameter is observed as a result of ligand binding, monitoring the changes can lead to binding information. The easiest way to monitor binding is to study changes in the relaxation rates of the solvent water and this is the method used most frequently (Chapter 11).

1.5.4. *The resonances of the cations and anions*

Many nuclei possess a spin quantum number, $I > \frac{1}{2}$, and a quadrupole moment is associated with such nuclei. The sensitivity of these nuclei is much less than that for hydrogens, and while the problems of assignment and resolution are fairly trivial, interpretation in many cases, particularly in the study of ^{23}Na and ^{39}K resonances, is ambiguous (Chapter 13). The study of the halide ion resonances has also received much attention but again the analysis of any changes in relaxation rates are difficult to quantify in terms of changes in macromolecular structure. However the study of such nuclei may well give additional useful information on the molecular motion of the bound ions, their exchange rates and binding constants.

1.6. Spin-label probes

We have noted above that perturbations in the n.m.r. spectrum can result from the presence of a paramagnetic centre, which is usually not covalently bound to the macromolecule. However, it is possible to modify chemically the macromolecule so that a paramagnetic probe becomes covalently attached. Such probes are called spin labels (Chapter 12). A spin-label probe is defined as a synthetic paramagnetic organic free radical with a chemical reactivity that results in its attachment to the macromolecule at a particular site, e.g. an SH group.

As before, the effects of the spin label on the relaxation rates of the various nuclear resonances can be monitored and again structural and kinetic information can be obtained. The experimental effects are usually smaller than those obtained with metal ions, partly because of the smaller magnetic moment of the spin label. The same concentrations as previously if not a little higher, will be required for observation of macromolecular resonances. Similarly, for experiments on the ligand resonances, spin-labelled enzyme concentrations of ca. 100 μM will be required with the ligand concentrations much as before.

There is an additional attraction in using spin-label probes in that one can observe the e.s.r. spectrum of the spin label itself. Because the vast majority of molecules in biochemical systems are diamagnetic, the paramagnetic (e.s.r.) spectrum is completely free from interference of other nuclei. The sensitivity of the e.s.r. spectral technique is much greater than that for n.m.r. Thus the present practical limits of detection, without special accessories, are in the range 10^{-7}–10^{-5} M in probe concentration with a sample volume which can be as little as 10 μl. (The actual concentration depends on the mobility of the probe.) The shape of the e.s.r. spectrum gives information on the mobility of the probe and the polarity of its environment. The main

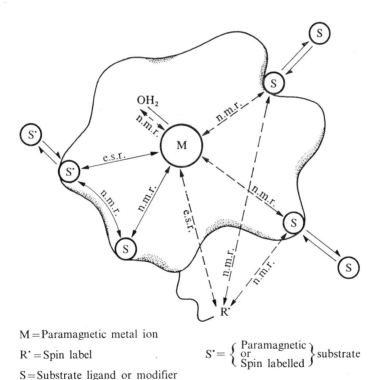

M = Paramagnetic metal ion

R˙ = Spin label

S = Substrate ligand or modifier

$S˙ = \left\{ \begin{array}{l} \text{Paramagnetic} \\ \text{or} \\ \text{Spin labelled} \end{array} \right\}$ substrate

FIG. 1.2. Some of the possible interactions, involving paramagnetic ions, that can be studied using magnetic resonance techniques. Note that the combination of two probes affords a means of triangulating on a particular substrate.

problems are in the exact quantification of these parameters, (i.e. the interpretation). The technique can also be used to monitor binding if the addition of a ligand causes a change in the e.s.r. spectrum and in some cases the mobility of the label can also be used as a conformational parameter.

The effect of a paramagnetic ion on the e.s.r. spectrum of the spin label may be to cause an apparent 'quenching'. This 'quenching' can be related to the distance between the probes.

Although we shall not deal specifically with e.s.r. in this book, it is worth noting that certain paramagnetic metal ions, e.g. Mn^{2+} also give observable e.s.r. spectra and in favourable cases it is possible to obtain information on the metal environment from the study of such spectra.

The structural information obtained by use of one paramagnetic probe (e.g. a metal ion) can be combined with that of a second paramagnetic probe (e.g. a spin label). Provided the distance between the two probes is known, triangulation procedures can be carried out and it becomes possible to 'map' out, in solution, the various nuclei, whose resonances are observed. Figure 1.2 summarizes the studies that can be carried out with paramagnetic probes.

1.7. Summary

Although we have indicated how some measurements can lead to useful information for the biochemist or biophysicist, we have also indicated that there may be difficulties in interpretation. We have stressed the importance of studies of chemical modification in helping in the interpretation. Paramagnetic probes too, in general, simplify the interpretation but, more importantly allow detailed structural information to be obtained. By use of such probes or chemical modifications, it is essential that the function of the biological molecule should not be disturbed to any marked degree. After all, the intention is to study the relationship between structure and function.

2

RELAXATION, CHEMICAL SHIFTS, SPIN–SPIN COUPLING CONSTANTS AND CHEMICAL EXCHANGE

2.1. Basic resonance theory

MANY atomic nuclei possess a spin with which a spin angular momentum is associated. The possession of both spin and charge confers a magnetic moment on the nucleus.

Quantum mechanical theory requires the spin angular momentum to be $\hbar\sqrt{(I(I+1))}$ where $\hbar$ is $h/2\pi$ and h is Planck's constant. I is the spin angular momentum quantum number and can take any of the values $0, \pm\frac{1}{2}, \pm1, \pm\frac{3}{2}...$ etc. depending on the mass number A and charge number Z of the particular nucleus.† The magnetic moment is given by

$$\mu_N = \gamma_I \hbar \sqrt{(I(I+1))} \tag{2.1}$$

where γ_I is called the magnetogyric ratio of the nucleus. The magnetic moment may also be expressed in terms of a dimensionless constant g_N (called the nuclear g factor) and β_N, the nuclear magneton. β_N is equal to $e\hbar/2Mc$ where e and M are the charge and mass of the hydrogen nucleus and c is the velocity of light.

$$\mu_N = g_N \beta_N \sqrt{(I(I+1))} \tag{2.2}$$

g_N and I are different for each nucleus considered.

If this nuclear magnet is placed in a strong uniform magnetic field, B_0, it experiences a torque, like a compass needle, and, at any particular orientation in the field, the energy of interaction E termed the nuclear Zeeman energy, is given by

$$E = -B_0 \times (\text{component of } \mu_N \text{ along } B_0).$$

Quantum theory demands that this energy is quantized so that the component of spin angular momentum along B_0 is $m\hbar$, where $m = I, (I-1), (I-2)...1, 0, 1, -2, -3...-I$. The *allowed* components of μ_N along B_0 are μ_m and are thus given by eqn (2.3). The maximum component of μ along the magnetic

$$\mu_m = g_N \beta_N m \tag{2.3}$$

field is that for which m is greatest, i.e. $m = I$. μ_m is then $g_N \beta_N B_0 I$ and it is this quantity which is usually *referred* to as the magnetic moment of a nucleus, μ. The energy of interaction corresponding to each value of μ_m is given by

---

† If A and Z are both even, $I = 0$. If A is even and Z is odd, I is integral while, if A is odd, I is half-integral.

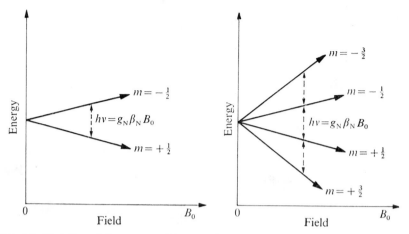

FIG. 2.1. Variation of the energy with magnetic field for nuclear spins $I = \frac{1}{2}$ and $I = 1$.

eqn (2.4). Typical values of $E(m)$ are 10^{-3} cal g^{-1} mol^{-1} which can be com-

$$E(m) = -g_N\beta_N B_0 m \qquad (2.4)$$

pared with the much larger energies involved in vibrational spectroscopy (ca. 10^3 cal g^{-1} mol^{-1}).

The energy-level diagrams for spin quantum number $I = \frac{1}{2}$ and $I = \frac{3}{2}$ are shown in Fig. 2.1. The difference in energy between two adjacent levels is given by:

$$\Delta E = g_N\beta_N B_0. \qquad (2.5)$$

As in every branch of absorption spectroscopy, we can cause transitions between these levels by applying a quantum of energy equal to the energy separation, i.e.

$$h\nu_I = \Delta E = g_N\beta_N B_0 \quad \text{or} \quad \nu_I = g_N\beta_N B_0/h. \qquad (2.6)$$

Such transitions only occur at the frequency, ν_I, and it is thus termed a resonant process; the phenomenon is known *magnetic resonance*. The selection rule governing transition is that $\Delta m = \pm 1$. Often it is convenient to express the resonance condition in terms of an angular frequency, ω_I. Since $\omega_I = 2\pi\nu_I$ and $\gamma_I = g_N\beta_N/\hbar$, we obtain eqn (2.7).

$$\omega_I = \gamma_I B_0 \qquad (2.7)$$

We have seen that there are $(2I+1)$ allowed values of m and thus $(2I+1)$ energy levels, each corresponding to a particular component of μ_N along B_0 and hence to a particular *orientation* of the nuclear magnet with respect to the direction of the applied magnetic field B_0. Now, when an object with spin angular momentum is subject to a torque, it precesses about the applied torque. In a similar manner the magnetic nuclei precess about the applied magnetic field with an orientation such that eqn (2.7) is valid. This is illustrated in Fig. 2.2 for nuclei with spin quantum numbers equal to $\frac{1}{2}$ and 1.

According to the Larmor theorem, the nucleus precesses about the magnetic field at an angular frequency, ω_I, given by eqn (2.8)

$$\omega_I = \gamma_I B_0 \qquad (2.8)$$

This equation describes the *motion of a nucleus in its particular energy level* and we thus see an important result. The angular frequency of the radiation required by the quantum theory to stimulate transitions of the nuclei from one energy level to another is equal to the Larmor frequency. When this condition is fulfilled experimentally then the system is said to be in resonance.

To investigate how to induce the transition, it is convenient to consider a nucleus for which $I = \frac{1}{2}$. We may visualize the situation as in Fig. 2.3. A transition corresponds to the magnetic moment vector passing from one orientation to another.

If we apply a second magnetic field, B_1, perpendicular to B_0, there will now be a torque tending to align the magnetic moment along the axis of B_1. However, if $B_1 \ll B_0$, this torque will be much smaller than that due to B_0. Furthermore, since the magnetic moment vector is changing its orientation with respect to B_1 as it precesses about B_0, the torque due to B_1 is relatively ineffective. To be continuously effective, B_1 must rotate in phase with the precession of the magnetic moment vector about B_0, i.e. B_1 must rotate at an angular frequency $\omega_I = \gamma_I B_0$. However the magnetic moment can only assume a finite number of orientations with respect to B_0. The same is true for its orientation with respect to B_1, so that it cannot precess simultaneously

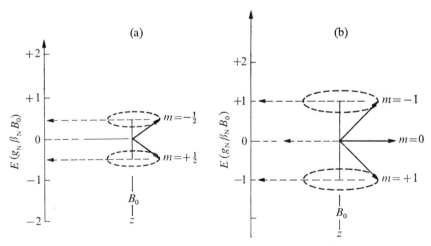

FIG. 2.2(a). Energy levels of a nucleus of spin $I = \frac{1}{2}$ in a field B_0. (b). Energy levels of a nucleus of spin $I = 1$ in a field B_0.

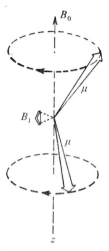

FIG. 2.3. Precession of the magnetic moment vector of nuclei in the two spin states ($I = \frac{1}{2}$). To induce transitions between the states B_1 must be perpendicular to B_0 and rotating at the Larmor frequency. (By convention the direction of the magnetic field is taken to be the z-direction.)

both about B_0 and B_1. The result is that B_1 then induces a transition from one orientation in the field B_0 to another, the energy for the transition coming of course from the field B_1.

Finally, we note that the resonance condition $\nu_I = g_N\beta_N B_0/h = \gamma_I B_0/2\pi$ can be approached either by keeping the field B_0 constant and varying the frequency of the applied field B_1 or *vice versa*. We see that the frequency required depends on the value of B_0 and also γ_I. For hydrogen nuclei $\gamma_I = 2\cdot 67519 \times 10^4$ rad G^{-1} s^{-1} but other nuclei have distinctly different values of γ_I. Thus at a given magnetic field their resonances occur at widely separated frequencies and in any experiment the n.m.r. spectrum of only one type of magnetic nucleus present in a molecule is recorded at any one time. Some relevant magnetic properties of important nuclei are given in Table 2.1.

2.2. Thermal equilibrium and spin–lattice relaxation

In the absence of a magnetic field the magnetic moments of a system of nuclei ($I = \frac{1}{2}$) will have random orientations, and the populations of the two nuclear energy levels must be equal. When the sample is placed in a magnetic field B_0 the nuclei must adopt one of two allowed orientations. If N_+ and N_- are the population of the lower and upper levels, respectively then at equilibrium according to the Maxwell–Boltzmann distribution law:

$$N_+/N_- = \exp(g_N\beta_N B_0/kT) \tag{2.9}$$

TABLE 2.1
Nuclear moments and spins

Nucleus	I	g_N	γ_I rad G^{-1} s^{-1}
^{1}H	$\frac{1}{2}$	5·585	26 753
^{13}C	$\frac{1}{2}$	1·405	6 728
^{19}F	$\frac{1}{2}$	5·257	25 179
^{31}P	$\frac{1}{2}$	2·263	10 840
^{2}D	1	0·857	4 107
^{7}Li	$\frac{3}{2}$	2·171	10 398
^{14}N	1	0·403	1 934
^{17}O	$\frac{5}{2}$	−0·757	−3 628
^{23}Na	$\frac{3}{2}$	1·478	7 081
^{35}Cl	$\frac{3}{2}$	0·548	2 624
^{37}Cl	$\frac{3}{2}$	0·456	2 184
^{39}K	$\frac{3}{2}$	0·261	1 250

At ordinary temperatures $g_N\beta_N \ll kT$ and the above equation becomes

$$\frac{N_+}{N_-} = 1 + \frac{g_N\beta_N B_0}{kT} \tag{2.10}$$

If $B_0 \approx 10^4$ G then the population of the lower energy level is in excess by ca. 1 in 10^5 spins.

This equilibrium is attained by a first-order rate process characterized by a rate constant $1/T_1$ where T_1 is known as the spin–lattice relaxation time because it is a measure of the time taken for the nuclei to reach equilibrium with their surroundings, namely, the molecular framework, which is known as the *lattice*, be it solid, liquid, or gas.

2.3. Resonance absorption

When radiation of suitable angular frequency, ω_I, is applied to the system in the appropriate way, transitions are induced between the energy levels; the probability of a quantum of radiation inducing an upward transition is the same as for a downward transition, so a net exchange of energy between the nuclear spins and the radiation can only occur if the populations of the two nuclear energy levels are unequal. If the lower energy level has an excess population (as it would have at thermal equilibrium) then energy is *absorbed* from the radiation, as spins in the lower level are promoted to the upper level (Fig. 2.4). The result of this net absorption of energy will be to reduce the excess population between the energy levels and so decrease the probability of further absorption.

To obtain continuous absorption it is thus necessary to have some mechanism by which the Boltzmann equilibrium population of spins is continually re-established.

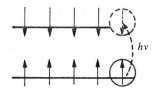

FIG. 2.4. Resonance absorption.

2.4. Mechanism of spin–lattice relaxation

In many branches of spectroscopy, molecules in excited energy states can relax to the ground state by collision with other molecules, in which process the extra energy of the excited states is lost as heat to the surroundings. Spontaneous emission is another cause of relaxation in which a molecule will make a spontaneous jump from one energy level to a lower one emitting a quantum of energy.

When dealing with nuclei, we have to remember that these are surrounded by electron clouds and therefore relatively isolated from the effects caused by collisions with other molecules. Further, the probability of spontaneous emission which is related to the energy difference between the various levels is, in n.m.r., vanishingly small. Thus neither of the above mechanisms occur in n.m.r.

For nuclei of spin $\frac{1}{2}$, the only mechanisms which can cause relaxation are magnetic in nature and must be due to fluctuating magnetic fields at the natural precession frequency of the nuclei† (i.e. the Larmor frequency, $\omega_I = \gamma_I B_0$). In general there are three requirements for a successful relaxation mechanism: (i) there must be some interaction which acts directly upon the spins; (ii) this interaction must fluctuate, i.e. it must be time dependent; and (iii) the interaction must have a suitable time scale. We now investigate the origin of these magnetic fields in the absence of an applied radio-frequency field.

2.5. Origin of magnetic relaxation

We shall be concerned mostly with liquids and we shall further confine our discussion to nuclei with spin $I = \frac{1}{2}$. In such liquids the interaction causing relaxation is generated by local dipolar magnetic fields. For example a nucleus may experience local fields from other nuclei moving past it. The magnetic moments of the nuclei in a liquid set up local fields B_μ in their environment which are proportional to μ/r^3, where μ is the nuclear magnetic moment and r is the distance from it to the point at which the field is measured.

† Theoretically, it also turns out that oscillations of local fields at $2\omega_I$ can also cause transitions.

Using a point-dipole simplification B_μ is given by

$$B_\mu = \pm\mu(3\cos^2\theta-1)/r^3 \qquad (2.11)$$

where θ is the angle between B_μ and the vector joining the nucleus to the point considered. The sign depends on the momentary orientation of the nucleus in the magnetic field.

These local magnetic fields fluctuate according to the molecular motion which may induce changes in r and/or θ and in the sign of the field B_μ. The frequency or variation of the interaction caused by molecular motion is referred to as *modulation* and is characterized by a *correlation time* τ_c which thus sets a time scale to the random fluctuations or noise. It correlates a given value of orientation or position of a nucleus at a particular instant with the value at some later time. If the random motions are rapid, τ_c is short for the nuclei lose memory (i.e. correlation) of the previous orientations very quickly.

2.6. Some examples of fluctuating local fields that can give rise to relaxation

We shall illustrate some of the types of motion that can modulate interactions between nuclei of spin $I = \frac{1}{2}$ in liquids. For example, *rotation* of a molecule will cause the local field at one nucleus, due to another in the same molecule, to fluctuate at the rotational frequency with a correlation time τ_R. The effect of rotation thus causes changes in θ. If one nucleus in a molecule is *relaxing very rapidly*, its alignment or spin state in the magnetic field is changing between its allowed orientations (i.e. with and against the magnetic field) and the local field this nucleus sets up at a neighbouring nucleus will change from $+$ve to $-$ve, at a rate equal to the relaxation rate. Alternatively, if an atom detaches itself from the molecule, as in *chemical exchange*, the atom which replaces it may have the orientation of its nuclear spin reversed, thus reversing the sign of the local field. The reversal will occur at a frequency $1/\tau_M$, which is the chemical exchange rate. Relative *diffusion* of molecules is also an example of fluctuations of local fields at the nuclei at one molecule due to another. In this case, however, changes in r and θ are produced. These motions are summarized in Fig. 2.5. Of course there are many other possibilities, but we shall mainly be concerned with these in this book.

2.7. The intensity of the randomly fluctuating fields—short and long correlation times

For any random motion there is a spectrum of frequencies of the fluctuations but the variation and analysis of the intensity of the fluctuations and their frequency distribution depends on the exact type of motion concerned. The intensity of the fluctuations is referred to as the *spectral density*, $J(\omega)$ and the form of $J(\omega)$ versus ω is shown in Fig. 2.6. In essence the Figure

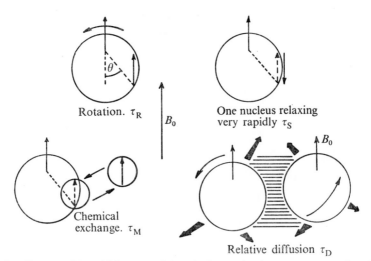

FIG. 2.5. Some motions which can modulate the interaction between two nuclei and cause relaxation.

shows every value of $J(\omega)$ which is initially independent of ω (the so-called *white-spectrum* region) but eventually there must be a limit to the frequency of these motions and obviously the intensity, $J(\omega)$ falls to zero as ω increases. The onset of the fall in $J(\omega)$ is called the *dispersion region* or non-white-spectrum region. The total 'area' of the spectral density curve is a constant for all types of motion at a given temperature. If the motion is slow (i.e. τ_c is long), there will be few high frequency components of the interaction. Consequently, the intensity (or spectral density) at the lower frequencies is increased. Conversely, if the motion is rapid (i.e. τ_c is short) the intensity is lower but extends over higher frequencies (Fig. 2.6).

The fall-off region in the curve of $J(\omega)$ versus ω, therefore, depends on the

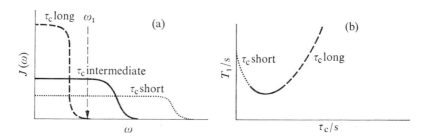

FIG. 2.6(a). Variation of spectral density function $J(\omega)$ with ω as a function of correlation time, τ_c. (b). Variation of T_1 with τ_c. Note the correspondence between solid and broken lines in the two diagrams.

value of τ_c and occurs when $\omega_I \approx 1/\tau_c$. Thus the 'white-spectrum' region is defined as $\omega_I \tau_c \ll 1$, and the non-white region as $\omega_I \tau_c \gg 1$.

2.8. Variation of $1/T_1$ with molecular motion

We have seen that it is only the fluctuations of the local magnetic fields at ω_I (and $2\omega_I$) which can stimulate spin–lattice relaxation. We therefore expect that the relaxation rate $1/T_1$ is proportional to the strength of the local fields, and to the value of $J(\omega)$ at ω_I which in turn depends on the value of the correlation time τ_c. Thus $1/T_1$ will depend on τ_c.

We consider a liquid, the molecular motion of which is characterized by a value of τ_c. If the value of τ_c is small so that $J(\omega)$ is very small near the Larmor frequency ω_I, (Fig. 2.6a) then spin–lattice relaxation will be inefficient and T_1 will be long. If the liquid is made more viscous (e.g. by lowering the temperature) the value of τ_c increases, $J(\omega)$ at ω_I increases and T_1 becomes shorter. However as τ_c increases still further, the 'spectral-density' curve is 'shifted' to lower frequencies such that there is now a decrease in the value of $J(\omega)$ at ω_I, and the value of T_1 becomes longer again (Fig. 2.6b).

A more quantitative description of this effect can be derived. For dipolar nuclei a convenient measure of the local field strength is given by the mean-square average of the local fields B_μ (eqn (2.11)). Therefore, we obtain

$$1/T_1 \propto \langle B_\mu^2 \rangle_{av}.(J(\omega_I)+J(2\omega_I)). \tag{2.12}$$

The first term $J(\omega_I)$ corresponds to fluctuations at ω_I, while $J(2\omega_I)$ corresponds to those at $2\omega_I$. To evaluate the equation, explicit forms of $J(\omega)$ must be known. An easy way to analyse fluctuations in terms of their frequency components is to use a Fourier Transform and $J(\omega)$ is then written

$$J(\omega) = \int_{-\infty}^{\infty} g(\tau_c)\exp(-i\omega\tau)\,d\tau_c. \tag{2.13}$$

where $g(\tau_c)$ is a correlation function. The exact form of this function depends on the type of motion. The correlation function is a measure of the persistence of the fluctuations. A particularly simple assumption is that the correlation function is an exponential with a time constant equal to the correlation time τ_c. This is physically reasonable for some types of motion, e.g. rotation. $J(\omega)$ is then given by an expression of the form:

$$J(\omega) = C\tau_c/(1+\omega^2\tau_c^2) \tag{2.14}$$

where C is a constant. From the functional form of $J(\omega)$, one can rationalize Fig. 2.6. At low frequencies $J(\omega)$ is independent of ω, but, when the region in which $\omega_I^2\tau_c^2 = 1$ is reached, the value of $J(\omega)$ falls rapidly to zero.

For other types of correlation functions (i.e. non-exponential) the curve in Fig. 2.6a is similar when $\omega_I \tau_c \ll 1$, but differs in the 'fall-off' region. More precise correlation functions can only be calculated on the assumption of a suitable model for the molecular motion. For example, in relaxation caused by the random translational diffusion between molecules, the diffusion equation does not predict that the orientation of a given molecule will change with time in an exponential manner. In this case, the values of $J(\omega)$ are given by a more complex equation. Another complication of particular importance in dealing with macromolecules, is the possibility of a distribution of correlation times. In other words, nuclei can exist in a range of environments, and experience different interactions, each of which is characterized by its own correlation time. The equations for the relaxation rate may be quite complex in both these cases. However, we mention them to point out some of the complications that can occur. In later sections, we will have reason to question again the validity of using single exponential correlation functions. For the present, we note that in many systems the use of such functions has been shown to be a good approximation, e.g. for rotational motion. If we consider a nucleus where $I = \frac{1}{2}$, experiencing a magnetic field from other similar spins at a distance r and that this motion is modulated by rotation, then eqn (2.12) becomes

$$\frac{1}{T_1} = \frac{3}{10} \frac{\gamma_I^4 \hbar^2}{r^6} . f_1(\tau_c) , \qquad (2.15)$$

where

$$f_1(\tau_c) = \frac{\tau_c}{1 + \omega_I^2 \tau_c^2} + \frac{4\tau_c}{1 + 4\omega_I^2 \tau_c^2}$$

Such an equation may apply to the rotation of a water molecule, in which one hydrogen nucleus experiences the field due to the second. In macromolecular systems, the rotational motion of methyl groups may be anisotropic and the equations have to be modified (see Chapter 6). Nevertheless, the same general form of the equation is applicable.

From the eqn (2.15), we can predict the variation of T_1 when the correlation time τ_c is altered (by varying the temperature of the liquid, for example). When $\omega_I^2 \tau_c^2 \ll 1$, T_1 is proportional to $1/\tau_c$. On the other hand when $\omega_I^2 \tau_c^2 \gg 1$, T_1 is proportional to τ_c. Thus T_1 passes through a minimum (which occurs when $\omega_I \tau_c \approx 0.6$). The functional form of $f_1(\tau_c)$ versus τ_c is shown in Fig. 2.7 and is calculated for two different values of ω_I. (Note that when $\omega_I^2 \tau_c^2 \gg 1$ the difference in values of T_1 for a given value of τ_c depends on the ratio of the square of ω_I.)

2.9. Classical versus quantum mechanical approach to n.m.r.

It is important to make clear to the reader the distinction between the classical and quantum mechanical (Q.M.) approaches to n.m.r. In a classical picture of the n.m.r. phenomena, we are only concerned with the net sample magnetization, while on the quantum mechanical picture we can discuss the

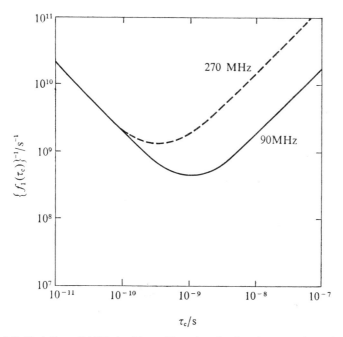

FIG. 2.7. Variation of $1/f_1(\tau_c)$ with τ_c. Note that the function goes through a minimum. $1/T_1 = (3\hbar^2\gamma_H^4/10r^6)f_1(\tau_c)$, where $f_1(\tau_c) = \{\tau_c/(1+\omega_0^2\tau_c^2)\}+\{4\tau_c/(1+\omega_0^2\tau_c^2)\}$. See eqn (2.15). NOTE. In accordance with Royal Society recommendations, axes on graphs have been labelled with the physical quantity represented divided by its units (or multiplied by their reciprocal) so that pure numbers only are plotted.

behaviour of the individual spins. It is dangerous and misleading to mix the two approaches; clearly each will have its advantages, but there will be several concepts which are better explained by one approach. In general we shall use the classical approach since it is more pictorial, but in some cases there may be difficulty in visualizing a particular concept and we may then invoke the Q.M. picture. After all, the failure to discuss some of the laws of physics classically led originally to the Q.M. approach. We shall mention if we need to change from one approach to the other.

2.10. The use of rotating co-ordinates

A system of rotating co-ordinates is a very convenient concept in describing many n.m.r. phenomena. In Fig. 2.8 an individual nuclear spin is shown as precessing at the Larmor frequency, $(\omega_I = \gamma_I B_0)$ about the z-axis which by convention, is the direction of the magnetic field. Suppose an observer now imagines himself in a frame of reference with co-ordinates $x'y'z'$ having the same z-axis but with the x'- and y'-co-ordinates rotating about z in the same sense and at the same frequency ω_I as the nucleus.

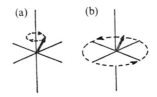

F IG. 2.8. (a) Nuclear precession. (b) Nucleus in a frame rotating at the Larmor frequency.

The nucleus will appear to the observer to be stationary. This transformation to rotating co-ordinates is entirely mathematical and, of course, we assume that none of the laws of physics are changed. For example if a nucleus is placed in a magnetic field it will precess, and conversely, if a nucleus precesses, it must be experiencing a magnetic field. In the rotating frame where the nucleus *appears* not to be precessing to the rotating observer, it effectively experiences a zero magnetic field (Fig. 2.8). This relationship can be expressed mathematically by

$$B_{\text{effective}} = B_0 - \omega/\gamma_I. \tag{2.16}$$

When $\omega = \omega_I$, the second term on the right hand side of eqn (2.16) becomes B_0, and the effective field in the rotating frame, $B_{\text{effective}}$, is zero.

2.11. Definition of nuclear relaxation times using the concept of a rotating frame

We start initially by considering the individual spins, i.e. a non-classical approach to develop two main points: at thermal equilibrium there is a net magnetization along the z-axis, and at equilibrium the net magnetization in the xy-plane is zero.

In Fig. 2.9 the individual nuclear spins are depicted by arrows which precess at the Larmor frequency ($\omega_I = \gamma_I B_0$) about the z-axis (by convention the direction of the magnetic field B_0). If thermal equilibrium has been

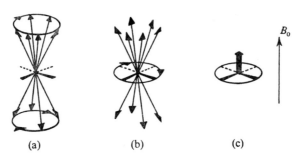

F IG. 2.9(a). Precession of nuclear spins about B_0; (b) spins stationary in the rotating frame; and (c) resultant magnetization.

established there will be slightly more nuclei in the $+z$-direction than in the $-z$-direction (i.e. more nuclei whose spins are aligned with the field) and thus there will be a resultant magnetization M_0 along the z-axis which appears stationary to the rotating observer.

A sample of finite size will contain many nuclei, all of which may experience identical magnetic fields, either because of local magnetic fields within the sample or because the external field B_0 is not completely uniform over the sample (i.e. magnetic field inhomogeneity). The observer will be effectively rotating at ω_I, which corresponds to the mean value of the magnetic field over the sample. Thus the observer will see many nuclei as stationary and some precessing slowly in either direction (corresponding to nuclei

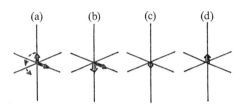

FIG. 2.10. (a), (b) 180° pulse of B_1; (c), (d) decay of nuclear magnetization.

experiencing magnetic fields $>$ or $<B_0$, i.e. a distribution of fields around B_0). The resultant magnetization in the $x'y'$-plane is zero, in general, since the nuclei are precessing around the z-axis with a *random symmetrical distribution of phases* with respect to one another. The resultant magnetization at thermal equilibrium is represented as shown in Fig. 2.9.

We now return to the classical approach where we must consider the resultant nuclear magnetization of a large number of nuclei for which $I = \frac{1}{2}$.

In the n.m.r. experiment, to induce transitions of the nuclei between their energy levels it is necessary to apply a small magnetic field B_1 perpendicular to the main field, e.g. along the x-axis. This field must rotate in the same sense and at the same frequency as the nuclear precession. To the rotating observer this field will appear stationary. The effect of this field on the spin system is such that the resultant magnetization M_0 rotates away from the z-axis in the zy-plane through an angle ζ (see eqn (2.17))

$$\zeta = \gamma_I B_1 t \tag{2.17}$$

where t is the length of time for which B_1 is applied. After a time $t_\pi = (\pi/\gamma_I B_1)$ s, the nuclear magnetization will have precessed through 180° so that the populations of the energy levels have been inverted, i.e. M_0 is now along the $-z$ axis. If B_1 is turned off, the spin system will return to thermal equilibrium, (Fig. 2.10) and we see, therefore, that $1/T_1$, the spin lattice relaxation rate, is the relaxation rate along the z-axis and for this reason it is sometimes called the *longitudinal relaxation rate*.

If the angle $\zeta = 90°$, the nuclear magnetization will be turned along the y-axis. If the stimulating radio-frequency field B_1 is then turned off the nuclear magnetization will now precess around the z-axis in the xy-plane at ω_I. *It is this rotating magnetization in the xy-plane that the n.m.r. spectrometer detects*, (i.e. the signal is detected in the xy-plane).

However the magnetization in the xy-plane is clearly not the equilibrium situation and it will therefore decay back to the equilibrium situation (which is with the magnetization along the z-axis). Classically, this decay is considered to be exponential with a characteristic time constant T_2 known as the *transverse relaxation time*. Thus the above relaxation times can be defined from the relationships†

$$\frac{dM_x}{dt} = -\frac{M_x}{T_2} \tag{2.18}$$

$$\frac{dM_y}{dt} = -\frac{M_y}{T_2} \tag{2.19}$$

$$\frac{dM_z}{dt} = \frac{M_0 - M_z}{T_1} \tag{2.20}$$

where M_x, M_y, and M_z are the components of M_0 along the axes. In terms of the quantum mechanical picture, T_2 can be visualized as the time taken for the individual nuclei to lose phase with each other in the xy-plane, when the net magnetization then decays to zero. At first sight it might be expected that the longitudinal and transverse relaxation times T_1 and T_2 are the same. After all, the magnetization in the xy-plane is decaying and being re-established along the z-axis. Any process which is involved therefore in the spin–lattice relaxation (T_1) processes will also be involved in transverse relaxation (T_2) processes. There are, however, additional mechanisms by which individual nuclei can lose phase in the xy-plane. We can discuss this in terms of the classical picture in which we suppose that the sample is divided up into a number of small identical elements such that all nuclei in the same element experience the same magnetic field. We then consider the resultant magnetization M of each element. Because of the small variation in magnetic field over the sample (from interactions between nuclei and from magnetic field inhomogeneities) not all these elements in the xy-plane precess at the same rate, but some will rotate faster and some slower than the reference frame. In the reference frame, the effect will be a 'fanning out' of the elements as indicated in Fig. 2.11 leading to a loss of phase in the xy-plane, and consequent decay of the n.m.r. signal. Also processes such as chemical exchange can contribute to this loss of phase. In this case, a nucleus

† These equations are a special case of the well-known Bloch equations.

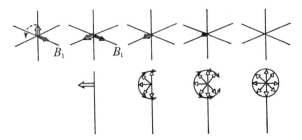

FIG. 2.11. Decay of transverse magnetization.

exchanges with one from a different environment, i.e. a different chemical shift. The precessional phase is, therefore, randomized because the nucleus comes from a site with a different precessional frequency. This contributes to a rapid decay of the transverse magnetization.

Another mechanism which occurs between like nuclei (i.e. same γ_I) is spin–spin exchange or as it is sometimes called, the *flip–flop* mechanism. In this mechanism nuclei in different energy levels exchange energy adiabatically, i.e. the downward transition (flip) of nucleus in the higher energy level is accompanied simultaneously by an upward transition (flop) of a nucleus in the lower energy level, or *vice versa*. This mechanism is particularly important in solids. Additionally, in solids the variations in the local magnetic field over the sample will be much greater than in liquids, the nuclei will lose phase more quickly and T_2 will be shorter. By contrast, T_1 will be long, since there is relatively little motion (and τ_c is very long). So in a solid interactions between the spins, (i.e. static and flip–flop) mainly account for the T_2 processes. Early relaxation-time studies were performed on solids and thus T_2 was called the spin–spin relaxation time. This term is still often used for liquids (where the contribution of spin–lattice relaxation processes to T_2 *is* significant). In fact, in a liquid, the molecules are moving rapidly and the magnetic fields effectively average out, leaving only the spin–lattice contribution to T_2. In these circumstances T_1 and T_2 become approximately equal.

2.12. Variation of T_2 with molecular motion

In view of the above discussion, we expect the general expression for $1/T_2$ to be similar to that for $1/T_1$ but to contain an extra term corresponding to the presence of the local static fields which cause a loss of phase in the xy-plane, i.e. the intensity of fields at zero frequency, represented by $J(0)$ in the terminology of p. 19. Thus the explicit form of $J(0)$ is then

$$J(0) = \text{Constant} \times \tau_c. \tag{2.21}$$

The full expression for $1/T_2$ can be shown to be

$$\frac{1}{T_2} = \frac{3}{20}\frac{\gamma_I^4\hbar^2}{r^6}f_2(\tau_c) \qquad (2.22)$$

where

$$f_2(\tau_c) = 3\tau_c + \frac{5\tau_c}{1+\omega_I^2\tau_c^2} + \frac{2\tau_c}{1+4\omega_I^2\tau_c^2} \qquad (2.23)$$

If the molecular motion is such that the fluctuations of the local field are rapid, the correlation time τ_c will be such that $\omega_I\tau_c \ll 1$. Comparison of the equations for $1/T_1$ and $1/T_2$ shows that

$$T_1 = T_2$$

as suggested in the previous section.

However, the behaviour of T_1 and T_2 is quite different when $\omega_I\tau_c \gg 1$ (see Fig. 2.12). We have seen that T_1 passes through a minimum, but this is not so for T_2, which continues to decrease linearly. It will be recalled that when $\omega_I\tau_c \gg 1$, T_1 increases (i.e. $1/T_1$ decreases) because the value of $J(\omega)$ at a given ω_I decreases as τ_c increases. In T_2 processes, we are also concerned with a term corresponding to $J(0)$ and, as τ_c becomes longer, it is this term

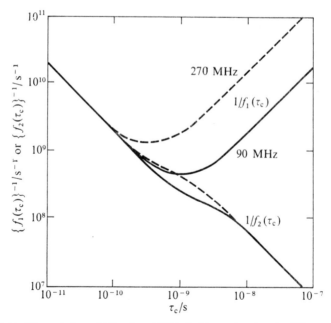

FIG. 2.12. The variation of $1/f_1(\tau_c)$ and $1/f_2(\tau_c)$ with τ_c, showing the difference in behaviour of the dipolar contributions to the relaxation times. See eqns (2.15) and (2.23) for definitions of $f_1(\tau_c)$ and $f_2(\tau_c)$.

which becomes increasingly important. Eventually a situation will be reached in the solid, where there is no motion whatsoever, the so-called rigid-lattice condition; T_2 will then remain constant.

2.13. Other nuclear relaxation mechanisms

We have considered dipole–dipole interaction between nuclei as the main source of the relaxation in liquids. Some other mechanisms are due to an anisotropic chemical shift, spin–rotational interactions, the presence of unpaired electron spins and electric quadrupole interactions.

2.13.1. *Anisotropic chemical shift*

Here the fluctuating local fields arise from changes in the chemical shielding of a nucleus as the molecule rotates. If the chemical shielding is not isotropic the secondary magnetic field generated by the motion of the electrons around a nucleus will not, in general, be parallel to the applied field and there will, therefore, be a component perpendicular to B_0 which will oscillate as the molecule rotates. This mechanism requires the presence of an external field B_0 (unlike most other mechanisms) and since the value of the secondary magnetic field depends on B_0, (Chapter 3) this relaxation mechanism will be proportional to B_0^2. The value of T_1 and T_2 are not equal, even under white spectrum conditions, when $T_1/T_2 = \frac{7}{6}$. Fluorine chemical shifts are often sufficiently anisotropic so that, in principle, this effect could be observed. However, dipole–dipole interactions between nuclei will dominate the relaxation in most cases, making the effect difficult to observe.

2.13.2. *Spin-rotational interactions*

The rotation of a molecule causes the electrons and nuclei to set up local magnetic fields which depend on the rotational quantum number, J. The correlation time for such fields is of the order of the time in which J changes, i.e. τ_J and it can be shown that

$$\tau_R \cdot \tau_J = I/6kT, \tag{2.24}$$

where I is the moment of inertia of the molecule, k is the Boltzmann constant, and T is the absolute temperature. As the temperature is increased, the molecules rotate faster, and τ_R will become shorter. Conversely, τ_J becomes *longer* as the temperature increases. For hydrogen nuclei this mechanism is not usually important, but for ^{31}P and ^{19}F it may often be significant.

We shall not discuss further the above mechanisms but we include them to help illustrate the general ideas of relaxation. By contrast we shall deal with the next two mechanisms in some detail.

2.13.3. *The presence of unpaired electron spins*

The nuclear relaxation times in a diamagnetic liquid are often dramatically reduced when very small concentrations of paramagnetic solutes are added.

The presence of dissolved oxygen, for example, in benzene, reduces the relaxation time of the hydrogen nuclei by a factor of 5. A simple explanation for this, is that the magnetic moments of unpaired electrons are about three orders of magnitude greater than nuclear moments, so that the local fields generated by them are correspondingly greater. This is discussed in Chapter 9.

2.13.4. *Electric quadrupole interactions*

Nuclei with spin $I > \frac{1}{2}$, possess a quadrupole moment which can interact with electric field gradients. The quadrupole moment defines a simple deviation of the positive charge distribution over the nucleus from spherical symmetry. The energy of a nucleus with an electric quadrupole moment depends on its orientation with respect to the electric field gradient. If the electric field gradients fluctuate, then there is an additional relaxation mechanism for the nuclei. This is discussed more fully in Chapter 12.

2.14. Chemical shifts

The resonance condition for a given nucleus depends not only on the nature of the nucleus (e.g. ^1H, ^{19}F) but also on its electronic environment. For nuclei of the same isotope, the differences in the resonance condition with chemical environment leads to the concept of a *chemical shift*.

The chemical shift has its origin in the magnetic 'shielding' or screening, of the nucleus produced by the electrons around it. The presence of an external magnetic field, B_0, induces an orbital motion of these electrons which thus produce a magnetic field, the magnitude of which is proportional to the applied field. Hence a particular nucleus experiences not the applied magnetic field, B_0, but one that is suitably modified by the 'shielding' of the electrons. Since the induced field is proportional to B_0, the local field at the nucleus B_{local} is given by

$$B_{\text{local}} = B_0(1-\sigma), \tag{2.24}$$

where σ is a dimensionless constant, the shielding or screening constant. It represents the summation of two terms:

$$\sigma = \sigma_D + \sigma_P. \tag{2.25}$$

The diamagnetic term, σ_D, is temperature independent and positive. It is, in effect, the local magnetic field that would be produced by the electrons around an isolated atom when placed in an external magnetic field. In a molecule, the electrons around the atom cannot circulate so freely as in a free atom (i.e. the motion becomes anisotropic) and this tends to reduce the local magnetic field. This reduction may be represented by a paramagnetic term σ_P which is negative. The distortion from spherical symmetry is really equivalent to the mixing of higher electronic states with the ground state. Thus the magnitude of σ_P depends on the energy separation between the ground and excited electronic states and also on the symmetry of the atomic orbitals of the circulating electrons. For example, for protons in *pure* ls orbitals this term vanishes, and indeed for protons in general σ_P is very small. On the other hand distribution of p-electrons in ^{19}F may vary from spherically symmetrical in the F^- ion ($\sigma_P = 0$) to asymmetric when fluorine is covalently bonded ($\sigma_P \ll 0$). Indeed, the term σ_P is usually more important than σ_D in the interpretation of ^{19}F chemical shifts.

So far we have only considered the effect of 'screening' from the circulation of electrons of the atom in question. The diamagnetic and paramagnetic contributions from other atoms must obviously be considered also. In general, the effects from neighbouring nuclei are important if the electrons on these atoms have a large anisotropic magnetic susceptibility as, for example, in $C\equiv C$, $C=O$, and $C=C$† bonds. Then too the effect is important if the electron cloud around the nucleus is perturbed by the neighbouring atom, for example when the bond to the nucleus is polar or a hydrogen bond.

There is one further contribution to the screening constant which is important for aromatic molecules, that is, interatomic currents which flow around closed conjugated loops. These currents are termed ring currents and are discussed in Chapter 3. In protein systems the effects of ring currents are important in interpreting shifts of nuclei which may be near to an aromatic residue.

2.14.1. *Range of chemical shifts of different nuclei*

Chemical shifts are measured usually with reference to some standard, either internal or external, and are expressed in units of magnetic field or frequency units. Since the chemical shift is proportional to the applied field, comparison at different magnetic field strengths is facilitated by expressing the chemical shift δ as a dimensionless unit. Conventionally, δ may then be defined as

$$\delta = 10^6 \times (\nu_{ref} - \nu_I)/\nu_0 \qquad (2.26)$$

where ν_I is the resonance frequency of the nucleus, ν_{ref} that of the reference compound and ν_0 the operating frequency of the spectrometer. δ is thus usually expressed in parts per million (p.p.m.).

Table 2.1 gives typical ranges of chemical shifts for the more important magnetic nuclei.

2.14.2. *Hydrogen chemical shifts*

In the majority of cases, we shall be dealing with hydrogen chemical shifts and we include a brief discussion of some of the main features which determine them. In general, hydrogen chemical shifts are determined by the magnitude of σ_D, from the local diamagnetic currents; the value of σ_D

† For a good discussion on this part see 'Nuclear Magnetic Resonance Spectroscopy' by R. M. Lyndon-Bell and R. K. Harris, 1969, Nelson.

TABLE 2.1

*Approximate range of chemical
shifts for some nuclei*

Nucleus	Approximate range (p.p.m.)
^1H	12
^{13}C	350
^{15}N	1000
^{17}O	650
^{19}F	500
^{31}P	700

depends on the electron density associated with the particular nucleus. To a first approximation, variations in electron density around a particular hydrogen nucleus may be correlated with the electronegativity of the group to which the hydrogen is bonded. Thus naively, when protons are attached to oxygen or nitrogen, the electron density around the proton will be reduced, with consequent deshielding, i.e. reduction of σ_D and hence the local field, B_{local}. At a fixed frequency, therefore, the magnetic field required to achieve the resonance condition is less than that for protons bonded to an aliphatic carbon atom. In other words, the —O*H* and —N*H* resonances occur at low field. The chemical shift values for representative types of chemically distinct protons are shown in Fig. 2.13.

The most commonly used *internal* standard for proton spectra is tetramethylsilane, $Si(Me)_4$, (TMS). It is chemically inert, volatile (B.p. $\sim$27 °C) and miscible with organic solvents. The hydrogen nuclei are all equivalent and highly shielded and therefore its hydrogen resonance occurs at high field. The convention adopted here will be that shifts to low field of TMS will be positive, and negative if at a higher magnetic field. Another standard, that is often used for biochemical work is hexamethyl disiloxane (HMS or HDMS) which has the advantage that it is less volatile. In aqueous systems the internal reference often used is $CH_3 \cdot Si(Me)_2 \cdot CH_2 \cdot CH_2 \cdot CH_2 \cdot SO_2^- Na^+$, often abbreviated as TSS or DSS. The methyl signals give a strong singlet resonance but the resonances of the methylene protons is a multiplet (due to spin–spin coupling) and this may interfere with the spectrum of the sample. Often with proteins an internal standard is not suitable and an *external* standard has to be used. In such a case, a correction to the chemical shift must be applied except when spherical sample tubes are used. The correction depends on the difference in *bulk* diamagnetic shielding (i.e. bulk susceptibility) between the solution containing the sample and the external reference and it is only possible to calculate the difference in chemical shifts if the bulk susceptibilities are known. With internal standards, of course, this complication does not arise, the bulk shielding is then identical for both the reference and sample nuclei.

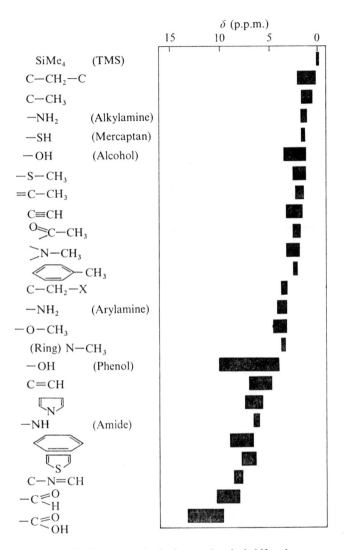

FIG. 2.13. Representative hydrogen chemical shift values.

2.15. Spin–spin coupling

In addition to the local fields experienced by a nucleus as a result of the motion of the electrons in the vicinity of the nucleus, we have to consider the local fields produced by the presence of other nuclei in the molecule which have a magnetic moment. The interaction between two nuclear spins is termed spin–spin coupling, and this results in the observation of multiplet structure in the n.m.r. spectrum.

A qualitative interpretation of the origin of spin–spin coupling comes from the following picture. Consider a molecule AX where the spin quantum numbers of A and X are both $\frac{1}{2}$. The spin of A interacts with the valence electron 'associated' with A. The two valence electrons in the A—X bond must, by the Pauli exclusion principle, have their spins anti-parallel. The other electron interacts weakly with the X nucleus. Therefore, there is an interaction path between the nuclei through the electrons, so that the X nucleus is sensitive to the spin of the A nucleus, but only as a result of a weak interaction with its associated electron. Any change in the spin state of A will result in a change in the energy level of the X nucleus and vice versa. The two spin states for A lead to a symmetrical splitting of the X resonance into two lines (Fig. 2.14). The separation between the two lines, J, is called the spin–spin coupling constant and is expressed in Hz. The splitting is generated by interaction between the nuclei and is not a function of the applied field, in contrast to the chemical shift. In general, the coupling of the spins of A and X tends to stabilize anti-parallel spins and by definition this is taken to imply a positive spin–spin coupling constant. The sign of J is usually positive for directly-bonded nuclei but some negative values (mostly small in magnitude) have been reported. In practice, coupling constants are reported as a number (i.e. their magnitude) without regard to their sign. The value of the coupling constant between two nuclei will be largest for directly-bonded nuclei being attenuated but not necessarily monotonically, as the number of intervening bonds increase. It will also depend on the nature of the bonds and the relative geometries of the nuclei.

Let us consider now three non-equivalent nuclei AMX again all with spin quantum number $= \frac{1}{2}$; we investigate the possibility of coupling between A—M and M—X (Fig. 2.14). The effect of the extra coupling is to split *all* the lines again. Note the effect is multiplicative rather than additive, i.e. the new nucleus splits all the lines of the other nucleus.

If, in fact, M is chemically identical and magnetically equivalent to X as, say, the hydrogens in a CH_2 group which undergoes free rotation, then Fig. 2.14 can be redrawn with the A—X and M—X couplings equal. The A resonance then appears as a triplet and the intensities of the lines is $1:2:1$. (The X resonance has its intensity doubled, since there are two nuclei, but remains a doublet.)

In general, for a nucleus coupled to n equivalent spins (for which $I = \frac{1}{2}$) there will be $(n + 1)$ lines of relative intensity given by a binomial distribution, e.g. $1:2:1$ for a triplet, $1:3:3:1$ for a quartet. This treatment can be easily extended to take into account other groups of equivalent nuclei.

The above discussion assumes that J is small compared with the chemical shift difference, between two given nuclei, or groups of equivalent nuclei. The spectrum obtained is called a first-order spectrum. When this condition is not fulfilled the spectrum is more complicated; the relative intensities of the

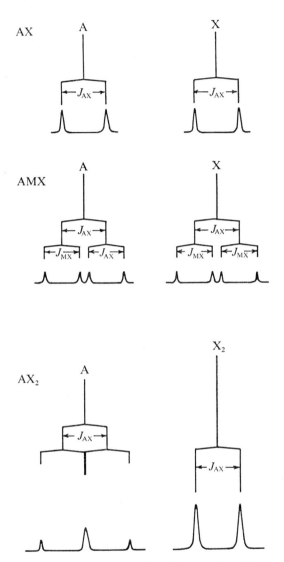

FIG. 2.14. Splitting of A and X resonance lines due to spin–spin interaction for nuclei
$I = \frac{1}{2}$ in AX (top), AMX, and AX$_2$ systems.

multiplet are not simple ratios and are no longer given by the binomial
coefficients and the various coupling constants are then found by computational analysis of the spectrum. We note, however, that it is possible to
simplify the spectrum of A by removing the splittings of X to A from the
spectrum, by strong irradiation of the X nuclei.

From the above discussion, it follows that n.m.r. spectra can be classified into two categories, those involving chemical shifts $\gg J$, and those involving chemical shifts of the order of J. Pople (1959) has introduced a nomenclature for these cases. When first-order spectra are observed, the nuclei are labelled with letters such as A, M, X, well separated in the alphabet (since the resonance signals are well separated) and the spectrum is termed an AMX spectrum. The symbols A, B, C are used for non-equivalent nuclei of the same species whose relative chemical shifts are approximately the same as the coupling constants between them. Equivalent nuclei are described by the same symbols, the subscript describing the number of equivalent nuclei in the group. Finally the symbol A is used for the nuclei resonating at lowest field e.g. HF is described as an AX spectrum; NH_3 as AX_3, the hydrogen spectrum of $CH_3 \cdot CHO$ as AX_3 (where A is the CHO proton). Finally, we note that the chemical shift is proportional to the applied field. If there are several resonances which overlap at one magnetic field strength, then, as the field strength is increased, the resonances appear as separate peaks. Apart from the advantages of greater sensitivity at the higher fields the simplification of a complex spectrum (e.g. ABC...) in such cases can be very important. It is possible, in theory, to separate the resonance lines sufficiently to obtain a first-order spectrum.

2.16. Conformational analysis using coupling constants

The importance of coupling constants is that they depend on intimate details of structure and can be used, in principle, to obtain geometric information. In practice, theoretical calculation of coupling constants is difficult because of the large number of electrons and geometrical parameters involved. However, one example where theory predicts quite accurately the observed behaviour is for vicinal coupling constants for hydrogens attached to carbon atoms. The value of the coupling constant, J, is given by an equation of the form:†

$$J(\phi) = A \cos^2\phi + B \cos \phi + C, \qquad (2.27)$$

where ϕ represents the dihedral angle formed by the H—C(1)—C(2) and C(1)—C(2)—H bond pairs and $A = 4 \cdot 22$, $B = -0 \cdot 5$, and $C = 4 \cdot 5$. This equation (the Karplus equation) is only strictly valid for systems containing tetrahedral carbon atoms to which hydrogen atoms are attached. Changes in hybridization of the bonds or the presence of electronegative substituents on the H—C—C—H′ fragment will clearly affect the bonding electrons and the coupling constants would change. Despite Karplus himself pointing out the inadequacies of this relationship and warning against its more general

---

† The equation is sometimes stated in the form

$$J = J_0 \cos^2\phi - 0 \cdot 28$$

where $J_0 = 8 \cdot 5$ for $0° \leqslant \phi \leqslant 90°$ and $9 \cdot 5$ for $90° \leqslant \phi \leqslant 180°$.

application, many groups of workers have chosen to ignore this (Karplus, 1963). Other groups, working on the assumption that the general form of the equation is correct, however, have modified it empirically, by adding other terms, derived from experiments which take into account the electronegativities of adjacent substituents. In practice it appears that, whilst the Karplus equation gives the form of the angular dependence of the coupling constants, different systems require different parameters to give consistent results. This makes the significance of the equation rather difficult to assess and must lead to some uncertainty in drawing conclusions based on its application.

In *cyclic* compounds, however, the dihedral angles will be determined by the mode and extent of buckling of the ring. Thus, in principle, measurement

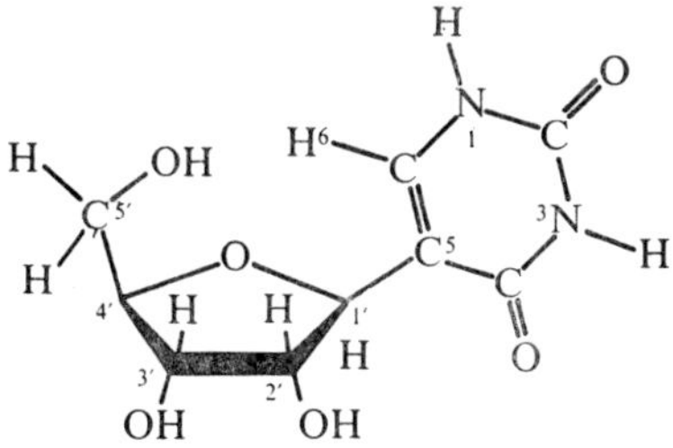

FIG. 2.15. Structural formula of β-pseudouridine.

of vicinal hydrogen coupling constants will allow the conformation of the ring in solution to be calculated, provided a 'suitable' equation can be used. The constants A, B, and C will then be best determined from model compounds closely related to the unknown compound. Experimentally, the vicinal coupling constants are found to depend on the orientation of the two nuclei and decrease in the order axial–axial > axial–equatorial, or equatorial–axial > equatorial–equatorial, if all other factors except ϕ are constant. This is true too for ^{19}F—C—C—^{1}H couplings, although it must be emphasized that the derivation of dihedral angles from fluorine–hydrogen vicinal coupling constants must be viewed with caution. Indeed such is also the case for dihedral angles determined from coupling constants such as ^{31}P—O—C—^{13}C, ^{31}P—O—C—^{1}H, and ^{1}H—N—C—^{1}H.

An actual example of a cyclic compound whose conformation has been analysed by measuring coupling constants is the furanose ring in β-pseudo-uridine (Fig. 2.15), (a minor constituent of tRNA). The coupling constants were determined using spin-decoupling techniques and by computer analysis of the spectrum (Hruska, Grey, and Smith, 1970).

In these studies the form of the Karplus equation used was $J = J_0 \cos^2\phi - 0.28$ with $J_0 = 9.27$ Hz for $0° < \phi < 90°$ and $J_0 = 10.36$ Hz for $90° < \phi < 180°$. The values of $J_{1'2'}$, $J_{2'3'}$, and $J_{3'4'}$ define, in principle, the conformation of the furanose ring. The values of $\phi_{1'2'}$, $\phi_{2'3'}$, and $\phi_{3'4}$

TABLE 2.2

Coupling constants of β-pseudouridine at 30 °C

	J(Hz)
$J_{1'2'}$	5·0
$J_{2'3'}$	5·0
$J_{3'4'}$	5·2
$J_{4'5'B}†$	3·2
$J_{4'5'C}†$	4·6
$J_{5'B5'C}†$	−12·7

† There are two protons on C(5′) labelled B and C.

calculated, *if* this equation is valid are approximately either 41° or 136°. However, values of $\phi_{1'2'}$ and $\phi_{3'4'}$ less than ca. 75° would require unreasonable buckling of the furanose ring and may be discarded. Similarly, values of $\phi_{2'3'} > 60°$ may be ignored since this would result in severe strain in the molecule since they require unfavourable rotation about the C(2′)—C(3′) bond. By making various models and comparing the calculated and observed coupling constants (Tables 2.2 and 2.3) it is obvious that none of the conventional conformations satisfy the requirement that $\phi_{1'2'} = \phi_{3'4'} = 136°$, and $\phi_{2'3'} = 41°$. By postulating that the observed values of J_{ij} and the observed chemical shifts are really time averages of two or more conformations then it becomes possible to explain this. Several equilibria involving a 'conventional' form, and (say) an eclipsed intermediate state can be visualized, e.g. (1) a C(2′)-endo ↔ C(3′)-endo conversion requiring $\phi_{1'2'}$ to vary from 105 to 165° (i.e. the average value ≈135°) predicting values 0·4 to 9·5 Hz for the coupling constants and thus (see Table 2.3) values between 4·3 and 9·0 Hz for $J_{2'3'}$. (2) a C(2′)-exo ↔ C(3′)-exo conversion requiring $\phi_{1'2'}$ and $\phi_{3'4'}$ to vary from 100 to 145° and the corresponding values of J to vary from 0 to 6·7 Hz and between 5·0 and 9·0 Hz for $J_{2'3'}$. Any proposed equilibrium must be

TABLE 2.3

Measured dihedral angles (φ) and calculated coupling constants for various furanose ring conformations

Atom out of plane†	$\phi_{1'2'}$, (deg)	$J_{1'2'}$, (Hz)	$\phi_{2'3'}$, (deg)	$J_{2'3'}$, (Hz)	$\phi_{3'4'}$, (deg)	$J_{3'4'}$, (Hz)
C(2′)-endo	165	9·5	45	4·3	105	0·4
C(3′)-exo	145	6·7	40	5·0	100	0
H(2′), H(2′) eclipsed	120	2·3	0	9·0	120	2·3
C(3′)-endo	105	0·4	45	4·3	165	9·5
C(2′)-exo	100	0	40	5·0	145	6·7
Observed		5·0		5·0		5·2

† endo means the atom is located on the same side of the plane defined by C(1′)O(1′) and C(4′), as the C(4′)—C(5′) bond. exo means that it is found on the opposite side. (Fig. 2.16).

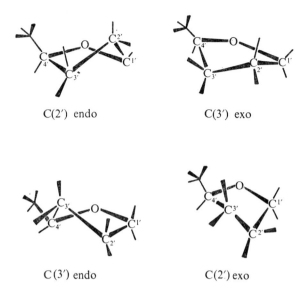

C(2') endo   C(3') exo

C(3') endo   C(2') exo

FIG. 2.16. Possible buckled conformations for the ribose ring of β-pseudouridine.

rapid on the n.m.r. time scale since in this case the entire spectrum of β-pseudouridine can be accounted for by a unique set of coupling constants and chemical shifts.

By addition of other suitable bases it is possible in principle to extend these kind of measurements to obtain conformations of the bases as they are stacked in tRNA. However it should be mentioned that any 'tolerances' or errors in the calculations of dihedral angles (and thus the conformations) obtained from coupling constants for each base would tend to become multiplicative as the number of bases increases. This could then lead to a calculated conformation that is totally different from that in reality. Although possible in principle such measurements must be undertaken with considerable caution.

2.17. Chemical exchange

2.17.1. General

In dealing with chemical exchange, the terms *slow* and *fast exchange* are frequently encountered in relation to nuclear relaxation times and/or chemical shifts. In this section we shall discuss chemical exchange between nuclei that can exist in two (or more) different environments with different chemical shifts.

We start by considering a ligand which exists in two different environments, A and M (for example a substrate *bound* to an enzyme and the substrate *free* in the bulk solution). If the rate of chemical exchange between these two environments is slower than their chemical shift difference, two separate resonances for the ligand nuclei are, in principle, observable (Fig. 2.17a). As

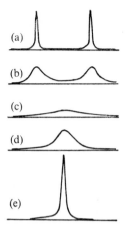

FIG. 2.17. Shapes of nuclear magnetic resonance peaks as a function of exchange rate.

chemical exchange of the ligand between the two environments becomes faster, there will be a distribution of nuclei in each environment, reflecting those that are characteristic of the environment, those which have just entered the environment (and still maintain the (phase) characteristics of their previous environment) and those that are intermediate in behaviour. This distribution is reflected in a broadening of the resonance signal (Fig. 2.17b). As the exchange becomes more rapid, the ligand nuclei will behave essentially in a manner that is characteristic of neither environment, (Fig. 2.17c and d) until the exchange rate is such that *all* the ligand nuclei behave in a manner which is an average of the characteristics of both environments (Fig. 2.17e) (the so called 'fast exchange' region). Since all the nuclei now behave in a like manner, a single sharp resonance is observed. In the fast-exchange region, the observed shift $\Delta\omega$ is given by

$$\Delta\omega = P_M q\, \Delta\omega_M + (1 - P_M q)\, \Delta\omega_A, \qquad (2.27)$$

where $\Delta\omega_M$ is the chemical shift between the two environments A and M, and $\Delta\omega_A$ the chemical shift of the A environment, measured with respect to some suitable standard. $P_M q$ is the fraction of the ligand in the M environment P_M and q are the mole fraction and number respectively of ligand nuclei in the bound site. If $P_M q \ll 1$ then:

$$\Delta\omega - \Delta\omega_A = P_M q\, \Delta\omega_M. \qquad (2.28)$$

If the shift is measured with respect to the pure A resonance then:

$$\Delta\omega = P_M q\, \Delta\omega_M \qquad (2.29)$$

and similar equations can be written in terms of the relaxation rates. However it is convenient to separate initially the effects of chemical shift and relaxation rates since the terms fast or slow exchange can have slightly different meanings in the two cases.

2.17.2. *The effects of chemical exchange on the chemical shift*

In many biochemical systems, however, the situation which we encounter most frequently is where the *concentration of ligand in one of the environments* (e.g. *bound* on an enzyme) *is very much smaller than in the other environment* (e.g. *free*, in the bulk solution). The relaxation times of the bound ligand also may be much shorter than those for the free ligand because the correlation times in the bound state are longer. In a case where both these conditions are met (i.e. very unequal populations and relaxation times) the chemical shift of the ligand in the *bulk* environment $\Delta\omega$ (measured with respect to its chemical shift in a solution in which there is only the A environment) is given by Swift and Connick (1962)

$$\Delta\omega = \frac{P_M q \, \Delta\omega_M}{\left(\dfrac{1}{T_{2M}}+\dfrac{1}{\tau_M}\right)^{2}+\tau_M^2 \, \Delta\omega_M^2} \tag{2.30}$$

where P_M is the fraction of *bound* ligand, $\Delta\omega_M$ the chemical shift between the two environments, τ_M is the life-time of a ligand nucleus, and T_{2M} is the spin–spin relaxation time in the bound environment. Thus the chemical shift depends on the exchange rate. An easy and convenient way of changing the exchange rate is to vary the temperature. The variation of $\Delta\omega$ with temperature is shown in Fig. 2.18. There is no chemical shift until τ_M becomes of the

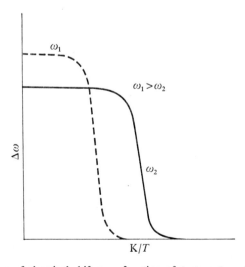

FIG. 2.18. Variation of chemical shift as a function of temperature (or exchange rate).

same order as $\Delta\omega_M$; then there is a sharp change in the chemical shift (the activation energy is approximately that for the rate of chemical exchange that is for $1/\tau_M$). As the resonances coalesce, the shift levels off in the fast-exchange region at the (weighted) average value for the two sites. Since we are primarily interested in the *bound* site, then of course the most profitable region in which to carry out experiments (if the bound site cannot be observed directly) is in the fast exchange region, for here the maximum effects of the bound site will be observed (as the weighted average).

Once the fast exchange region obtains, however, the equation becomes

$$\Delta\omega = P_M q \, \Delta\omega_M,$$

which is identical to eqn (2.29).

If the magnetic field is increased, then the chemical shift $\Delta\omega_M$ will increase and the fast-exchange condition $\tau_M \, \Delta\omega_M \ll 1$ will occur at a higher temperature. (Dotted curve in Fig. 2.18.) Incidentally in Fig. 2.18, the chemical shift in the fast exchange region is shown as being temperature independent. In many cases, however, $\Delta\omega_M$ does depend on temperature as, for example, when the *bound* environment is a paramagnetic site (see Chapter 3).

TABLE 2.4

Exchange-rate region	Definition	Spectrum
Slow	$\tau_M \, \Delta\omega_M \gg 1$	Two resonances
Intermediate	$\tau_M \, \Delta\omega_M \approx 1$	Broad single resonance
Fast	$\tau_M \, \Delta\omega_M \ll 1$	Single narrow resonance, the shift of which is the weighted average of the two environments

The above discussion now allows us to define slow intermediate, or fast exchange in terms of the chemical shift difference between the two sites as shown in Table 2.4.

2.17.3. *The effects of chemical exchange on the relaxation rates*

As in the previous section, we consider initially that the nuclei may exist in either the bulk solvent or bound to the macromolecule. The sites are denoted by A and M. We saw qualitatively that the line-widths (i.e. $1/T_2$) can alter as a consequence of chemical exchange between two sites. In this section we attempt a brief quantitative explanation. Again we shall only consider the case when a small fraction of the nuclei are bound. We start by discussing each site in turn.

2.17.3.1. *Unbound site, no exchange.*
The relaxation rates of the unbound nuclei are $1/T_{1,A}$ and $1/T_{2,A}$. Each of these consists of two contributions to the

relaxation caused by inter- and intra-molecular interactions. To derive information on molecular motion from measured relaxation rates, it is necessary to separate these contributions. By considering various models for the molecular motion, it is possible to obtain explicit equations for the relaxation rates. However we are not concerned with such treatment here, we merely note that in principle it can be carried out in a fairly straightforward manner when the molecular motion is well defined.

2.17.3.2. *Bound site.* The relaxation times in the bound site are denoted $T_{1,M}$ and $T_{2,M}$. In general, we are considering the case when the population of the bound site is small. Thus intermolecular interactions between like molecules are absent, although the motions of the solvent molecules can still cause relaxation if their nuclei possess a magnetic moment. If the bound site is on a macromolecule, there are many different magnetic nuclei in a variety of environments. The motions of these nuclei can cause relaxation of the ligand nuclei, and this may often dominate any intramolecular contribution to the relaxation rate. An added complication is that the motions of the nuclei on the macromolecule may be anisotropic and hence analysis is difficult and not unambiguous.

2.17.3.3. *Unbound site with chemical exchange.* In the presence of chemical exchange between the two sites, the relaxation rates in each site may be significantly changed. We shall only consider the bulk site, since this is easier to measure experimentally because of its large concentration.

If the observed relaxation rates are $1/T_{1,obs}$ and $1/T_{2,obs}$ then: (Luz and Meiboom, 1964, Swift and Connick, 1962):

$$\frac{1}{T_{1,obs}} = \frac{1}{T_{1,A}} + \frac{P_M q}{T_{1,M} + \tau_M} \tag{2.31}$$

and

$$\frac{1}{T_{2,obs}} = \frac{1}{T_{2,A}} + \frac{P_M q}{\tau_M} \left[\frac{\frac{1}{T_{2,M}}\left(\frac{1}{T_{2,M}} + \frac{1}{\tau_M}\right) + \Delta\omega_M^2}{\left(\frac{1}{T_{2,M}} + \frac{1}{\tau_M}\right)^2 + \Delta\omega_M^2} \right], \tag{2.32}$$

where $P_M q$ is the fraction of nuclei in the bound site. P_M and q are (of course) the mole fraction and number, respectively, of ligand nuclei bound to the macromolecule. $1/\tau_M$ is the rate of chemical exchange between the two sites and $\Delta\omega_M$ the chemical shift between them.

2.17.4. *The dependence of the observed relaxation rates on chemical exchange*

For convenience, we shall consider the case where the exchange rate is altered by changing the temperature. Similar considerations would apply if the exchange rate is altered by, for example, a change in pH. The theoretical temperature dependence of eqns (2.31) and (2.32), is illustrated in Fig. 2.19. Three distinct regions of the graph are apparent. Some physical insight into these regions may be gained by considering *three limiting forms* of eqns (2.6) and (2.7). Initially we shall analyse the behaviour of the spin–spin relaxation rates.

Region I, is the 'slow exchange' region. Here $\Delta\omega_M^2 \gg 1/T_{2,M}^2$, $1/\tau_M^2$ and eqn (2.32) would, in the limiting case, simplify to

$$\frac{1}{T_{2,obs}} - \frac{1}{T_{2,A}} = \frac{P_M q}{\tau_M} \tag{2.33}$$

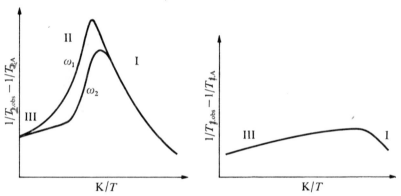

FIG. 2.19. Illustration of the temperature dependence of the relaxation rates for the case when only a small fraction of nuclei are bound. I, II, and III are regions of 'slow', 'intermediate', and 'fast' exchange, respectively; ω_1 and ω_2 represent two different n.m.r. frequencies ($\omega_1 > \omega_2$). (Strictly speaking the relaxation rates in region III are also dependent on the value of ω.)

that is, the relaxation rate is governed by the rate of chemical exchange between the bulk and bound sites. Alternatively, if for some reason the relaxation rate in the bound site is very rapid, i.e. $1/T_{2M}^2 \gg 1/\tau_M^2$ and $\Delta\omega_M^2$ the same equation obtains. The temperature dependence of τ_M is given by the Eyring relationship (eqn (2.34))

$$\frac{1}{\tau_M} = \frac{kT}{h} \exp\left(-\frac{\Delta H^{\ddagger}}{RT} + \frac{\Delta S^{\ddagger}}{R}\right) \tag{2.34}$$

where k in the Boltzmann constant, T the temperature (K) and $\Delta H^{\ddagger}$ and $\Delta S^{\ddagger}$ are the enthalpy and entropy of activation, respectively, for the first-order reaction of chemical exchange. Thus, by measuring the relaxation rates as a function of temperature in Region I, kinetic information can be obtained. In practice $1/T_{2,obs}$ is often measured from the line width and so, if conditions of *slow exchange* obtain, *increasing the temperature leads to an increase in the measured line width.*

In Region II, the exchange rate is increased and we notice that the line width of a resonance would now decrease with increasing temperature. In this region $1/\tau_M^2 \gg \Delta\omega_M^2 \gg 1/T_{2,M}\tau_M$ when the following equation is obtained

$$\frac{1}{T_{2,\text{obs}}} = \frac{1}{T_{2,A}} + P_M q \tau_M \Delta\omega_M^2 \qquad (2.35)$$

$1/T_{2,\text{obs}}$ in Region II depends on the chemical shift between the two sites $\Delta\omega_M$, which itself is dependent upon the (measuring frequency or) magnetic field. So measurements at different magnetic-field strengths will give different values of $1/T_{2,\text{obs}}$ as indicated in Fig. 2.19, for two field strengths (frequencies). (When eqn (2.35) is valid, the ratio of $(1/T_{2,\text{obs}} - 1/T_{2,A})$ at two different magnetic fields will be equal to the ratios of the square of the magnetic fields.)

Region III is that of 'fast exchange' where now the observed relaxation rates are essentially a weighted average of those of the two sites, i.e. $1/(T_{2,M}\tau_M) \gg 1/T_{2,M}^2, \Delta\omega_M^2$ leading to

$$\frac{1}{T_{2,\text{obs}}} - \frac{1}{T_{2,A}} = \frac{P_M q}{T_{2,M}}. \qquad (2.36)$$

In this region, in principle, information concerning molecular motion of the *bound* site can thus be obtained, if explicit equations can be written for $T_{2,M}$.

In Region III, the relaxation rate usually decreases with increasing temperature as shown in Fig. 2.19. In general, the relaxation rate will also decrease with increasing magnetic field, but we shall not consider this complication here.

Turning now to the behaviour of $1/T_{1,\text{obs}}$ in Region I where there is 'slow-exchange' behaviour, i.e. $1/T_{1,M} \gg 1/\tau_M$ eqn (2.31) becomes

$$\frac{1}{T_{1,\text{obs}}} - \frac{1}{T_{1,A}} = \frac{P_M q}{\tau_M} \qquad (2.37)$$

i.e. similar to that for $1/T_{2,\text{obs}}$. The plot of $1/T_{1,\text{obs}}$ versus temperature initially has the same form as that for $1/T_{2,\text{obs}}$. When the reverse condition of the above holds, i.e. $1/\tau_M \gg 1/T_{1,M}$ the relevant equation becomes:

$$\frac{1}{T_{1,\text{obs}}} - \frac{1}{T_{1,A}} = \frac{P_M q}{T_{1,M}}. \qquad (2.38)$$

Note that there is no region for $1/T_{1,\text{obs}}$ corresponding to Region II for $1/T_{2,\text{obs}}$, that is, $1/T_1$ does not depend on the chemical shift between the two sites. The difference in behaviour between the relaxation rates when chemical exchange is important may be a decisive factor in the interpretation if only a limited range of the temperature dependence of the relaxation rates is experimentally observable.

In Region III, $1/T_{1,M}$ and $1/T_{2,M}$ may or may not be equal. The explicit equations for $1/T_{1,M}$ and $1/T_{2,M}$ in general, are dealt with briefly in Section 2.17.5.

The definition of slow and fast exchange as applied to $1/T_2$ (the line-width) is the same as that already mentioned under the section on chemical shifts above, i.e. $\tau_M \Delta\omega_M \gg 1$ for slow exchange etc. For $1/T_1$, however, the explicit equation, relating to the observed behaviour in the presence of chemical exchange, does not depend on $\Delta\omega_M$. In this instance *slow exchange is defined as* $\tau_M \gg T_{1,M}$ while fast exchange is simply the converse (see eqns (2.37) and (2.38)).

If we now consider the case when there is *no* chemical shift between the two sites, the general equation for $1/T_{2,obs}$ becomes

$$\frac{1}{T_{2,obs}} = \frac{1}{T_{2,A}} + \frac{P_M q}{(T_{2,M} + \tau_M)}, \tag{2.39}$$

which is analogous to that for $1/T_{1,obs}$. The definition of slow and fast exchange is then also analogous, i.e. $\tau_M \gg T_{2,M}$ for slow exchange.

The above equations relating to chemical exchange are only applicable when the *bulk* resonances are being observed and then only when a small fraction of the nuclei are bound. (In practice it turns out that the equations are reasonably accurate if $P_M q < 0.3$, i.e. 30 percent or less of the nuclei are bound to the macromolecule.) The simplest case is when the bulk site consists of only one equivalent set of nuclei. If the bulk site contains many different nuclei then the equations must be applied to each different nucleus. Conditions of fast or slow exchange then refer to *each* nucleus. It is not difficult to envisage a situation in which the relaxation rates and shifts of one set of nuclei are such that the fast-exchange condition obtains (i.e. $\tau_M \Delta\omega_{M,1} \ll 1$) while those for a second set of nuclei in the *same molecule* obey the slow-exchange condition ($\tau_M \Delta\omega_{M,2} \gg 1$). Alternatively, the first set of nuclei may have a chemical shift between the two environments, i.e. $\Delta\omega_{M,1}$, while the second set may have no chemical shift. In the latter case fast or slow exchange is then defined with respect to the relaxation times. In many systems, where the high-resolution spectrum of a ligand is observed, on binding to a macromolecule, differential 'broadenings' (i.e. $1/T_{2,obs} - 1/T_{2,A}$) of the nuclear resonances are often interpreted in terms of different mobilities of the nuclei on the macromolecule (i.e. Region III). Such an interpretation assumes fast exchange for each nucleus, despite the possibility of a variety of exchange conditions. Where possible other measurements should be performed to help confirm the interpretation and, if possible, measurements should be carried out at different field strengths, and at different temperatures.

Measurements of $1/T_{1,\text{obs}}$ will also help to choose between various possibilities.

The above treatment only considers chemical exchange between two sites. If there are more than two sites or environments for a particular nucleus, then the treatment of exchange phenomena becomes rather more complex. Each site may have a distinctive population, chemical shift and/or relaxation time. In such cases exchange rates are usually extracted by the use of a computer programme which can simulate magnetic resonance line shapes and which take into account the different parameters for all the sites.

2.17.5. The explicit form of $\Delta\omega_M$, $T_{1,M}$, and $T_{2,M}$

A complete analysis of a system exhibiting chemical exchange behaviour as discussed in this chapter, necessitates explicit equations for $\Delta\omega_M$, $T_{1,M}$, and $T_{2,M}$. These parameters depend on the nature of the system and in this book we shall encounter three main types, which may be classified as *dipolar*, *paramagnetic*, and *quadrupolar*. We can illustrate these as follows. Let us consider a ligand exchanging between a bulk site and a site bound to an enzyme. In the first example the relaxation rates in the bound site result from dipole–dipole interactions between nuclei either on the ligand itself and/or on the enzyme. In general, however, the increase in the correlation time characterizing the dipolar interaction in the bound site, is the most important factor (see Chapter 6). The presence of a paramagnetic centre results in very efficient relaxation and because of the importance of paramagnetic probes, this case is dealt with in some detail in Chapters 9 and 10. The final example involves chemical exchange of nuclei possessing quadrupole moments. Here the change in correlation time and electric field gradients around such nuclei when bound to an enzyme can result in very rapid relaxation in the bound 'site'. This example is discussed in Chapter 13.

The explicit equations for $1/T_{1,M}$ and $1/T_{2,M}$ for the diamagnetic and quadrupolar cases are generally obtained by adding the subscript M to the usual equations for relaxation. Thus for instance in the diamagnetic case, we simply use the same equations as quoted in this chapter (but note of course that the values of r and τ_c may be very different in the two sites). The situation is somewhat more complex for relaxation resulting from the presence of a paramagnetic centre and the explicit forms of $1/T_{1,M}$ and $1/T_{2,M}$ are given in Chapter 9.

The mechanism responsible for the chemical shift difference between the two sites, $\Delta\omega_M$, is not always known. Some important interactions resulting in chemical shifts are discussed in Chapter 3, but for dipolar and quadrupolar systems, they involve parameters that can not easily, in general, be determined. The origin of chemical shifts (hyperfine shifts) resulting from the presence of a paramagnetic centre is more easily quantified. The explicit equations are given in Chapter 3.

Note on nomenclature. Throughout this book we shall use the subscript M to denote the bound site. The units of $\Delta\omega_M$ are rad s^{-1}. Often books quote these shifts in Hz and there is the obvious relationship

$$2\pi \cdot \Delta\nu_M \text{ (Hz)} = \Delta\omega_M \text{ (rad s}^{-1}\text{)}.$$

The symbol $\Delta\omega$ (or $\Delta\nu$) represents the observed shift of the bulk resonance but it is possible that some confusion could arise since $\Delta\nu$ is also used to denote line-width. The line-width of a resonance is related to the spin-spin relaxation rate by the general equation

$$\frac{1}{T_2} = \pi \, \Delta\nu$$

where $\Delta\nu$ is the line-width (in Hz) at half height of the resonance signal. If one is referring to the bound site the suffix M is then added and although there is the possibility of some ambiguity in symbols, the meaning is generally obvious from the text.

References

HRUSKA, F. E., GREY, A. A. and SMITH, I. C. P. (1970), *J. Am. chem. Soc.*, **92**, 4088.
KARPLUS, M. (1963), *J. Am. chem. Soc.* **85**, 2870.
LUZ, Z. and MEIBOOM, S. (1964). *J. chem. Phys.* **40**, 2686.
POPLE, J. A., SCHNEIDER, W. G. and BERNSTEIN, H. J. (1959). *High resolution nuclear magnetic resonance.* McGraw-Hill, New York.
SWIFT, T. J. and CONNICK, R. E. (1962). *J. chem. Phys.* **37**, 307.

Bibliography

As we indicated in the preface, there are many comprehensive text books and articles on the topics covered in this chapter. Below we list a few of these. Without doubt, Abragam's book must be the standard work on the subject. However it is rather mathematical (as is Slichter's book) and the biochemist is advised to consult some of the others for a simpler approach.

BECKER, E. D. (1969). *High resolution n.m.r.* Academic Press, New York.
BOVEY, F. A. (1969). *Nuclear magnetic resonance spectroscopy.* Academic Press, New York.
BOVEY, F. A. (1972). *High resolution n.m.r. of macromolecules.* Academic Press, New York and London.
CARRINGTON, A. and McLACHLAN, A. D. (1967). *Introduction to magnetic resonance.* Harper International Edition, Harper and Row, (New York and London).
EMSLEY, J. W., FEENEY, J. and SUTCLIFFE, L. H. (1965). *High resolution nuclear magnetic resonance spectroscopy.* Pergamon Press.
FARRAR, T. C. and BECKER, E. D. (1971). *Pulse and Fourier Transform n.m.r.* Academic Press, New York and London.
JACKMAN, L. M. and STERNHELL, S. (1969). *Applications of nuclear magnetic Resonance Spectroscopy in Organic Chemistry.* Academic Press, New York.
POPLE, J. A., SCHNEIDER, W. G. and BERNSTEIN, H. J. (1959). *High resolution nuclear magnetic resonance.* McGraw-Hill Book Co., New York, Toronto, and London.

RICHARDS, R. E. (1961). *Advances in Spectroscopy* Vol. II—(ed. H. W. Thompson) Interscience, New York and London.

ROBERTS, J. D. (1961). *An Introduction to the analysis of spin–spin splitting in high resolution nuclear magnetic resonance spectra.* Benjamin, New York.

More mathematical texts

ABRAGAM A. (1961). *The principles of nuclear magnetism.* Clarendon Press, Oxford.

SLICHTER, C. P. (1963). *Principles of magnetic resonance.* Harper and Row, New York.

3

SOME IMPORTANT INTERACTIONS LEADING TO RESONANCE SHIFTS IN HIGH RESOLUTION SPECTRA OF PROTEINS

3.1. Introduction

IN an earlier section, we saw that the chemical shift was a consequence of different screening constants for nuclei in chemically different environments. In principle it is possible to predict theoretically the observed chemical shifts, but in practice, in all but the simplest systems, there are difficulties. In proteins, therefore, because of the large number of possible contributions to the screening constant from neighbouring atoms the task of predicting the chemical shifts of individual nuclei would be complex. Then too, this can only be done when the crystal structure is known. Many perturbations can lead to changes in the screening constant and therefore in the chemical shifts. For example, if an enzyme is activated by Ca^{2+}, then addition of Ca^{2+} to a solution of the enzyme could result in changes in its proton high-resolution spectrum. At best, however, such changes can be interpreted qualitatively.

Shifts resulting from hyperfine interactions due to the presence of a paramagnetic ion or shifts resulting from 'ring currents' present in aromatic systems can, to a large extent, be interpreted quantitatively. In many enzymes these effects are particularly important and further it is possible to introduce specific 'probes' which may cause shifts as a result of these effects. In this chapter we consider these effects with appropriate examples.

3.2. Ring-current shifts

A simple physical explanation for ring current shifts is based on the fact that, when an aromatic ring is placed in a magnetic field, a current is set up which arises from the circulation of the delocalized π-electrons. This current, termed a ring current, produces a local magnetic field which opposes the externally applied field in the areas above and below the plane of the aromatic ring, but reinforces it otherwise (Fig. 3.1 and 3.2). At least qualitatively one can easily see that nuclei either near to or located inside the aromatic ring, on or near to the sixfold axis normal to the ring plane, are partially shielded from the applied field. Therefore *higher* external fields are required to achieve the resonance condition. On the other hand nuclei outside the ring which are in, or nearly in, the plane of the aromatic ring have their resonances shifted to *low* field by ring currents.

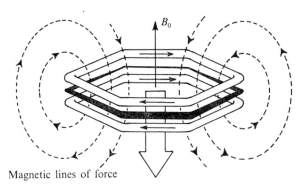

FIG. 3.1. Current and magnetic lines of force induced in benzene by an external field.

Quantitatively, however, it is not easy to express the values of the perturbations on the chemical shifts of nuclei next to an aromatic region. Several treatments have been tried; for example, Johnson and Bovey (1958) have *calculated* the chemical shift of a hydrogen, placed at any point in space, which arises from the influence of a planar benzene ring by assuming a current ring situated above and below the plane of the ring. This model correctly predicts a positive shielding inside the ring and, of course, a negative shielding outside it. The calculations show that these ring currents may be quite large and extend over considerable distances. The model is useful for semi-quantitative estimations of the shielding effect of the benzene molecule and a contour diagram of the ring-current shifts (in p.p.m.) is shown in Fig. 3.3. We see from Fig. 3.3 that the resonance of an 'in-plane'

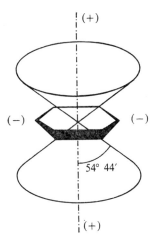

FIG. 3.2. Benzene ring and shielding envelope. The cone separates the shielding (+) and the deshielding (−) regions.

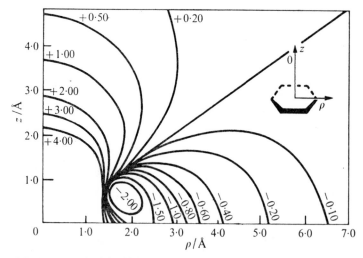

FIG. 3.3. Contour plot of shift in the n.m.r. shielding values which will be experienced by hydrogens as a result of the ring currents associated with the benzene ring. The z-direction is along the hexagonal axis of the benzene ring while ρ is the direction in the plane of the carbon atoms (From Johnson and Bovey, 1958).

hydrogen is still 0·1 p.p.m. to low field even at 6·5 Å, while a proton on the sixfold (z) axis is shifted 0·5 p.p.m. to high field at 4·8 Å. A methyl group has a minimum approach to a benzene ring of ca. 3 Å along the sixfold axis and will experience a shift of ca. 2 p.p.m. to high field. The theory, though, becomes increasingly inaccurate as the nucleus approaches the benzene ring.

Another approach, which can be used in cases where the nuclei are sufficiently far from the perturbing groups, is to calculate the anisotropic contribution to the shielding constant σ'_{ani} of a neighbouring cylindrically-symmetrical unsaturated bond (e.g. —C≡C—).†

For an axially symmetric bond

$$\sigma'_{\text{ani}} = \frac{\chi_\| - \chi_\perp}{12\pi r^3}\,(1 - 3\cos^2\theta). \tag{6.1}$$

Otherwise:

$$\sigma''_{\text{ani}} = \frac{1}{12\pi r^3}\,((\chi'_\| - \chi'_\perp)(3\cos^2\theta'_1 - 1) + (\chi''_\| - \chi''_\perp)(3\cos^2\theta''_1 - 1)). \tag{6.2}$$

In these equations θ_i represents the angle between the principal axis and the distance vector r, and χ_i represents the anisotropic susceptibility in the different directions.

† It is often assumed that C=O and C=C too may be treated as having axial symmetry and though this is not strictly true, it is often a good approximation. For C=O and C=C $\chi_\| - \chi_\perp$ is *positive* resulting in deshielding along the bond direction. However for C≡C, $\chi_\| - \chi_\perp$ is negative.

The geometric form of the equations correspond to cones whose axes coincide with the direction of the principal symmetry axis. For the special case of axial symmetry, e.g. benzene, the cone is spherical with half angle 54°44′. (The same as the angle the dividing line makes with the x-axis in Fig. 3.3; this shows the equivalence of the two treatments.)

One important property of these shifts is that they are *intrinsically independent of the temperature*. This is often a very important means of distinguishing them from other types of shifts (see later).

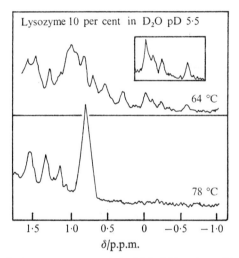

FIG. 3.4. 220 MHz hydrogen spectrum of the high-field region of lysozyme (64° native; 78° random coil form) (From McDonald and Phillips, 1967 © 1967, Amer. Chem. Soc.).

In aromatic amino acid residues, the sizes of the ring currents are not well studied and they have been assumed to be similar to those in benzene. There are, however, considerable difficulties in making these estimates quantitative and extension to other aromatic ring systems such as those of histidine and tryptophan must be viewed with considerable caution in the absence of suitable models to test the theory.

Lysozyme catalyses the break down of the polysaccharide components of the bacterial cell wall. It is widely distributed in many secretions and is particularly plentiful in egg whites. Its molecular weight is ca. 14 000. There are 129 amino acid residues in a single chain containing four disulphide bridges. Its X-ray crystal structure has been determined.

Figure 3.4 provides an illustration of the effects of ring currents for lysozyme. Lysozyme denatures reversibly at ca. 68 °C. In the random-coil form (78 °C) the methyl groups of the six valine, eight leucine, and six isoleucine residues resonances are all at −0·85 p.p.m. However, in the native

conformation (64 °C) the degeneracy is removed and methyl groups of apolar
residues of lysozyme give way to several peaks in the range −0·7 to 0·7 p.p.m.
From the known crystal structure of the enzyme it is possible to suggest that
in the native (folded) form some of these resonances are perturbed by the
proximity of aromatic rings. It must be stressed that assignment of the
various high-field resonances to specific interactions between hydrogen

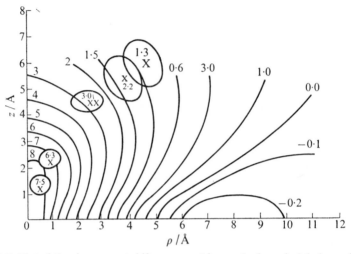

FIG. 3.5. Plot of the ring-current shifts measured in porphyrin and phthalocyanine complexes where the positions can be deduced from the chemical formulae. The estimated uncertainties in the positions are indicated by the ellipses and are larger than the errors obtained in deriving the ring-current shifts from measured positions. The co-ordinates z and ρ
refer to the distances perpendicular to the plane of the ring and outward in the plane of the
ring, respectively. The lines show the contours of equal ring-current shifts. (From Shulman,
Wütrich, Yamane, Patel, and Blumberg, 1970.)

nuclei and aromatic ring current is only possible since the crystal structure is
known. In this case, by treating heteronuclear aromatic rings as phenyl rings
and tryptophans as two fused phenyl rings each making an independent
contribution to the ring current shift, it was possible to predict the high-field
region of the spectrum. Since the calculated and observed values agree this
suggests that the conformation in solution is identical to that in the crystal.
(McDonald and Phillips, 1967; Sternlicht and Wilson, 1967.)

The ring-current shifts associated with porphyrin rings are, in general,
much larger than those of other aromatic ring systems. Figure 3.5 shows a
plot of the ring-current shifts measured in some model porphyrin and
phthalocyanine complexes where the position of hydrogens with respect to
the plane of the ring can be estimated from stereochemical considerations.
Using these results it is possible to modify (empirically) the expression which

Johnson and Bovey (1958) had used in their calculation of the ring-current shifts due to a benzene ring (Fig. 3.3). When this is done, the contour lines of equal ring-current shifts shown in Fig. 3.5 are obtained.

An important illustration of the ring-current shifts in porphyrin complexes is in the study of conformational changes of myoglobin upon oxygenation. In deoxymyoglobin, the iron is five co-ordinate; the sixth position is occupied upon oxygenation. The low-field n.m.r. spectrum of the two forms are shown in Fig. 3.6. There are many differences between the two forms and some of

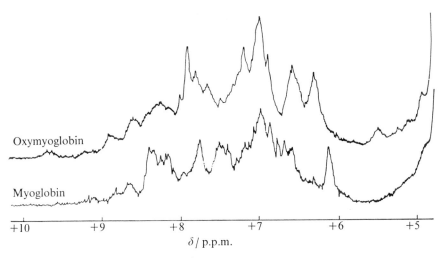

FIG. 3.6. A comparison of the aromatic regions of sperm-whale myoglobin and oxymyo-globin at 25 °C and 220 MHz (From Shulman *et al.*, 1970).

these might arise from the paramagnetism of myoglobin. The temperature independence of many of the resonance shifts implies however that the changes in chemical shifts result from a change in the conformation of the protein, and probably from changes due to ring-current shifts. For example, the resonance at $-6\cdot13$ p.p.m. in myoglobin corresponds to approximately three hydrogens and shows no temperature dependence. On the basis of the known amino acid sequence, the crystal structure, an empirical expression for ring-current shifts (as discussed above) and a suitable line-shape function, it is possible to predict the observed spectra. The resonance at $-6\cdot13$ p.p.m. has a component which is assigned to the *meta*-hydrogens of a particular phenylalanine (CD-1).† This resonance is shifted upfield by ca. 1 p.p.m. from the normal diamagnetic region (ca. $7\cdot3$ p.p.m.) largely because of ring

† In the myoglobin structure the residues are identified by letters and a number. A single letter refers to one of the helical residues A–H; a double letter to an irregular region between two helicies.

currents from the porphyrin ring. On oxygenation there is a further change of 0·2 p.p.m. (to ca. 6·37 p.p.m.). From inspection of Fig. 3.5, we note that changes perpendicular to the 1·0 p.p.m. line of *several tenths of an ångstrom* would change the ring current shifts by the required amount (ca. 0·2 p.p.m.).

Although at present it is difficult to make meaningful quantitative deductions from differences of this order since so many other factors could contribute to this shift, the example does illustrate the type of experiment that can be done.

We may summarize as follows:

(1) Measurable ring-current shifts may occur over quite large distances (ca. 6 Å) but no rigorous theories for these shifts have as yet been established. For these, we need to know not only the dependence on direction and distance of the perturbing fields but also the position of the perturbing group. In large rings this is not always well defined.

(2) In proteins, ring-currents arise not only from the aromatic rings of phenylalanine, tyrosine, tryptophan, and histidine residues but also from prosthetic groups such as porphyrins and flavins.

(3) Assignment of these interactions relies on a prior determination of the crystal structure the assumption that the structure of the molecule is relatively unchanged in solution.

(4) In the absence of environmental changes these interactions are temperature independent.

3.3. Hyperfine shifts resulting from the presence of a paramagnetic centre

The magnetic resonances of nuclei in paramagnetic complexes often show large shifts from their corresponding positions in diamagnetic complexes. Figure 3.7 shows the ^1H n.m.r. spectrum of cyanoferriporphin compared with that of the diamagnetic zinc porphin complex. We note first the effect of 'ring currents' of the porphin which give a low-field resonance of the hydrogen nuclei. However, in the paramagnetic complex, resonances a and b are shifted upfield, that of the meso protons (b) by ca. 9 p.p.m. and that for the hydrogen 1–8 (a), by ca. 24 p.p.m. These shifts result from interaction with the unpaired electron and in general may, be of two types.

(1) *Contact shifts* which result from the delocalization of unpaired electron spin density at the resonating nucleus. This electron density is usually transmitted through chemical bonds although other mechanisms such as hyperconjugation have been suggested in some cases.

(2) *Pseudo-contact or dipolar shifts* which result from the dipolar interaction between the electron spin and the nuclear spin. This shift is *only* observed if the magnetic field produced by the unpaired electron does not average to zero i.e. it is anisotropic.

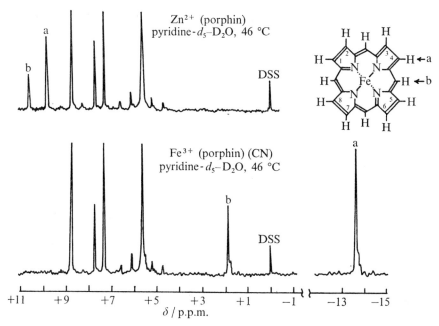

FIG. 3.7. ^{1}H n.m.r. spectra at 220 MHz of cyanoferriporphin and porphin zinc(II). The lines between +4.5 and +9·5 p.p.m. correspond to the solvent resonances and their spinning side-bands (From Wüthrich, 1970).

The presence of a paramagnetic ion in solution will generally increase the relaxation rate of the solvent and substrate nuclei, i.e. in the high-resolution spectrum the lines may become broader. Clearly for any hyperfine shifts to be easily detectable, any effects of the broadening must be less than those of the shift. In general paramagnetic ions which give rise to easily measurable shifts have short electron-spin relation times e.g. Co(II), Ni(II) low-spin Fe(III) and most of the lanthanide ion complexes. (We note in passing that if the electron spin relaxation time is short, the e.s.r. line will be very broad and not easily observable.)

3.3.1. *Contact shifts*

This is sometimes referred to as the Fermi or isotropic contact interaction. The interaction is generally transmitted through chemical bonds to the appropriate nucleus and the shift is given by (Bloembergen, 1957)

$$\frac{\Delta\omega_M}{\omega_I} = -\left(\frac{A}{\hbar}\right)\frac{g\beta S(S+1)}{3kT\gamma_I} \tag{3.3}$$

for the first-row transition metals and where $A/\hbar$ is the hyperfine coupling constant in rad s^{-1}, g is the Lande g factor, T is the absolute temperature,

γ_I is the magnetogyric ratio of the hydrogen nuclei in rad s^{-1} G^{-1}, β is the Bohr magneton, S is the total electron spin, $\Delta\omega_M$ is the shift from the corresponding position in the diamagnetic complex, and ω_I is the irradiating frequency.

A shift to high field, at fixed frequency, is considered a positive shift (i.e. $A/\hbar$) is negative. We note the dependence of the shift on $1/T$, in contrast to ring-current shifts which are temperature independent. Finally, we note that it is possible to relate the value of the hyperfine coupling constant to a particular nucleus to the spin density at, or near to, this nucleus. We now consider some examples of this.

3.3.2. Calculation of spin densities

It has been shown that the hyperfine coupling constants of *aromatic hydrogen nuclei* can be related to the spin density in the π-orbital of the ring carbon, ρ_c^π, next to the observed protons by (McConnell, 1956)

$$A/h = Q\rho_c^\pi \tag{3.4}$$

in which ρ_c^π is the π-electron density expressed as the fraction of one unpaired electron and $Q = -6.3 \times 10^7$ Hz (McConnell and Chesnut, 1958). For atoms or groups, other than hydrogen atoms, attached to an aromatic ring, the value of Q is different. For example, for a methyl group attached to aromatic ring compounds, $Q^{Me} = +7.5 \times 10^7$ Hz, but in general Q^{Me} has been found to differ for different compounds and it has to be obtained by an essentially empirical approach. For example, in low spin ferric haemes an approximate value of $Q^{Me} = 3.0 \times 10^7$ Hz was obtained from a comparison of the hyperfine shifts of the ring methyl hydrogens and the 2,4 hydrogens in cyanoferrideuterioporphyrin IX (Fig. 3.8). The shift, taken as the difference between the positions of corresponding resonances in the paramagnetic and diamagnet deuterioporphyrin–metal complexes, is assumed to be entirely contact in origin. The similarity of the shifts for the four ring methyls as well as the 2,4-hydrogens (f) was taken to indicate a symmetrical distribution of the spin density in the porphoryin ring. Thus it was assumed that the average of the spin densities on the ring carbon atoms next to the four methyl groups is approximately equal to the average of the spin densities next to the 2,4-protons. The calculated values of A/h for the 2,4-hydrogens were -0.99 and -0.95 MHz and those for the methyl protons 0.32, 0.37, 0.37, and 0.45. Using $Q = -6.3 \times 10^7$ Hz and the values of A/h in eqn (3.4) we obtain, $Q^{Me} \approx 3.0 \times 10^7$ Hz. Note that for a given spin density centred on a carbon atom, the shifts in the two cases cited here differ in both sign and magnitude. Theoretical calculations of spin density using molecular orbital methods can be performed quantitatively if a single mechanism operates, e.g. via the σ- or π-system or by a hyperconjugation mechanism. By a comparison with the observed values one may obtain information on the type of bonding.

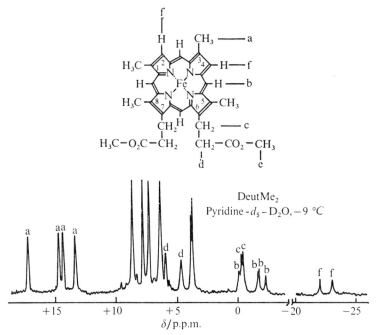

FIG. 3.8. ¹H N.m.r. spectrum at 220 MHz of cyanoferriduterioporphyrin IX dimethyl-ester (Deut Me₂). The letters refer to the resonance assignments of Fig. 3.7. The distinction between the resonances c and d is arbitrary (From Wüthrich, Wyluda, and Caughey, 1969).

Of course, some generalizations can be made, such as that the transfer of spin density through the π-system will occur over larger distances from the metal ion than transfer through the localized σ-system, the effects of which will rapidly be attenuated as the number of bonds between the metal and nucleus increases. Also, in the case of π-delocalization, there is an alternation of the sign of the shift.

In biochemical systems, the calculation of spin densities could be of use in, for example, tracing the path of the electron in the electron-transport chain of cytochromes. For instance, because of the distances involved it is extremely likely that electron transfer between Fe(II) and Fe(III) occurs via the intervening chemical bonds. Observation of contact shifts for the resonances of the intervening nuclei can, in principle, then lead to calculation of spin densities and hence the path taken by the electron. In the oxygen-carrying proteins, calculation of spin densities could well lead to the definition of any conformational changes associated with the reversible combination with oxygen.

3.3.3. *Some notes relating to the contact-shift equation*

Equation (3.3) is only valid for systems with isotropic g-values. If the g-value is anisotropic, then the equation has to be modified. The reader is referred to the papers by Jesson (1967), Kurland and McGarvey (1970), and Horrocks (1970) for further details. Here we need only note that certain paramagnetic ion complexes, such as those of octahedral Co(II), tetrahedral Ni(II), and octahedral low-spin Fe(III), may have several thermally-populated excited electronic states. The 'mixing' of these states with the ground state results in a temperature dependence much different from the T^{-1} predicted by eqn (3.3). One simple way of interpreting this, often used for the haeme proteins, is to assume that the hyperfine coupling constant itself is temperature dependent (Wüthrich, 1970).

Another example of a temperature dependence different from T^{-1} arises when there is an equilibrium involving two states of the paramagnetic ion. For example, consider the case where an equilibrium exists between two states A and B, each of which is paramagnetic. The equilibrium is

$$\underset{(1-\alpha)}{A} \rightleftharpoons \underset{\alpha}{B}$$

where $(1-\alpha)$ and α are the relative concentrations of A and B. The equilibrium constant is given by

$$K = \alpha/(1-\alpha). \tag{3.5}$$

If we assume that the rate of chemical exchange is such that only a single resonance is observed then, at a particular temperature T the observed mean hyperfine shift of a nucleus $\Delta\omega_T$, is given by

$$\Delta\omega_T = (1-\alpha)\,\Delta\omega_A + \alpha\,\Delta\omega_B. \tag{3.6}$$

where $\Delta\omega_A$ and $\Delta\omega_B$ are the hyperfine shifts of the two forms given by eqn (3.3). Combining eqns (3.5) and (3.6) and using the relationship: $\Delta G^\circ = -RT \ln K$ leads to:

$$\Delta\omega_T = \frac{\Delta\omega_B}{\exp(\Delta G^\circ/RT)+1} + \frac{\Delta\omega_A}{\exp(-\Delta G^\circ/RT)+1}. \tag{3.7}$$

An example in which both A and B represent paramagnetic forms would be in an equilibrium between a 'high-spin' and 'low-spin' species. In such a system there will in general be large deviations from a linear dependence of the contact shifts on $1/T$.

If A and B represent a system undergoing a diamagnetic–paramagnetic conversion, then the term in $\Delta\omega_A$ in eqn (3.7) is omitted. Such an equation has been used in the nickel(II) aminotroponeiminates systems (Eaton and Phillips, 1965) in which there is a rapid equilibrium between the square-planar diamagnetic and tetrahedral paramagnetic forms of the molecules. The tetrahedral form has a very short electron spin relaxation time which leads to the observation of sharp n.m.r. resonances.

3.4. Pseudo-contact shifts

The failure of the dipolar interaction between the unpaired electron spin and a given nucleus to average to zero gives dipolar or pseudo-contact shifts. For transition metals the magnitude of this shift is given by

$$\frac{\Delta\omega_M}{\omega_I} = \frac{\beta^2 S(S+1)}{6kT} \times r^{-3} F' \tag{3.8}$$

where $F' = (g_z^2 - g^2)(3\cos^2\theta - 1) + (g_x^2 - g_y^2)\sin^2\theta \cos 2\phi$ in which

$$g^2 = \tfrac{1}{3}(g_x^2 + g_y^2 + g_z^2).$$

θ and ϕ are defined from the relationships $\cos\theta = z/r$, $\sin\theta\cos\phi = x/r$, $\sin\theta\sin\phi = y/r$. For axial symmetry $g_z = g_\parallel$ and $g_x = g_\perp^2$ whence

$$F' = \tfrac{2}{3}(g_\parallel^2 - g_\perp^2)(3\cos^2\theta - 1).$$

In some cases there may be interactions between the various electronic energy levels arising from the presence of ligand-field interactions. In such a case there is an *extra* contribution to the induced shift which is given by (Bleaney, 1972; Kurland and McGarvey, 1970)

$$\frac{\Delta\omega_M}{\omega_I} = \beta^2 \frac{S(S+1)(2S-1)(2S+3)}{60(kT)^2} \times r^{-3} F'' \tag{3.9}$$

$F'' = \tfrac{2}{3} g_z^2 D_z (3\cos^2\theta - 1) + (g_x^2 D_x - g_y^2 D_y)(\sin^2\theta\cos 2\phi)$ and where D_x, D_y, and D_z represent the components of the ligand field. We note the presence of a term in $1/T^2$.

The assumptions in the above equation are that the ligand field splitting $\ll kT$ and that the tumbling time of the complex is less than the relaxation time of the spin of the electron causing this shift. Other cases also have been derived which depend on the relative magnitudes of the Zeeman anisotropy, the tumbling time of the complex τ_R and the electron spin relaxation time $\tau_{1,s}$ (see Kurland and McGarvey, 1970). However, Vega and Fiat (1972) have shown that the pseudo-contact shift is independent of the magnitude of τ_R and $\tau_{1,s}$.

The most frequently encountered form of eqn (3.8) is that for the case of axial symmetry. In solution, many possible conformations of a molecule exist and thus the values of the parameters θ and r will be the result of averaging of the motions of the molecule which are rapid on the n.m.r. time scale. For this reason, the equation for the case of axial symmetry is written as

$$\frac{\Delta\omega_M}{\omega_I} = D \left\langle \frac{3\cos^2\theta - 1}{r^3} \right\rangle_{av}. \tag{3.10}$$

3.5. Separation of contact and pseudo-contact shifts

One of the difficulties in assigning shifts to a pseudo-contact mechanism is that it is important to show that contact shifts are absent, or, if present, what contribution they make to the total shift. Unfortunately the temperature dependence of the shift does not necessarily yield any extra information since both contact and pseudo-contact shifts may have similar dependences. One way that has been used in a series of six-coordinate Ni(II) and Co(II) complexes is the 'ratio method'. (Horrocks, 1970.) Here it is assumed that the Ni(II) atom is in an octahedral environment and there is then no anisotropic dipolar interaction between the unpaired Ni(II) electrons and the ligands, and the pseudo-contact interaction is zero. For the Co(II) complex both kinds of shift are possible. With the assumption that the modes of delocalization of spin densities, as well as the geometric factors, are identical in each complex it is possible to estimate the pseudo contact contribution.

In dealing with larger molecules some comparative approaches have been used but few quantitative estimates have as yet been done. If spin densities

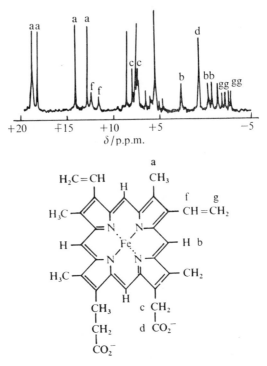

FIG. 3.9. ^{1}H N.m.r. spectrum at 220 MHz of cyanoferriprotoporphyrin IX (Proto CN). The distinction between resonances c and d is arbitrary. In the resonances of the 2,4-vinyl groups (f and g) the fine-structure from hydrogen–hydrogen spin–spin coupling is partially resolved. (From Wüthrich, 1970.)

are to be calculated, then it is important to be able to estimate the pseudo-contact contribution to the shift. For example, in the case of cyanoferrideuterioporphyrin 1X (Deut CN) and cyanoferriprotoporphyrin (Proto CN) (Figs. 3.8 and 3.9) the similarity of the resonance positions of the meso-hydrogens (b) and the ring methyl hydrogens (a) seem to indicate that the electronic structures are quite similar in the two molecules. However the resonances of the 2,4-vinyl hydrogens (f) in Deut CN and the α-hydrogens of the 2,4-vinyl groups in Proto CN are 35 p.p.m. apart and the shifts are of different sign. The positions of these hydrogen signals relative to the haeme iron is almost identical so that the dipolar shift would be almost identical. Thus, these shifts are assumed to arise from contact interactions with different values of the hyperfine coupling constant.

In a large molecule, there are a great number of nuclei not directly involved in the binding of a metal ion, which will be in close proximity to it. These nuclei should experience pseudo-contact shifts. If only a few resonances shifted from the normal diamagnetic range are observed, then it is likely that the shifts are contact ones, otherwise, as we have noted, a considerable number of the hydrogen atoms in the polypeptide chain would experience the pseudo-contact shifts giving rise to a large number of shifted resonances.

References

BLEANEY, B. (1972). *J. mag. Res.* **8**, 91.

BLOEMBERGEN, N. (1957). *J. chem. Phys.*, **27**, 595.

EATON, D. R. and PHILLIPS, W. D. (1965). In *Advances in Magnetic Resonance* (ed. J. S. Waugh) p. 103. Academic Press, New York and London.

HORROCKS, W. DE W. (1970). *Inorg. Chem.* **9**, 690.

JESSON, J. P. (1967). *J. chem. Phys.* **47**, 579.

JOHNSON, C. E. and BOVEY, F. A. (1958). *J. chem. Phys.* **29**, 1012.

KURLAND, R. J. and MCGARVEY, B. R. (1970). *J. mag. Res.* **2**, 286.

MCCONNELL, H. M. (1956). *J. chem. Phys.* **24**, 764.

——, and CHESNUT, D. B. (1958). *J. chem. Phys.* **28**, 107.

MCDONALD, C. C. and PHILLIPS, W. D. (1967). *J. Am. chem. Soc.* **85**, 3736.

SHULMAN, R. G., WÜTHRICH, K., YAMANE, T., PATEL, D. J. and BLUMBERG, W. E. (1970). *J. Mol. Biol.* **53**, 143.

STERNLICHT, H. and WILSON, D. (1967). *Biochemistry*, **6**, 143.

VEGA, A. J. and FIAT, D. (1972). *Pure appl. Chem.* **32**, 307.

WÜTHRICH, K. (1970). *Structure and Bonding* **8**, 53.

——, WYLUDA, B. J. and CAUGHEY, W. S. (1969). *Proc. natn. Acad. Sci. U.S.* **62**, 636.

4

EXTRINSIC SHIFT PROBES

4.1. Introduction

PROBES can be defined as small molecules or ions which can be attached to specific regions of another (usually larger) molecule and which, through their physical or chemical properties, are able to report on some features of the structure at, or close to, their site of binding. There are essentially two types of probes: detecting and perturbing. Of course some probes can act as both detecting and perturbing probes. Such is the case for spin-label probes and we shall return to these in Chapter 12. Perturbing shift probes causes shifts in the molecule being examined while a detecting shift probe is one in which the chemical shifts of the probe itself are monitored. In dealing with bio-chemical systems we shall often have recourse to use probe nuclei and in order to indicate the type of information that can be obtained from probes we now consider some examples of each category.

4.2. Detecting probes

There are a large variety of potential probes that can be used for magnetic resonance experiments ranging from ionic to covalent. (The use of covalent probes is of course, restricted to systems where chemical modifications are possible.) Ideally, the probe should be introduced into the system at strategic sites without changing the biological function or activity of the system. Clearly the choice of probe depends on the system and the magnitude of any effects the probe is sensitive to. For example Tl^+ is a good probe for K^+ as it resembles it chemically quite closely, has a higher n.m.r. sensitivity (easier to detect) and exhibits larger chemical shifts. (Additionally, because of the higher value of its magnetogyric ratio, Tl^+ will be more sensitive to dipolar interactions with paramagnetic ions, see p. 239.) Similarly, fluorine can be used as probe for hydrogen for it has almost the same n.m.r. sensitivity, it is not too different in size, and undergoes a far greater range of chemical shifts. Examples of the use of detecting probes will be found in the general text and in Chapters 7 and 8.

4.3. Perturbing probes

There are many kinds of perturbing probes which, for a variety of reasons, will displace the resonance of a nucleus from its 'normal' position. For example, the addition of a diamagnetic metal-ion probe to a molecule may result in a change in the electron distribution around a particular nucleus with a consequent chemical shift. Shifts, as we have seen, may also result when perturbations arise from ring-current effects or from hyperfine interactions (contact and pseudo-contact) with a paramagnetic metal ion. In general, not every nucleus is equally affected by these perturbations and the resulting shifts may lead to a considerable simplification of the spectrum.

The magnitude of the shift induced by any probe can be estimated quantitatively if the equations relating to the perturbation are known. This is so in only a few cases mainly when perturbations result from ring-current shifts or hyperfine interactions. We note, however, that in contrast to the equation for contact interactions the equations for ring-current or pseudo-contact interactions contain terms which depend on the distance and direction of the perturbed nucleus from the probe. In such cases, the position of the perturbed nuclei with respect to the probe can be obtained. However the 'position' of the probe or its perturbing field is not always well defined as, for example, the perturbing fields due to ring currents. Thus, although many probes may cause a simplification of the spectrum, only those paramagnetic probes causing pseudo-contact shifts will be of general use in obtaining structural information. Some of the more common paramagnetic probes which are used for this purpose, together with some systems that have been studied are listed in Table 4.1, and we note that there are both charged and uncharged probes to cater for the different systems.

TABLE 4.1

Paramagnetic shift probes

	Diagmagnetic control	Shift probe (Example)	System (Example)	Solution
Lanthanides aqueous ions pH < 6	La(III), Lu(III)	Eu(III), Ho(III), Nd(III), Dy(III)	(a) pH < 6 AMP, ApA, ApAp Lysozyme	Water DMSO
Lanthanides metal buffer† at pH < 12		Eu(III), Ho(III), Nd(III), Dy(III)	(b) pH < 12 thyroxine, peptides	Water
Transition metal cations (Acid pH)	Mg(II)	Ni(II), Co(II)	All anions	Water
Porphyrin complexes	Zn(II) Ni(II)	Co(II)	Metal porphyrin–sterol complexes	CHCl₃
[M(CN)₆]³⁻ complexes	Co(III)	Fe(III)	Positively charged molecules, e.g. acetylcholine	Water
Ln(AcAc)₃	La(III)	Eu(III)	Steroids etc. water in membranes	Organic solvents

† The metal buffer anion is oxydiacetate, or tiron.

6

Several interesting points arise from inspection of Table 4.1. Firstly, we note the many applications of the lanthanide ions and, further, that by making various complexes of them it is possible to extend their use considerably. For example, the dipyridine adduct of trisdipivalomethanato-europium(III)/(Eu(DPM)$_3$·2py) when added to a solution of cholesterol in

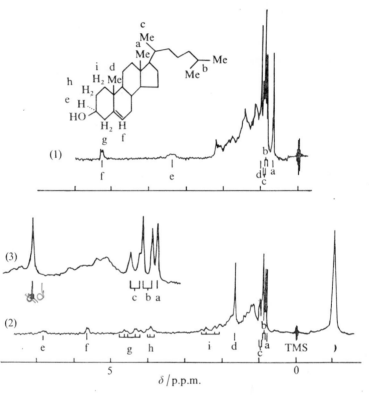

FIG. 4.1. 100 MHz spectra of cholesterol monohydrate (0·1 M in CCl$_4$). Spectrum (1): no Eu(III), spectrum (2): +0·05 M Eu(DPM)$_3$.2py. Spectrum (3) is an expansion of the methyl region of spectrum 2. (From Hinckley, 1969 © 1969 Amer. Chem. Soc.).

carbon tetrachloride gives considerable simplification of the spectrum (Fig. 4.1). The (pseudo-contact) shifts result from the binding of this complex to cholesterol and will, therefore, depend on the concentration of the metal complex in a manner that reflects the binding constant. Sometimes it is possible to combine two types of perturbations into one probe. For instance the Zn(II)–porphyrin probe forms loose molecular complexes with organic molecules; the ring currents from the porphyrin may give shifts in some of the resonances of the organic molecule. The substitution of Zn(II) by a paramagnetic metal-ion probe which causes hyperfine interactions will result in further chemical shifts, which may be easier to monitor as a result of any simplification caused by the ring currents. Since the hyperfine interactions in these loose complexes will be essentially pseudo-contact (contact shifts are transmitted through chemical bonds) structural information can be derived.

One final note on Table 4.1 is that, since geometric information can only be obtained if the shifts result from pseudo-contact interactions, it is important to show that contact interactions are absent. In the next section we shall see that, for the lanthanide probes, it is possible to carry out controls but this is not usually so for the first-row transition metals. With Co(II) in particular, there appears to be a significant contribution to the shift from contact interactions in many of the systems studied. Thus it would appear that the lanthanide probes are probably the most useful for obtaining structural information from chemical-shift data. We shall now consider them in more detail.

4.4. The lanthanide ions as shift probes

The magnetic properties of the lanthanides are well understood and, in general, the ground state is specified by a quantum number J, resulting from strong spin–orbit couplings. This is, of course, in marked contrast to the first row transition metal ions where spin–orbit coupling is relatively unimportant and the ground state is characterized by a spin quantum number S. (Readers unfamiliar with this terminology will find a useful account in Phillips and Williams, 1966.) In the lanthanides, the value of J ranges from $\frac{5}{2}$ to 8 and in a magnetic field each energy level is split into $2J+1$ levels. In a complex, these $2J+1$ levels are further split by the ligand field but the splitting is small ($\ll kT$ at room temperature), since the $4f$-electrons are well shielded from the ligand. The result is that the hyperfine levels arising from the ligand field are, to a first approximation, all equally populated and the environment of the ion is thus symmetric and g is isotropic. The contact shift $\Delta\omega_M$ is then given by

$$\frac{\Delta\omega_M}{\omega_I} = -(A/\hbar)\frac{\beta J(J+1)g}{3kT\gamma_I}. \tag{4.1}$$

To a first approximation, there are no pseudo-contact shifts since g is isotropic, i.e. the contribution to the shift described by equations similar to 3.8 will be zero. However, to a second approximation, pseudo-contact shifts are observed when the symmetry of the ligand field is less than cubic. The magnitude of this shift is (Bleaney, 1972)

$$\frac{\Delta\omega_M}{\omega_I} = \frac{g^2\beta^2 J(J+1)(2J-1)(2J+3)}{60kT^2}\times r^{-3}F \tag{4.2}$$

where $F = D_z(3\cos^2\theta-1)+(D_x-D_y)\sin^2\theta\cos 2\phi$ and D_x, D_y, and D_z are the components of the ligand field. θ and ϕ are defined as previously (p. 59).

In general, however, we note that the temperature dependence of the shift is proportional to $1/T^2$ as long as the above equation is valid. *For axial symmetry*, such as is often the case for the lanthanides, $D_x = D_y$, and the second term in F disappears.

For Eu^{3+} and Sm^{3+} the situation will be slightly more complex since there are terms from excited states with a different J value that can contribute to the shift.

TABLE 4.2

Relative n.m.r. shifts of various ligand nuclei bound to lanthanides at room temperature

Ln^{3+}	Theory (p.p.m.)	Expt.† (p.p.m.)
La	0	0
Ce	−6·3	
Pr	−11·0	−8·2
Nd	−4·2	−3·0
Pm	+2·0	
Sm	−0·7	−0·2
Eu	+3·5	+5·1
Gd	0	
Tb	−86	−50
Dy	−100	−100
Ho	−39	−42
Er	+33	+16
Tm	+53	(+10)
Yb	+22	+12

† Scale to −100 for Dy^{3+}. A minus sign indicates a downfield shift. (From Bleaney, Dobson, Levine, Martin, Williams, and Xavier, 1972.)

Table 4.2 shows the degree of agreement between the experimental and the theoretical values for the shifts of the protons in the lanthanide–cytidine monophosphate complexes which are given relative to those in the dysprosium complex. The theoretical calculation, using equation 4.2, assumes axial symmetry and that the value of D_z is constant throughout the series.

Of course, in this example, it is important to show that the contribution of the contact term to the hyperfine shift is negligible. This can be demonstrated by measuring the ratio of the shifts of the same two nuclei for the different lanthanide ions. Since the lanthanide ions are chemically very similar they are assumed to bind identically in each case. If contact shifts are important, then this ratio will be expected to change for the different lanthanide ions since they have differing numbers of unpaired electrons and the contact shift would be expected to reflect this. If contact shifts are unimportant, this ratio should be constant.

The lanthanides have many advantages as probes because of their chemical properties. For example, isomorphous substitution makes it possible to choose probes that shift resonances either upfield or downfield. Their chemistry, and particularly that of Eu^{2+}, strongly resembles that of Ca^{2+} so that they may be useful in studying the structural role of Ca^{2+} in biochemical systems. Also, the chemistry of the lanthanides is similar to several of the actinides, particularly Uranium(II), which is often used as a heavy-atom probe in X-ray crystallography. (Incidentally, as the lanthanides are heavy atoms they can also be used in X-ray crystallography.) Often the position of UO_2^{2+} in a molecule is known from X-ray work and it is then possible to 'place' the lanthanide probes since these would be expected to bind in much the same position.

4.5. An example of the use of lanthanide ion probes to obtain the conformation of the Ln(III)–cyclic AMP molecule

4.5.1. General

The method essentially involves using n.m.r. pseudo-contact-shift data as an elegant 'filter' of the many possible conformations obtained using a computer programme (Barry et al., 1971). The programme calculates all the possible conformations of a molecule or complex by allowing free rotation about every bond,† but using the constraints imposed by Van der Waal's contacts.

The magnitude of the pseudo-contact shifts of a nucleus, i, depends on its distance r_i from the metal and the angle θ_i between the vector r_i, and the principal symmetry axis of the complex averaged over motions which are rapid on the n.m.r. time scale. For the lanthanides, assuming axial symmetry, we have seen that it is sufficient to write

$$\frac{(\Delta\omega_M)_i}{\omega_I} = D\left\langle\frac{3\cos^2\theta_i - 1}{r_i^3}\right\rangle_{av}. \tag{4.3}$$

In general, the value of D is unknown and it is more convenient to measure ratios of the shifts of two nuclei in the same molecule, i.e.

$$R_i = \frac{(\Delta\omega_M)_1}{(\Delta\omega_M)_2} = \left\langle\frac{(3\cos^2\theta_1 - 1)}{r_1^3}\right\rangle_{av}\bigg/\left\langle\frac{3\cos^2\theta_2 - 1}{r_2^3}\right\rangle_{av}. \tag{4.4}$$

To proceed further however the principal symmetry axis must be known but this is not usually the case.

By 'placing' the metal ion on the complex, and choosing a symmetry axis, it is possible to compute the ratios of the observed shifts for each conformation. This can be done for any number of metal co-ordination sites and principal symmetry axes. The solution is obtained when the predicted and experimentally observed 'ratios' agree. In practice, several 'families' of

† If there are four rotatable bonds and each one is allowed to rotate in a steps of 4°, the number of possible conformations is ~64 000 000!

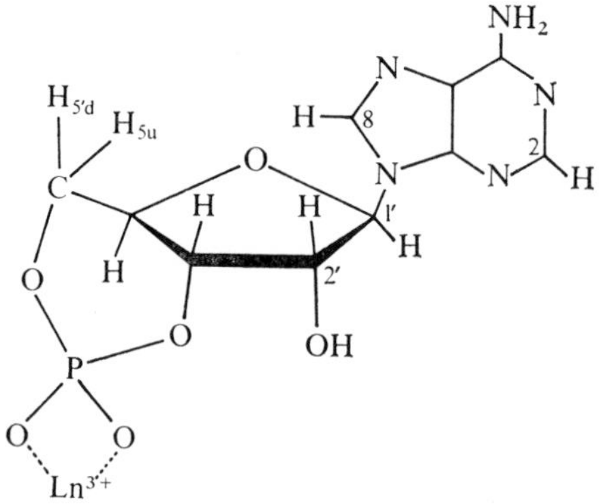

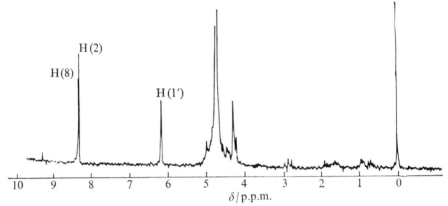

FIG. 4.2. Ln(III)–Cyclic AMP.

FIG. 4.3. 'Unperturbed' spectrum of cyclic AMP (25 mM) at pH = 2·0 and 25 °C. (From Barry, Martin, Williams and Xavier, 1973.)

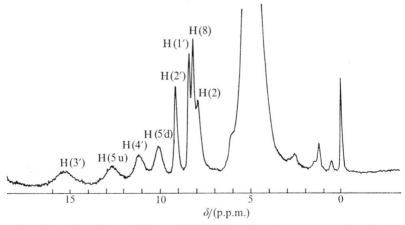

FIG. 4.4. The effect of the addition of 0·04 M Dy(III) on the spectrum of cyclic AMP (see Fig. 4.2). (From Barry et al., 1973.)

possible conformations are obtained, and it is then necessary to use some extra measurable parameter, such as the ratio of relaxation times of the two nuclei, to limit the number of possibilities.

In the lanthanides, we have concentrated on those metal ions which have very short electron relaxation times and cause little broadening (or change in relaxation times) of the observed resonances. However we shall see later (Chapter 9) that several of the lanthanides, notably Gd^{3+}, have relatively long electron relaxation times and their interaction (even at very low concentration) with nuclei results in a marked change in the relaxation times of the nucleus under observation. These relaxation times can provide a second filter since the ratio of the relaxation rates R'_i is given by (see Chapter 9)

$$R'_i = \langle r_0^6 \rangle / \langle r_i^6 \rangle. \tag{4.5}$$

Several points are worth noting:

(1) If the position of the metal is known, the number of possible solutions is dramatically reduced. Sometimes simple tests can be carried out to 'place' the metal. For example, in the binding of lanthanide(III) to AMP, the metal is assumed to bind to the phosphate since in control experiments pseudo-contact shifts in the hydrogen resonances were obtained for ribose monophosphate but not adenosine. (Barry, North, Glasel, Williams, and Xavier, 1971).

(2) The more ratios that are measured the more single valued will be the solution. Thus in the lanthanide(III)–AMP complexes, shifts of the ^1H, the ^{31}P, and ^{13}C nuclei can be observed.

(3) The use of a different probe as an extra filter implies that this probe binds in an identical manner to the first probe and the case is thus restricted to elements where isomorphous replacement is possible.

4.5.2. Method

The method involves measurements of chemical shifts and line widths (or relaxation times) as a function of the added lanthanide ions. The experiments are usually run as titrations using a constant concentration of the molecule under study and adding increasing amounts of the probes. Figures 4.3 and 4.4 show the simplification of the spectrum of cyclic AMP (see Fig. 4.2) produced by the addition of Dy(III). The titration curve which reflects the binding of Dy(III) to cyclic AMP is shown in Fig. 4.5 for various hydrogen nuclei. At each point on the titration curve the shift ratios R_i for each hydrogen to some given hydrogen shift of the molecule are calculated (in this case $H(5u)$). These ratios are then extrapolated to zero lanthanide(III) concentration (Fig. 4.6). This allows for any effects caused by variations in ionic strength or from weaker (secondary) binding sites, for the extrapolation effectively gives the value of the ratios at infinite dilution of the metal ion. In fact, Fig. 4.7 shows that if the

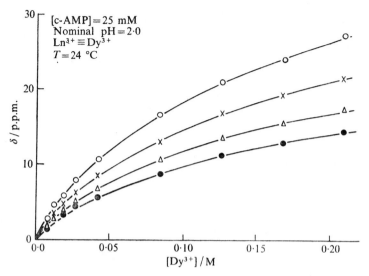

FIG. 4.5. Shifts for Dy(III)–cyclic AMP corrected for the diamagnetic (La(III)) shift. (From Barry *et al.*, 1973.)

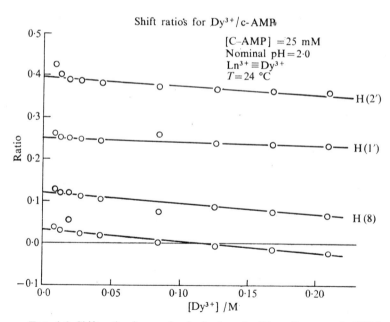

FIG. 4.6. Shift ratios for varying ionic strength. (From Barry *et al.*, 1973.)

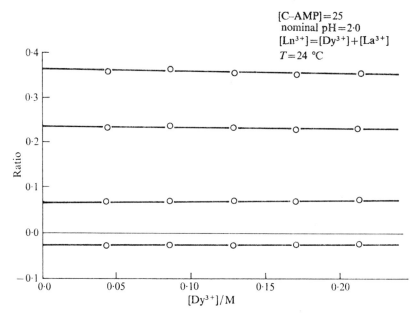

FIG. 4.7. Shift ratios for constant lanthanide titration. (From Barry *et al.*, 1973.)

ionic strength is constant the ratio of the shifts of the hydrogen nuclei R_i is constant.

The next stage is to demonstrate that there is no contact contribution to the observed shift. This can be done by measuring the ratio of shifts R_i for the *same* two nuclei for different lanthanide ions (which may cause shifts in different directions). If there is a contact-shift contribution it would be expected to vary with the value of J for the metal ion (eqn (4.1)). For this case the ratios of the shifts of H(8), H(2), and H(1′) relative to that of H(5u) are within experimental error independent of the metal ion for Eu(III), Nd(III), and Ho(III).

The final stage is the addition of the second relaxation probe, e.g. Gd(III) (Fig. 4.8). If the ratios with this relaxation probe are required for each hydrogen nucleus, it will be often necessary to 'shift' the resonances first into a region where they can be easily observed. This involves using the two probes simultaneously. (Since small amounts of the relaxation probe can cause quite dramatic changes in the line widths (or relaxation times) an extrapolation procedure such as that mentioned above is usually unnecessary.)

The computed conformations of the Ln(III)–cyclic AMP complex, which are consistent with the experimental results, are shown in Fig. 4.9. The two possible alternatives were differentiated by irradiating H(2′). In one case (see Fig. 4.9) a nuclear Overhauser effect was observed. (see Chapter 14).

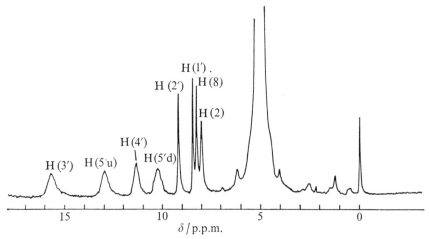

FIG. 4.8. The broadening of the resonances of cyclic AMP from Fig. 4.4 by the addition of 10^{-4}M Gd(III). (From Barry *et al.*, 1973.)

4.5.3. *Some difficulties in using this method*

The use of this method for determination of structures has several assumptions. Below we list some of the points that should be kept in mind if the method is to be generally used.

(i) For nuclei *other than hydrogen*, there is a greater possibility of a contact contribution to the shift. It will be recalled that the presence of a contact term for a given nucleus results from some delocalization of the unpaired electron density of the paramagnetic ion to the particular nucleus. The magnitude of the hyperfine coupling constant $A/\hbar$ reflects the interaction between an electron and a nucleus. More specifically, it reflects the interaction between the nucleus and an electron in an *s*-orbital since only an *s*-orbital has a non-zero electron density at a nucleus. Several postulates have been put forward for the mechanism by which the unpaired electron density is transmitted to the *s*-orbital but it is sufficient here to state that this interaction is smallest for an electron in a 1*s*-orbital and increases thereafter for the 2*s*, 3*s*, and 4*s*, etc. orbitals. For example, calculations have shown that for *one* electron in an isolated 1*s* orbital of a ^{1}H atom the coupling constant to the nucleus is ca. 1500 MHz while the corresponding value in an isolated 2*s*-orbital (e.g. in a ^{19}F atom) is ca. 50 000 MHz. With nuclei other than protons (1*s*-orbitals), there will be contributions from the higher *s*-orbitals so that contact shifts are likely to be more important. If contact shifts are important, then it may be possible to allow for their contribution using, for example, a 'ratio method' (see Chapter 3). Gd(III) has a half-filled 4*f*-shell and is not subject to any ligand field interaction and is thus expected to be spherically symmetrical under all conditions. Thus unlike the other lanthanide(III) ions, Gd(III) will only induce contact rather than pseudo-contact shifts. Hence the value of $A/\hbar$ can be evaluated and it is then possible to estimate this contribution for all the other lanthanides. There are still difficulties, for Gd(III) will tend to broaden (see page 211) the resonances as well as to shift them and so accurate shift determinations may be difficult. If the observed shifts induced by the lanthanide(III) ions have a significant contact contribution, then shift errors in estimating this contribution could

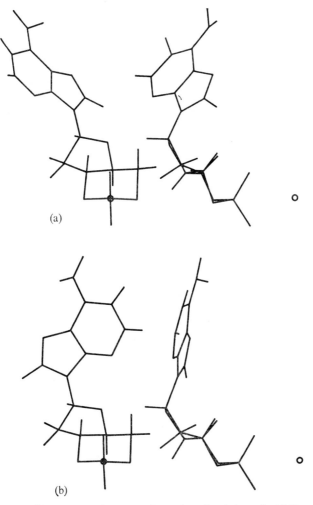

FIG. 4.9. Two possible structures for the conformation of La(III)–cyclic AMP. ○ represents the metal. The two orthogonal views show that in Fig. 4.9(a) H(8) is nearer to the ribose ring. Irradiation of H(2′) and H(3′) results in a nuclear Overhauser effect on H(8) indicating that (a) is the correct structure. (From Barry *et al.*, 1973.)

lead to large errors in the ratio R_i. An alternative method that has been used (Dobson, Williams, and Xavier, 1973) to separate the contact and pseudo-contact contributions to the shifts is to write the observed shift as a sum of these, viz:

$$\text{observed shift} = D \text{ (pseudo-contact shift)} + C \text{ (contact shift)} \quad (4.6)$$

where A and B are constants. Dobson *et al.* (1973) used this method to analyse the shifts of the ^{31}P resonances of Ln(III)–CMP complexes. They assumed that the protons only experience pseudo-contact shifts. If D represents the *ratio* of the ^{31}P shift to a given ^{1}H shift then the ^{31}P pseudo-contact shift is simply D times the observed ^{1}H shift. The contact term is given by eqn (4.1) and $A/\hbar$ is assumed to be constant throughout the series. Thus for each lanthanide ion, an equation of the form (4.6) is obtained. These equations then can be solved graphically or otherwise to obtain D. This ratio may then be used as discussed earlier.

(ii) The shifts of resonances other than hydrogen, are often very sensitive to the electronic environment. Thus lanthanides with different numbers of unpaired electrons could lead to 'orbital mixing' (i.e. the paramagnetic term σ_p in eqn (2.25) becomes more important) with consequent chemical shift. Unfortunately if this occurs it is difficult to allow for, but it is probably not very important.

(iii) The shifts induced by the lanthanide to any other probe may be such that conditions of slow, intermediate, or fast exchange apply. In the case of slow exchange the resonances of the complex may be observed directly and the induced shifts ($\Delta\omega_M$) measured. In the case of fast exchange the observed shift is of course given by

$$\Delta\omega = P_M q \,\Delta\omega_M \quad (4.7)$$

where $P_M q$ is the fraction bound and $\Delta\omega_M$ is the shift measured with respect to the appropriate 'control'. Thus the geometric parameters can still be obtained since, for any two nuclei, the ratio of the shifts is still given by eqn (4.4). It is possible that intermediate exchange conditions ($\tau_M \,\Delta\omega_M \simeq 1$) apply for some of the resonances. From Fig. 2.18, we may see that this means that $\Delta\omega$ the measured shift is too small and thus the values of R_i will not be a reflection of the true conformation. The lanthanide pseudo-contact shifts decreases as $1/T^2$ so by changing the temperature it is possible to achieve fast-exchange conditions. Unfortunately changing the temperature may alter conformation and this may be inconvenient if comparisons of structure at one particular temperature are required for a number of compounds. One way of overcoming this problem is to use a different lanthanide ion which induces small shifts (so that the condition $\tau_M \,\Delta\omega_M \ll 1$ is applicable). Additionally, if the values of R_i for different lanthanide ions are the same, it is highly probable that fast-exchange conditions do apply.

(iv) With probes like Gd(III), which alter the relaxation rates of the nuclei (causing broadening of the resonances), it is often experimentally difficult to measure the relaxation rates of the *bound* nuclei since they are usually very short (i.e. very broad resonances). Information on these nuclei has thus to be obtained under conditions of fast exchange.†

(v) In cases where there is no axial symmetry eqn (4.2) for the pseudo-contact shift has to be used. The computation is then much more complex. As a general 'rule of thumb' we may note that if the 'ratios' R_i of different lanthanides are the same, it is a reasonable indication of axial symmetry.

The difficulties that we have encountered can almost all be overcome by doing the same experiment with different lanthanide ions. Herein lies the power of these ions as probes and some exciting developments should follow in the next few years.

References

BARRY, C. D., NORTH, A. C. T., GLASEL, J. A., WILLIAMS, R. J. P. and XAVIER, A. V. (1971). *Nature (Lond.)* **232**, 236.

——, MARTIN, D. R., WILLIAMS, R. J. P., and XAVIER, A. V. (1973). to be published.

BLEANEY, B. (1972). *J. mag. reson.* **8**, 91.

——, DOBSON, C. M., LEVINE, B. A., MARTIN, R. B., WILLIAMS, R. J. P., and XAVIER, A. V. (1972). *Chem. Commun.* 791.

DOBSON, C. M., WILLIAMS, R. J. P., and XAVIER, A. V. (1973). to be published.

HINCKLEY, C. C. (1969). *J. Am. chem. Soc.* **91**, 5760.

PHILLIPS, C. S. G. and WILLIAMS, R. J. P. (1966), Inorganic Chemistry, Vol. 2, O.U.P., Oxford.

† Fast exchange for relaxation probes may be defined differently from that for shift probes—see Chapter 2.

5

PROTON HIGH RESOLUTION SPECTRA
OF PROTEINS

5.1. General

THE ^{1}H resonances of most diamagnetic compounds occur in the relatively narrow spectral range between δ 0 and 12. In general, the appearance of an n.m.r. spectrum of a compound is governed by three factors: the chemical shifts of the various nuclei; the spin–spin coupling constants between the nuclei; and the width of the resonance.

In proteins there will be a large number of hydrogen nuclei in slightly different environments with different chemical shifts, with consequent overlap of many resonances. Also the slow motions (large values of the correlation times, τ_c) associated with proteins will lead to relatively broad lines. Since the line-width may be greater than the spin–spin coupling constants then these will not be resolved. The result is to produce a broad envelope for the spectrum.

Some simplification may result from the use of high magnetic fields. Since the chemical shifts are proportional to the field strength the near equivalence of many of the protons may be lessened. This is illustrated for ribonuclease in Fig. 5.1, in which the spectra at 60 and 220 MHz are compared.

For proteins such as ribonuclease and lysozyme (Mol. wt. $\approx$ 15 000) the average line widths are ca. 15 Hz. Thus at 220 MHz it has been possible to resolve large regions of their spectra. However as the molecular weight of the protein increases, the individual resonance lines will become more difficult to resolve, but there may be the possibility of local motional freedom leading to some sharper resonance lines.

Such sharp lines are expected in the random-coil form of a protein where there is considerable segmental motion of the side chains. For example, the spectrum of the random coil form of ribonuclease is shown in Fig. 5.1b and from the difference from Fig. 5.1a we note that the n.m.r. spectrum of the *native* form is influenced by the tertiary structure and *in principle* it should be possible to obtain from the spectrum the complete structure of a protein in solution. The difficulties are such, however, that this has not yet been achieved and, of course, the larger the protein the more difficult the problem becomes. However in *random coil* proteins, the segmental motion of the side chains is not a function of the molecular weight of the protein. This means that, to a good approximation, the line width of any particular resonance is also independent of molecular weight in random coil proteins (Bradbury and King, 1969). Of course if a native protein of large molecular weight has

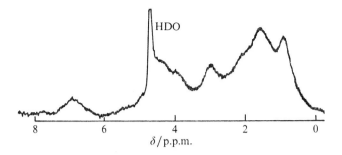

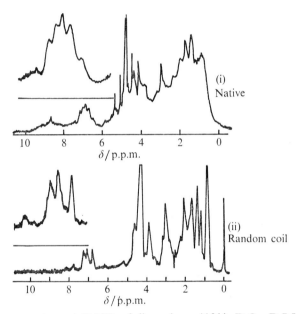

FIG. 5.1(a). ¹H N.m.r. spectrum at 60 MHz of ribonuclease, 11% in D_2O, pD 7·5, 38 °C. (From McDonald and Phillips, 1967.) (b). ¹H N.m.r. spectrum at 220 MHz of ribonuclease, 11% in D_2O, pD 6·8 at (i) 22 °C (native form) and (ii) 72·5 °C (random coil form). (From McDonald and Phillips, 1967. © 1967 Amer. Chem. Soc.)

flexible portions within it, these too should be observable by magnetic resonance.

Before discussing the protein spectra it is useful to list the various intrinsic factors which can contribute to the position of the resonance of a particular nucleus. These include: the atom to which the nucleus is bonded; ring currents; hydrogen bonding; local electrostatic effects; different degrees of protonation; and paramagnetic centres. Various combinations of these factors can lead to regions of the spectrum which are often quite simplified

and thus easy to study. For example the N*H* region is at very low field
(δ 8–10 p.p.m.) due to the electronegativity of nitrogen. Similarly the C(2)
hydrogens of histidine are markedly deshielded by the two adjacent nitrogens
and also occur at low field. The presence of a paramagnetic centre, as we have
already seen for some of the haeme proteins, can result in large hyperfine
shifts and groups of protons may then be shifted right outside the spectral
envelope.

One of the difficulties of estimating the effects of intrinsic perturbations
such as ring currents etc. is having suitable standards with which to compare
the resonance positions in the native protein. Sometimes the random-coil
form may provide a partial answer, as we have seen in the case of lysozyme
where ring-current effects cause significant shifts in some of the resonances
of the native form compared with the random-coil form. Also, the intrinsic
paramagnetism of the iron in some of the haeme proteins causes large
hyperfine shifts of many of the hydrogen nuclei of the haeme group. Assignment of the resonances in this case is made on the basis of relative integration
and comparison with suitable models.

These difficulties are often overcome by the use of extrinsic perturbations
which have the advantage that there is a control, i.e. the unperturbed protein.
It has the disadvantage that the perturbation might cause changes in the
protein structure, and, of course where possible suitable tests should be
carried out to try to assess any changes. The obvious extrinsic perturbations
are: pD or pH; temperature; addition of substrate or inhibitors; addition of
paramagnetic ions; and chemical modification of a select group or groups. A
slightly different approach is the use of isotopic substitution e.g. replacement
of hydrogens by deuterium for which can cause simplification of the ^{1}H
spectrum.

In the following an attempt is made to give some indication of the problems
and procedures in assignment. We start first with the much simpler case of
random-coil proteins.

5.2. Random-coil proteins

Some idea of how the native protein is folded may come from a study of
the unfolding process. Thus the spectrum of the random-coil form has to be
known. In the random-coil protein the amino acid side chains are in an
aqueous environment and, as a first approximation, it is to be expected that
their resonances will sharpen relative to the native form (as the segmental
motion of many residues is increased with consequent decrease in correlation
time) and that their positions will be similar to those of the individual amino
acids. However protons bonded to α-carbon atoms are dependent to a small
extent on the nature of the nearest-neighbour residues and if necessary
this can be allowed for empirically. Also other hydrogen nuclei such as those
of a methyl group, which may have been in different environments, (and

therefore non-equivalent) in the native protein, become equivalent in the random-coil form, when they are assumed to rotate freely. There will, of course, be other effects to be considered, such as the effect of hydrogen bonding, pH, solvent, and temperature. In different solvents, for example, different empirical rules will have to be derived. Also, effects of ring currents, and other local electrostatic effects may well have to be assessed. Some idea of the progress that has been made is illustrated by the work of McDonald and Phillips (1969a) using a 220 MHz spectrometer. From the spectra of the L-amino acids commonly occurring in proteins and a number of small peptides (in D_2O at pH 7) they obtained the various resonance positions (Appendix 5.1). By modifying these assignments to agree with the random coil spectra of lysozyme and ribonuclease they obtained the 'random coil' resonance positions for most of the different types of hydrogen atom in proteins. From observations of polypeptides and random-coil proteins the resonance width at half-height was ca. 10 Hz. It was assumed (by McDonald and Phillips 1969a) that the line shape was adequately represented by a triangle of height equal to the resonance height and base equal to twice the resonance width at half-height. Spin–spin couplings were allowed for by assuming that the individual components again have a resonance width at half-height of ca. 10 Hz, and by calculating the 'apparent half-width'. The resonance is then represented by a triangle with a base that is twice the value of this 'apparent half-width'. The height is correspondingly decreased in proportion to the amount by which the base has been increased from the normal width of 20 Hz. This maintains a constant area to represent the intensity of a single hydrogen nucleus. If there are a number of chemically equivalent hydrogens in a residue, the height of the respective triangle is then scaled up proportionately. There is a further scaling if there are other equivalent residues.

For example, the histidine C(2) hydrogen is a singlet and is represented in the appropriate position by a triangle of base 20 Hz and height 10·0 units (arbitrary). On the other hand, the three methyl protons of alanine are equivalent, but their resonance is split into a doublet by the α-hydrogen ($J \approx 8$ Hz).

If we consider *one* of the protons of the methyl group, then we have the doublet represented as two triangles with apexes ca. 8 Hz apart. The resultant triangle then has an 'apparent half-width' of 18 Hz and a base-width of 36 Hz. To keep the area constant the height has to be reduced to $\frac{20}{36}$th of its previous value, i.e. to 5·56 units. This complete methyl group is then characterized by a triangle of base-width 36 Hz and height 16·6 units. A list of the amino acid residues for which the chemical shifts and line shapes have been obtained is to be found in Appendix 5.1. Use of these is made in Fig. 5.2 which shows the actual and computed ^1H n.m.r. spectrum of pepsin (which has a molecular weight of ca. 35 000). The computed and actual spectrum agree quite well throughout the high-field region, but in the low-field region,

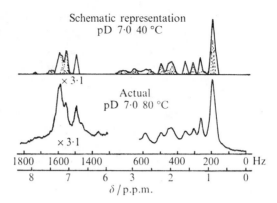

FIG. 5.2. ¹H N.m.r. spectra of porcine pepsin at 220 MHz. (From McDonald and Phillips, 1969a.)

the resonance of phenylalanine at ca. 1600 Hz is somewhat stronger than expected. Of course more sophisticated curve fitting procedures will undoubtedly be used, but this method appears quite good as a first step. More recently procedures for spectrum simulation have been developed which make use of an improved line shape function derived from measured spectra (Bradbury and Rattle, 1970). This has been done from a study of histone spectra and may be more applicable since it is known that histones adopt a random coil form in water.

We may note some of the general characteristics of random coil protein resonances from Appendix 5.1. The least shielded protons are those of the histidine C(2), and tryptophan indole. The aromatic region which includes hydrogens of tryptophan, tyrosine, phenylaline, and the C(4) histidine hydrogens appears at $\delta \approx 7$ p.p.m. The α-C hydrogens occur at higher field and generally appear as a broad envelope which may be partly obscured by the solvent peak. The 'methyl region' usually occurs at the highest field region at $\delta \approx 1$ p.p.m. and consists of the methyl groups of the aliphatic side chains of valine, leucine, and isoleucine.

The concentrations of protein for which spectra can be obtained are usually ca. 1 mM; the concentration of water when used as a solvent is 55·5 M. Thus at the high gains necessary to observe the protein resonances, the majority will be obscured by the large solvent peak. For this reason, spectra are usually observed in D_2O, and in most spectra an HOD peak is observed arising from proton exchange of any residual water in the system with deuterium in D_2O. In general, in D_2O solution, most of the NH resonances will also disappear because of chemical exchange and even in H_2O, these protons may broaden or disappear for the same reason.

Finally, we note that the process of denaturation can be followed by observation of the n.m.r. spectrum as a function of temperature, pH or concentration of added denaturant etc. In principle every hydrogen resonance can be observed and it ought to be possible to see which resonances appear to 'unfold' first. For example, in lysozyme Bradbury and King (1971) showed that the NH resonances of arginine sharpen before those of many other hydrogens and concluded that three clusters of arginine side chains, involving seven arginine residues unfolded before the bulk of the molecule. Information on the nature of any intermediate states can also be deduced from analysis of the spectrum (see e.g. Bradbury and King, 1971, 1972). Clearly this type of work will be extended in the future.

5.3. Native proteins

5.3.1. *Assignments*

Once a spectrum has been obtained, it becomes necessary to assign each resonance not only to an amino acid type but to a specific residue. This obviously requires that the protein amino acid sequence is known. In general there are several approaches which can be used for identifying and studying the resonances: (1) Comparison with spectra of pure amino acids and oligo peptides. (2) By comparison with the random coil form it is possible to assign *some* of the resonances. For example, we have already noted the effects of ring currents in lysozyme in which a *knowledge of the crystal structure* allowed a prediction of which resonances in the native form would experience ring-current shifts. (3) Analysis of resonances shifted from the main spectral envelope e.g. histidine C(2) and tryptophan NH at low field, and the resonances of porphoryin rings in haeme resulting from hyperfine shifts with the paramagnetic centre. (4) Isotopic substitution, which leads to a simplification of the spectra. (5) The use of extrinsic perturbations. (6) Difference spectroscopy.

We now discuss several examples to illustrate some of these points, some of the problems encountered in assigning the resonances, and the kind of information that can be obtained from studying these resonances under a variety of conditions.

5.3.2. *Haeme proteins*

5.3.2.1. *Myoglobin.* Myoglobin was the first protein whose structure was determined by X-ray diffraction. Its molecular weight is ca. 18 000 and its function is to bind oxygen and store it in the muscle until required for metabolic oxidation. Each molecule contains 153 amino acid residues and a haeme group. This haeme group is not covalently bound to the protein, and can be removed and replaced by other kinds of haeme units. In myoglobin and haemoglobin, the haeme group is a protoporphyrin iron(III) and proto-haeme 1X unit. The iron is bound to four nitrogen atoms (e.g. see Fig. 5.3) and one of the axial ligands is a histidine residue of the polypeptide chain. The sixth co-ordinate position of the iron varies with the state and function

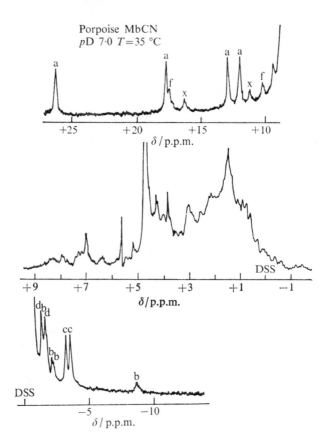

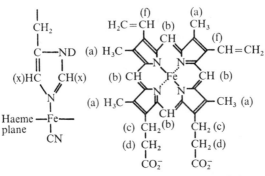

FIG. 5.3. ¹H N.m.r. spectrum at 220 MHz of cyanoferrimyoglobin (MbCN). Different horizontal and vertical scales were used for the spectral regions $\delta-1$ to $+9$ p.p.m., and $\delta-1$ to -10 and $\delta+10$ to $+30$ p.p.m. The sharp lines between $\delta+3\cdot8$ and $+6$ p.p.m. correspond to the resonance of HDO and its first and second spinning side band. The structure of protohaeme IX and of the axial ligands of the haeme iron in cyanoferrimyoglobin are shown. (From Wüthrich *et al.*, 1970.)

of the protein and constitutes the 'active site'. For example, in haemoglobin and myoglobin this position may be filled by O_2, CO, and NO (for the Fe(II) state) and H_2O, CN^-, N_3^-, OH^- (for the Fe(III) state).

The ^1H n.m.r. spectrum at 220 MHz of cyanoferrimyoglobin is shown in Fig. 5.3. The δ 0–10 p.p.m. range contains the usual diamagnetic resonances of ca. 1000 hydrogens of the protein part of the enzyme. Outside this range there are a number of resolved resonances corresponding in intensity to one–three hydrogens. These resonances have been shifted from their usual positions by intrinsic perturbations (ring currents or hyperfine shifts). The high-field part of the spectrum δ 0– -5 p.p.m. contains ca. nine–eleven resonances corresponding to groups of one–three hydrogens each. Between δ $+10$ and $+27$ p.p.m. nine resonances are visible which again correspond to one–three hydrogens each. In the haeme group there are 14 groups of hydrogens and two single hydrogens on the imidazole group co-ordinated to Fe(III) which may experience interactions with the metal ions. This number agrees quite well with the number of resonances found and their intensities.

The shift of the low-field lines is inversely proportional to the reciprocal of the absolute temperature (Fig. 5.4) as expected for resonances which are shifted because of contact interactions. The assignments have been made on the basis of their intensities, by comparison with the spectra of reconstituted myoglobins having haeme units containing hydrogens and ethyl groups in place of the vinyl groups (e.g. see Fig. 3.8), and by observation of the effects of the inert molecules, xenon and cyclopropane, on the spectrum (Shulman, Peisach, and Wyluda, 1970).

In the up-field part of the spectrum there are six distinguishable resonances close to δ 0 p.p.m. The shift is temperature independent and this suggests that the resonances are due to peptide methyl groups or other amino acid side-chain hydrogens shifted by ring currents.

One problem is apparent from close inspection of Fig. 5.4. The hyperfine shift is not strictly proportional to $1/T$. This may result if the hyperfine coupling constant is temperature dependent (see Chapter 3).

5.3.2.2. *Haemoglobin.*   The molecular weight of haemoglobin is about four times greater than that of myoglobin. The intrinsic width of most resonances is thus about four times greater and the resonances in the diamagnetic region overlap even more strongly than in myoglobin. The assignments are complicated further since haemoglobin consists of two α-chains and two β-chains. Figure 5.5 shows the 220 MHz spectrum. The only peaks definitely assigned are the two ring methyl resonances of each haeme group at δ $+21$ to $+22$ p.p.m. and δ $+15$ to $+16$ p.p.m. The temperature dependence indicates that these result from contact interactions. Not all the hydrogens expected (in comparison with myoglobin) can be seen probably because some are not shifted sufficiently.

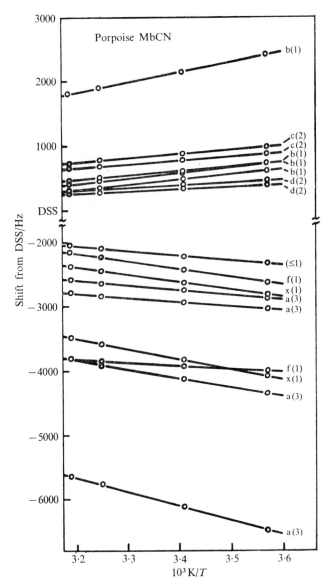

FIG. 5.4. Dependence on the reciprocal of temperature of the positions (220 Hz = 1 p.p.m.) of the hyperfine-shifted resonances of porpoise cyanoferrimyoglobin (MbCN). The letters and numbers on the right refer to the assignments of the resonances (Fig. 5.3), and the number of hydrogen nuclei corresponding to their intensities. All of the other resolved resonances of the spectrum in Fig. 5.3 were found to be independent of temperature. (From Wüthrich *et al.*, 1970.)

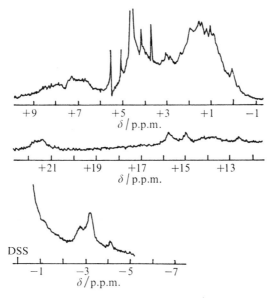

FIG. 5.5. ^{1}H N.m.r. spectrum at 220 MHz of human cyanoferrihaemoglobin at 36 °C. Different scales were used for the different spectral regions. (From Wüthrich, 1970.)

By making partially-oxidized haemoglobin such as $\alpha_2(Fe^{2+}O_2)\beta_2(Fe^{3+}CN^-)$ and the complimentary compounds in which the α-chains are paramagnetic, it has been possible to distinguish the resonances of the α- and β-chains. For example, a comparison of the spectra of $\alpha(Fe^{3+}CN^-)$ and $\beta(Fe^{3+}CN^-)$ shows definite differences between them in the mixed oxidation state samples while $\alpha(Fe^{3+}CN^-)\beta(Fe^{3+}CN^-)$ has a spectrum which is an exact superposition of the two chains. Thus it is possible to distinguish the resonances of the α- and β-chain. This offers the possibility of studying allosteric behaviour in these compounds and is likely to increase our understanding of subunit inter-actions, and thus of the mechanism for oxygen uptake in haemoglobins.

5.3.3. Ribonuclease A (RNase A)

RNase A has a molecular weight about 14 000 and 124 amino acid residues. It catalyses the hydrolysis of 3',5'-phosphodiester linkages of ribonucleic acids or nucleotide esters at the 5'-ester bond, in a two-step reaction. In the first step, the 2'-OH group attacks one of the phosphodiester bonds to form a 2',3'-cyclic phosphate and this is followed by hydrolysis of this to give a terminal 3'-phosphate. The three-dimensional structure, determined from X-ray studies, indicates that the shape of the enzyme is approximately spheroidal with a large cleft in one side to accommodate the substrate. His-12 and His-119, and Lys-7 are grouped together on one side of this cleft and Lys-41 is on the other. From chemical studies these four residues have been shown to be at the active site.

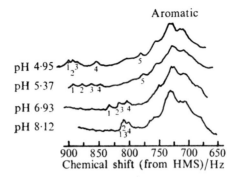

FIG. 5.6. 100 MHz N.m.r. spectra of the aromatic region of 0·012 M RNase A in deuterio-acetate buffer (pH 4·95, 100 CAT scans; pH 5·37, 14 CAT scans; pH 6·93, 51 CAT scans; pH 8·12, 57 CAT scans). Peaks 1–4 are C(2) imidazole peaks of the four histidine residues. Peak 5 is a C(4)H imidazole resonance. The envelope labelled 'aromatic' includes three other C(4)H imidazole peaks as well as peaks from six tyrosine and three phenylalanine residues (CAT = computer of average transients—see Chapter 14). (From Meadows, Markley, Cohen, and Jardetzky, 1967.)

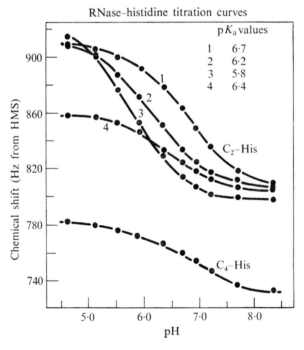

FIG. 5.7. Titration curves of C(2)H peaks of histidine residues of RNase A. Curves 1–4 correspond to peaks 1–4 of Fig. 5.6. Approximate pK_a values are: curve 1, 6·3; curve 2, 5·9; curve 3, 5·6; curve 4, 6·1. (From Meadows et al., 1967.)

RNase A contains four His residues at positions 12, 48, 105, and 119 in the primary sequence. As His-12 and His-119 are known to be at the active site, the observation of the resonances of these compounds offers an attractive intrinsic probe to study the enzyme in solution. The classic work initiated by Jardetzky and his co-workers illustrates well the complexities and subtleties in assigning the four histidine $C(2)H$ resonances.

We have already noted that the histidine ring hydrogens absorb to very low field because of the deshielding effect of the nitrogen atoms in the imidazole ring. On protonation, the ring becomes 'aromatic' and the ring currents produce a further deshielding. By varying the degree of protonation (pH) it is possible to obtain a titration curve for the imidazole $C(2)$ and $C(4)$ hydrogen resonances.

The $C(2)H$ resonances at 100 MHz are shown in Fig. 5.6, and the pH dependence of these resonances, with that of the $C(4)H$ of one of these residues are shown in Fig. 5.7. The four residues have pK_a values 6·7, 6·2, 5·8, and 6·4, as indicated. The single $C(4)H$ exhibits a pK_a value of 6·7, identical to the value for one of the $C(2)H$ resonances and thus these two resonances are from the same residue. Note in Fig. 5.7 that, at low pH, peak 4 has an 'abnormal' chemical shift. Since this resonance had a greater line-width than the others it was assumed to reflect a lower mobility than the other three histidines. Peak 4 was thus assigned to His-48 which is known from X-ray studies to be the 'buried' histidine of RNase A. The complete identification proceeded in two stages; chemical modification and specific residue deuteration.

5.3.3.1. *Chemical modification.*   Iodoacetate reacts with RNase A to form two derivatives, the 1-carboxymethyl His-119 and the 3-carboxymethyl His-12. As above, the pH dependence of the chemical shifts of the histidine resonances were measured. The titration curves of either alkylated derivative produced changes in the pK_a values of the same two histidines irrespective of which alkylated derivative was used (see Figs. 5.8 and 5.9). The conclusion was that either a conformational change had occurred or that two histidine residues were close together in the three-dimensional structure of the enzyme. If the latter is true, then it would be expected that the titration curves would be the same irrespective of which alkylated derivative is used since the size of the carboxymethyl group is such that it would perturb both the histidines. Using the nomenclature in Fig. 5.7 peaks 2 and 3 are assigned to histidines-12 and -119 and peaks 1 and 4 to histidines-48 and -105.

Next use was made of the fact that the RNase A could be cleaved by the enzyme subtilisin at the 20–21 peptide bond. Earlier work had shown that RNase S formed by bringing together, without covalent linkage, the two components i.e. the S-peptide (residues 1–20) and the S-protein (residues 21–124) had almost the same enzymatic activity and specificity as the parent

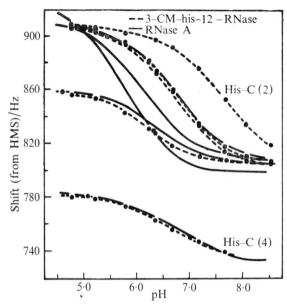

FIG. 5.8. Titration curves of the histidine residues of ribonuclease A (solid curve) and 3-carboxymethyl(3 CM-His-12 RNase)-ribonuclease His-12 (broken curve). The ordinate is the chemical shift of the C(2)*H* or C(4)*H* peaks, in Hz (at 100 MHz) downfield from hexamethyldisiloxane (HMS). (From Meadows, Jardetzky, Epand, Rüterjans, and Scheraga, 1968.)

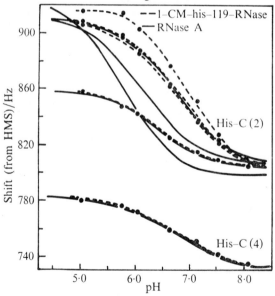

FIG. 5.9. Titration curves of the histidine residues of ribonuclease A (solid curve) and 1-carboxymethyl His-119 ribonuclease (broken curve). The ordinate is the chemical shift of the C(2)H or C(4)H peaks, in Hz (at 100 MHz) downfield from hexamethyldisiloxane (HMS). (From Meadows *et al.*, 1968.)

RNase A. At pH = 5·0, three of the histidine resonances of RNase S had identical chemical shifts as in the parent RNase A, but the fourth (His-48) was shifted ca. 0·20–0·30 p.p.m. downfield and at lower pH values now approached more closely the chemical shifts of the other histidines (Fig. 5.10). Since His-48 is in the vicinity of the peptide bond between residues 20 and 21 this may arise from it becoming less 'buried' when this bond is cleaved.

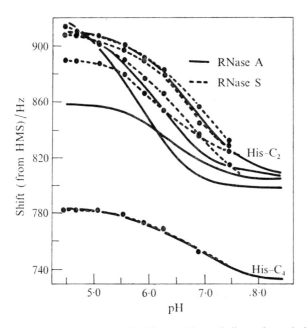

FIG. 5.10. Titration curves of the histidine residues of ribonuclease A (solid curve) and ribonuclease S (broken curve). The ordinate is the chemical shift of the C(2)H or C(4)H peaks, in Hz (at 100 MHz) downfield from hexamethyldisiloxane (HMS). (From Meadows *et al.*, 1968.)

5.3.3.2. *Specific residue deuteriation.*   The specific deuteriation experiments were carried out by separating the S-peptide (residues 1–20) from the S-protein and incubating the S-peptide at 40 °C with D_2O at pH 7 for 5 days. The C(2)H of His-12 is thus exchanged for deuterium. The two components were recombined to give RNase S' and the new spectrum contained only three histidine peaks (Fig. 5.11). A comparison of the titration curves of RNase S' and RNase S indicated that either peak 1 or 2 was absent in RNase S'. The experiments in Section 5.3.3.1 imply that peaks 2 and 3 are histidine-12 and -119 and thus the exchanged hydrogen must be that of His-12. The complete assignment in RNase S is given in Table 5.1

The pK_a values of the histidines in ribonuclease are dependent on the ionic strength in the solution (Rüterjans and Witzel, 1969). For example, in

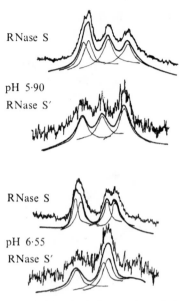

FIG. 5.11. Comparisons of the imidazole C(2)-H region of the n.m.r. spectrum of ribo-nuclease S and ribonuclease S′ (in which the C(2)-*H* of histidine-12 has been exchanged for deuterium). The curves below the spectra were obtained using a curve resolver, and show the decomposition of each spectrum into three or four Lorentzian peaks of equal area. (From Meadows *et al.*, 1968.)

Table 5.2 the pK_a values of His-119, -12, and -105 in three solutions of different NaCl solutions are given. It is seen that the three pK_a values increase with increasing ionic strength.

It is noteworthy that the extension of the assignments of RNase S to RNase A rests on the assumption that the titration curves for His-12 and His-119 both move to lower pH values by about the same extent (ca. 0·5 pH unit see Fig. 5.10) on going from RNase S to RNase A, i.e. the pK_a of His-12 falls from 6·7 to 6·2 and His-119 from 6·3 to 5·8. An alternative possibility is that the pK_a of His-12 drops from 6·7 to 5·8 and that of His-119 from 6·3 to 6·2. That this possibility is not the case has been demonstrated by King and

TABLE 5.1

Peak	pK_a†	His-residue
1	6·7	105
2	6.2	12
3	5·8	119
4	6·4	48

† 32 °C, 0·2 M sodium acetate buffer.

TABLE 5.2

pK_a *values of the histidine*-119, -12, *and* -105 *of ribonuclease in solutions of different* NaCl *concentrations. The concentration of ribonuclease is* 0·01 M

NaCl (M)	pK_a		
	His-105	His-12	His-119
<0·1	6·4	5·6	5·1
0·2	6·5	6·1	5 8
0·4	6·6	6·3	6 1

Bradbury (1971) by the use of *n.m.r. difference spectroscopy*, (see p. 99) in which the C(2) and C(4) hydrogens of all the histidine resonances in RNase A can be 'resolved'. Figure 5.12 shows the titration curves (at low ionic strength) obtained for the C(2) and C(4) hydrogens in RNase A which give apparent pK_a values of 5·2, 5·8, and 6·4 for the resonances assigned to His-119, -12, and -105 respectively. These are in good agreement with those in Table 5.2. In Fig. 5.12 we see that the C(2) hydrogen resonance of His-48 is not visible at pH 6, although the C(4)H resonance is visible throughout the titration range (Fig. 5.13). The apparent discontinuity shown by the dotted line in Fig. 5.13 may be associated with a conformational change.

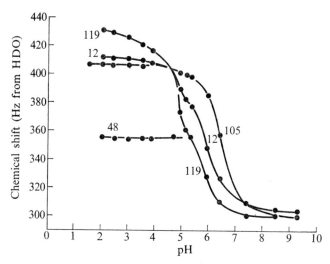

FIG. 5.12. Histidine C(2) hydrogen titration curves of ribonuclease A. The numbers refer to the residues in ribonuclease A as previously assigned. (From King and Bradbury, 1971.)

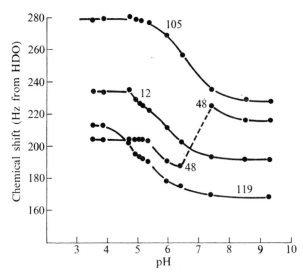

FIG. 5.13. Histidine C(4) hydrogen titration curves of ribonuclease A. The numbers represent assignments of resonances to specific residues in the protein. (From King and Bradbury, 1971.)

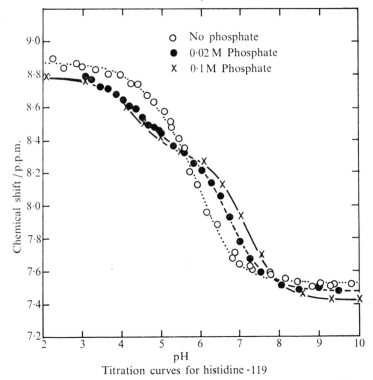

Titration curves for histidine -119

FIG. 5.14. Titration curves of the histidine C(2)-H resonance of His-119 of ribonuclease as a function of phosphate concentration. The lines are computed assuming an interacting group with a pK_a ≈ 4·5. The shift is measured relative to external TMS. (From Cohen, Griffin, and Schechter, 1973.)

Another reason for the slight displacement in the titration curves between RNase A and RNase S could be the presence of phosphate in the RNase S solutions (Cohen *et al.*, 1973). Figure 5.14 shows the effect of increasing phosphate concentration on one of the histidine $C(2)H$ (His-119) titration curves. In these it is seen that the effect of phosphate is to 'shift' the curves to higher pH values. If RNase S is desalted then the curves are identical with those of RNase A (Cohen *et al.*, 1973).

Examination of the titration curves of His-12 and His-119 indicate that they are asymmetric. Such asymmetry has been observed in histidine model compounds and has been shown to result from the titration of amino and carboxyl groups. The titration curves have been analysed by a mathematical model (details of which are included in Appendix 5.2.) in which it is assumed that there is an interaction between the two adjacent titrating groups. For RNase A, an analysis of the His-12 and His-119 titration curves suggests that the asymmetries result from interaction with groups having pK_a values 4·5 and 8·4, respectively. The X-ray data suggests that these groups may be Asp-121 and Lys-41.

Imidazole–imidazole interaction is ruled out on the basis that a 'good fit' to the experimental points of both histidine resonances is *not* obtained for these (Fig. 5.15). This is particularly important in discussing the mechanism of RNase A. For example, it tends to rule out a mechanism involving

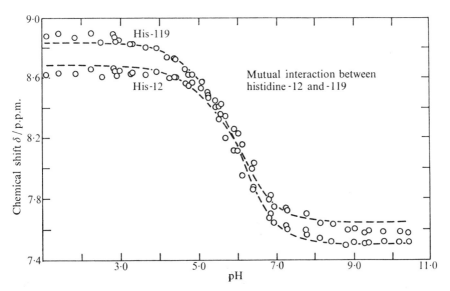

FIG. 5.15. The pH dependence of the chemical shift (relative to external TMS at 220 MHz) of His-19 and His-12 in RNase A. The dotted line indicates the computer fit that would be expected for two interacting histidines. ($T = 22$ °C, 0·1 M, NaCl). (From Schechter *et al.*, 1972.)

hydrogen bonding between the two imidazoles at the active site (Schechter, Sachs, Heller, Shrager and Cohen, 1972).

The shift inflexions in Fig. 5.14 arise from groups with pK_a values of ca. 4·6 and ca. 6·8. One of these pK_a values might arise from interaction with the phosphate itself. Phosphate is an inhibitor of RNase and has two pK_a values of ca. 2 and 7·2. The binding of phosphate to a positively-charged group (histidine) could possibly result in a *lowering* of the pK_a values. Although this is unlikely the pK_a of 7·2 could be reduced to ca. 4·6. (His-119 has a pK_a value of 6·8.) In the presence of 0·1 M sulphate, two titratable groups are again observed with pK_a values of 3·45 and 6·31. Since the pK_a of SO_4^{2-} is ca. 2, and interaction with a positive charge decreases the pK_a, the pK_a of ~3 to 4 in the titration curves does not arise from the phosphate and must arise from another group close by on the enzyme.

Summing up, the RNase A system illustrates that the buffer, the pH, the temperature, the added salts are all important in determining the exact histidine spectra that will be observed. These are obviously important lessons in extrapolating to other systems.

5.4. Use of isotopic substitution in assignments

5.4.1. *Deuteriation of exchangeable NH and OH hydrogens*

Most of the observations of ^1H spectra of proteins are carried out in D_2O solution in order to suppress the very strong H_2O resonance that can essentially 'blot out' most of the protein resonances. If some NH and OH hydrogens are exchangeable then this gives simplification of the spectra, but there may well be a loss of information in this process.

For example, if the indole NH hydrogen of a tryptophan residue is thought to participate in substrate binding then observation of the NH resonance would be particularly important. Of course, if the NH hydrogens are too labile, then their resonance would not be observable since exchange with the hydrogen nuclei of the solvent, water, would lead to the merging of the two resonances (see Chapter 2). In general the —NH resonances occur at very low field and in favourable cases, they can be observed in H_2O solutions, well removed from the main spectral envelope and therefore an attractive region to study.

Hen egg white (HEW) lysozyme contains six tryptophan residues. The NH hydrogens in the denatured form appear as a single peak indicating that they are equivalent with respect to the primary amino acid sequence. In the native form, five separate NH resonance positions are resolved (the sixth being displaced perhaps into a different region). The spectra are shown in Fig. 5.16. As previously, extrinsic perturbations, namely chemical modification, inhibitor perturbation, and differential deuterium exchange rates, have been used to help in the assignment of the tryptophan residues in lysozyme. For example, two of the resonances exchanged slowly in D_2O and

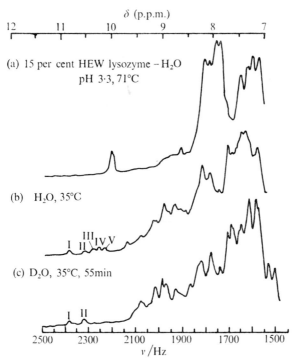

FIG. 5.16. 220 MHz ^{1}H n.m.r. spectra of low-field resonances of HEW lysozyme: (a) thermally denatured in H_2O, (b) native in H_2O, (c) native after 55-min, exchange in D_2O at 5 °C. (From Glickson, McDonald, and Phillips, 1969.)

this suggests that they do not have free access to the solvent. Examination of the X-ray structure indicates that these may be tryptophan residues 28 and 108 since these are involved in intramolecular hydrogen bonding.

The amide N*H* hydrogens involved in the secondary and tertiary structure of proteins are expected to have a rather more complicated spectrum than the single peaks obtained for the N*H* hydrogen of tryptophan, since there are obviously very many more different environments for these hydrogens.

5.4.2. *Selective amino acid residue deuteriation*

One way of simplifying the ^{1}H n.m.r. spectrum of a protein is to prepare proteins that have hydrogens at specific residues replaced by ^{2}H. For example, in the assignment of the four histidine residues in RNase S, deuteriation of the S-peptide (residues 1–20) removed the C(2)*H* of His-12. The procedure can be extended to the entire protein (Bradbury and Chapman, 1972) and the rate of disappearance of the resonances can be monitored. Under the conditions used by Bradbury and Chapman (37 °C, in D_2O at pH 8·5) the C(2)*H* resonances of His-105 and His-12 disappeared almost completely.

A dramatic example of simplification of a spectrum resulting from selective

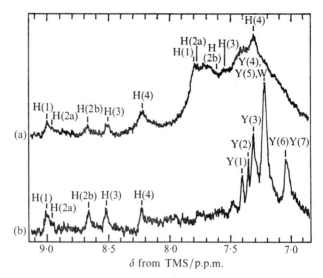

F<small>IG</small>. 5.17. Comparison of the aromatic region of the n.m.r. spectra of staphylococcal nuclease (Nase) and as electively deuteriated analogue (Nase-D4). Assignments: His, H(1), H(2a), H(2b), H(3), and H4 (the low-field peaks are C(2) and the high-field peaks C(4) imidazole hydrogens; Tyr, Y(1–7); Trp, W(C(2) ring proton). (a) Nase, pH 6·0, (b) Nase-D4, pH 6·0. (From Markley (1969).)

deuteriation is shown in Fig. 5.17 which compares the spectra of staphylococcal nuclease and a selectively deuterated analogue. In this case use was made of the fact the bacteria, Staphylococcus aureus from which the nuclease is made requires all amino acids in its growth medium. By using a mixture of deuteriated amino acids and only selected ^{1}H amino acids it is possible to achieve the selective deuteriation.

This particular strain (Foggi) of staphylococcal nuclease contains four histidine, seven tyrosine, one tryptophan, and three phenylalanine residues. In the selectively deuteriated analogue, the phenylalanines were fully deuteriated, the histidines deuteriated at C(4) and the tyrosines at C(3) and C(5) (removing any spin–spin coupling in the tyrosine spectrum and transforming the resonance of each individual tyrosine to a singlet). The problems of assignment may still be far from easy and other perturbations, such as the addition of inhibitor or metal ion, then have to be used to help the assignment. Of course, in principle, it is possible to prepare a protein with all but one or two types of amino acid residues deuteriated or, more simply, to have only one type of amino acid residue deuteriated. In the second case, comparison with the spectra of the analogue containing only hydrogen nuclei may allow identification of the resonance position of the 'missing' resonances.

This selective deuteriation technique can only be effected successfully under conditions where the necessary components in the growth medium are deuteriated. These conditions are only easily fulfilled where growth of micro-organisms is concerned. The most elegant results will be obtained when the protein of interest is an induced enzyme and can be obtained in large quantities. For non-bacterial proteins, the difficulties of deuteriation are very real, and it is to be hoped that advances in experimental ingenuity will provide a solution.

5.5. Use of paramagnetic shift probes in assignment

We have seen that for proteins where a n.m.r. spectrum can be observed, some simplification may result by using high magnetic fields which lessens the near equivalence of many of the chemical shifts. In some cases even greater simplification may result from the use of a shift probe, and, with a suitable choice of probe, it is possible to calculate the distance of each affected residue from this probe and thus map out the probe binding site. Even if the geometry can not be obtained, then shift probes can be used in a qualitative way to help assignments as, for example, binding of Co(II) to hen egg white (HEW) lysozyme. In general, Co(II) induces both contact and pseudo-contact shifts. Some representative changes in the positions of resonances in the −100 Hz to −200 Hz region (at 220 MHz) are shown in Fig. 5.18. The degeneracy between resonance lines marked 7 and 8, 5 and 6,

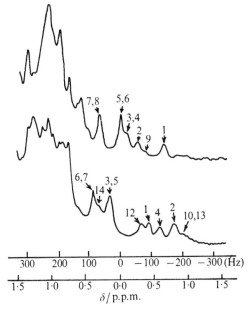

FIG. 5.18. Perturbation of the 220 MHz ^{1}H n.m.r. (300 to −300 Hz region—relative to DSS) spectrum of HEW lysozyme by Co(II). Lysozyme 7×10^{-3} M in D_2O, pD 5·5, 55 °C. Co(II) concentrations: upper trace 0·0, lower trace 0·154 M. (From McDonald and Phillips 1969b)

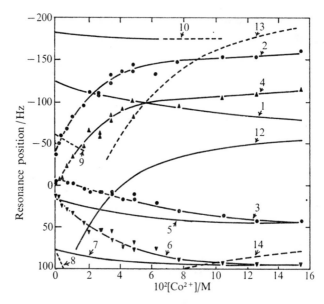

FIG. 5.19. Co(II)-induced shifts of HEW lysozyme resonances versus Co(II) concentration. Lysozyme 7×10^{-3} M in D_2O, pD 5·5, at 55 °C. Referred to internal DSS at 220 MHz. (From McDonald and Phillips 1969b.)

and 3 and 4 is removed on addition of Co(II). Integration of the 'perturbed' resonances in the spectrum then suggests that resonances 1–7 have intensities corresponding to methyl groups. The change in shift as a function of the concentration of added Co(II) will reflect the binding of Co(II) to lysozyme and is shown in Fig. 5.19 for various resonances. From this 'binding curve,' the dissociation constant of the Co(II)–protein complex can be calculated to be ca. 17·5 mM at pD = 5·5 and 55 °C with the assumption of 1:1 binding. By measuring the dependence of the shift of each resonance as a function of pD, it is possible to obtain a titration curve which gives an apparent pK_a value. For those resonances associated with the strong binding site (e.g. 2, 4, 6...) pK_a values of ca. 5·2 were obtained, but the pK_a varied between 5 and 6 depending on the Co(II) concentration, decreasing with increasing [Co(II)]. This decrease in apparent pK_a was attributed to competition between Co(II) and H^+ ions for the same site. Binding may occur at one or more of the carboxyl groups of aspartic-52 and glutamic-35 acids, since Co(II) inhibits the enzyme and these acids are known to be at the active site.

   In this example it is not possible to obtain geometric information about the Co(II) site because it is not clear what contribution to the shifts are the result of contact interactions. The lanthanide ions have, however, been shown to bind very close to the active site and with these it ought to be possible to map out the residues around active site by using them in an analogous manner (see Section 5.7).

However, if quantitative information is to be obtained, it is important to know if there is more than one metal binding site which could cause the shifts. For example, inspection of Fig. 5.19 shows that, while most of the resonances are 'saturated' at 0·15 M Co(II), those marked 1 and 3 are still increasing and may therefore be associated with Co(II) binding at a second weaker site. If the binding sites, though, are sufficiently different the chemical shifts arising from the stronger site alone can be obtained by suitable extrapolation techniques similar to those described previously (see page 69).

5.6. Simplification of ¹H n.m.r. spectra by difference spectroscopy

A good example of this method is the identification of the His-C(4)H resonances in RNase A. In the normal n.m.r. spectrum only one of the C(4) hydrogen resonances, that arising from His-105, can be distinguished from the overlapping aromatic resonances of phenylalanine and tyrosine. As the C(2) hydrogen resonances are pH dependent, the behaviour of C(4) hydrogen resonances should be similar. We use this fact to observe them, since the other aromatic hydrogen resonances are pH independent between ca. 4·5 and 7·5. Thus Fig. 5.20 shows the n.m.r. spectra of the aromatic region of RNase A at pH 5·96 and 5·35, together with the difference spectrum

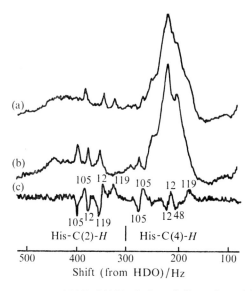

FIG. 5.20. 100 MHz spectra at 34 °C of 10% solution of ribonuclease A in D₂O. pH values are meter readings uncorrected for deuterium isotope effects. (a) pH = 5·96, (b) pH = 5·35, (c) is the difference spectrum (a)−(b). The numbers refer to hydrogen resonances of the histidine residues. (From King and Bradbury 1971.)

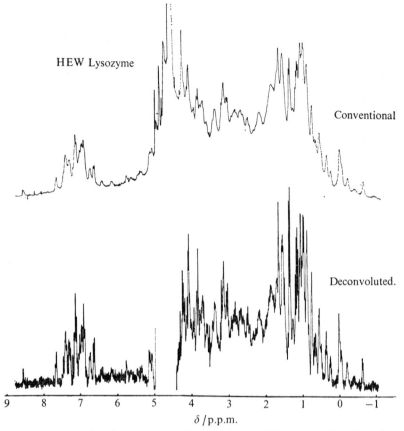

FIG. 5.21. Conventional Fourier transform spectrum of lysozyme (5 mM) and corresponding deconvoluted spectrum. (From Campbell *et al.*, 1973.)

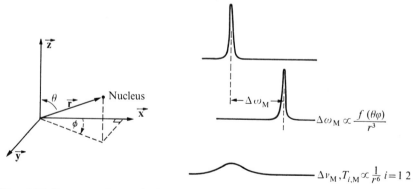

FIG. 5.22. Summary of perturbations caused by paramagnetic lanthanide ions. $\Delta\omega_{\text{M}}$ is the pseudo-contact shift, $\Delta\nu_{\text{M}}$ is the linewidth and $T_{i,\text{M}}$ are the relaxation times.

obtained by subtracting one from the other. A computer was used on-line with the n.m.r. spectrometer to increase the signal-to-noise ratio by averaging spectra (see Chapter 14). The difference spectrum in Fig. 5.20 clearly separates those peaks whose chemical shifts are dependent on pH from those which are not. Several subtractions may be required for each spectrum, because of occasional cancellation of coincident peaks. For example, in Fig. 5.20 the C(4)H peak of His-48 at pH 5·96 has the same chemical shift as the C(4)H peak of His-119 at pH 5·35, and so these two peaks are absent in the difference spectrum. To reveal these peaks, subtractions are then necessary from spectra at other pH values.

5.7. The possibility of quantitative determination of the structure of proteins in solution illustrated for lysozyme

One major future field in the application of n.m.r. spectroscopy to bio-chemistry will come from the combination of the use paramagnetic ions and new n.m.r. instrumental techniques. For example by using Fourier transform spectroscopy in conjunction with a deconvolution technique (see Chapter 14) new line shapes can be produced giving much greater effective resolution. For example, Fig. 5.21 shows the Fourier transform spectrum of lysozyme, which after deconvolution reveals several multiplet structures. This means that use of spin-decoupling experiments (normally inapplicable to proteins), can help in the assignment of resonances to types of amino acids.

Geometric information on the tertiary structure of lysozyme can be obtained from the use of selective perturbations of the resonances by para-magnetic ions, such as we have already discussed for Co(II). In this example, Campbell, Dobson, Williams, and Xavier (1973) used the lanthanide ions. We have already discussed the information obtainable from the use of these (Chapter 4) and for simplicity we summarise this in Fig. 5.22. The *induced shift* of a resonance depends on its distance, r from a paramagnetic centre and two angles† while the *broadening* of a resonance only depends on its distance.‡ The lanthanide ions bind lysozyme between Asp-52 and Glu-35. Thus if the protons of the amino acid residues can be placed in space *relative* to the metal, then the structure of lysozyme in solution can be determined.

Using lysozyme as a test case, because the crystal structure is known and because its structure is expected to be very similar in solution, Campbell *et al.* (1973) made use of difference spectroscopy to determine which amino acid residues were near to the metal binding site. The addition of Gd(III) selectively *broadens* the resonances near to it. By taking the *difference spectrum* of a solution with and without Gd(III) only the difference spectrum of those *resonances near to the metal* will be observed.

Figure 5.23 illustrates this effect for the methyl region of lysozyme. Four

† Axial symmetry is *not* assumed here.

‡ For further discussions of effects of chemical exchange, concentrations of paramagnetic ions and correlation times see Chapters 2, 4, and 9.

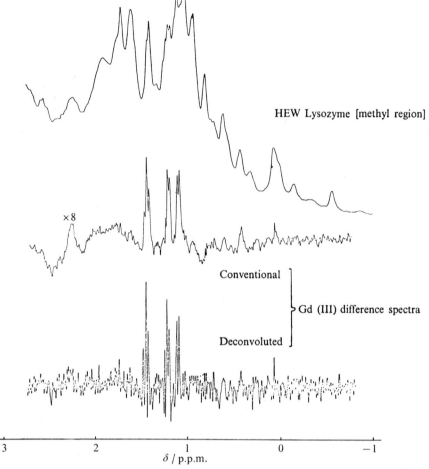

FIG. 5.23. Methyl region of hen egg-white lysozyme and Gd(III) difference spectra·
(From Campbell *et al.*, 1973.)

methyl peaks (two overlap) are 'resolved' by this technique. Additionally
Campbell *et al.* (1973) report that *one* peak only is observed in the aromatic
region.

Figure 5.24 shows a two-dimensional representation of part of the structure
of lysozyme based on the co-ordinates of the crystal structure. There are
indeed three methyl groups and one aromatic proton (from Trp-108) that
appear much nearer than the rest. The methyl peaks in the spectrum are then
in the positions expected for Val(2), Ala(1), and Trp(1) if these residues are
unaffected by ring currents. By the use of higher concentrations of Gd(III),
the technique can be extended to the other residues.

From the use of Gd(III), information on the distances between the residues

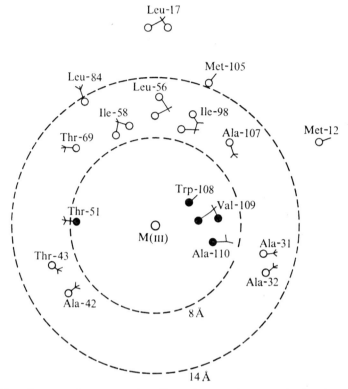

FIG. 5.24. Two-dimensional representation (from the crystal structure) of amino acid residues around an active metal site in lysozyme. (From Campbell *et al.*, 1973.)

and the metal can be obtained (see Chapters 9 and 10). Angular information is obtained as described by observing the *shifts* of these resonances. This is done using both Gd(III) and a *shift* probe at the same time (as illustrated for the example of cyclic-AMP discussed in Chapter 4). The Gd(III) difference spectrum is first obtained and then the shift of the resonances observed on addition of a second lanthanide. Figure 5.25 shows the Gd(III) difference spectra in combination with two different lanthanides.

The preliminary work by Campbell *et al.* (1973) suggests that the structure in solution is indeed very similar to that expected on the basis of the crystal structure.

Although initial work has relied on the crystal structure, this may not always be necessary. Computer programmes used in small molecules can be adapted to obtain fits for a given structure. Obviously the limitation of the technique is in the initial resolution of the protein resonances. However this remains potentially one of the most exciting developments in n.m.r. spectroscopy in the last decade.

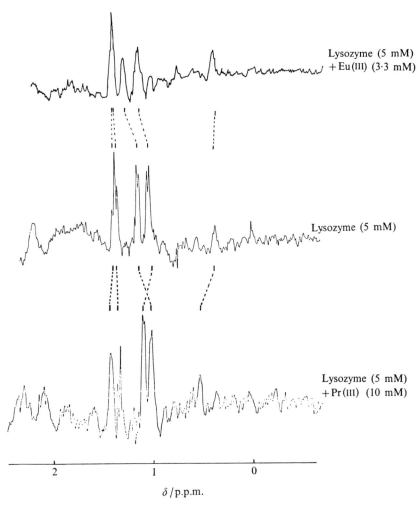

FIG. 5.25. Gd(III) difference spectra of methyl region of lysozyme. Note the pseudo-contact shifts that arise from Pr(III) and Eu(III). (From Campbell *et al.*, 1973.)

Appendix 5.1

Chemical shifts and line-widths of ^1H resonances of amino acids and peptides in neutral D_2O at 40 °C. (From McDonald and Phillips 1969a.)

Hydrogen type	Equivalent hydrogens per residue	Compound	Resonance position, (Hz)†	Apparent half-width, (Hz)‡
Leucine CH_3	6	L-Leucine	208	15
		Glycyl-L-leucine	197	13
		L-Histidyl-L-leucine	193	20
β-CH_2+γ-CH	3	L-Leucine	374	20
		Glycyl-L-leucine	350	16
		L-Histidyl-L-leucine	345	25
α-CH	1	L-Leucine	813	14
		Glycyl-L-leucine	920	15
		L-Histydyl-L-leucine	ca. 905	
Isoleucine CH_3	3	L-Isoleucine	202	16
CH_3	3		217	17
CH_2	1		273	29
CH_2	1		319	29
β-CH	1		429	26
α-CH	1		800	14
Valine CH_3	3	L-Valine	214	17
CH_3	3		225	17
β-CH	1		494	24
α-CH	1		789	15
Alanine CH_3	3	L-Alanine	322	18
		Glycyl-L-alanine	294	18
		L-Tyrosyl-L-alanine	292	
		Glycyl-L-phenylalanyl-L-alanine	288	
		L-Tryptophanyl-L-alanine	300	
α-CH	1	L-Alanine	827	18
		Glycyl-L-alanine	916	18
Threonine CH_3	3	L-Threonine	288	16
β-CH	1		928	20
α-CH	1		772	15
Glycine α-CH_2	2	Glycine	776	10
		Glycylglycine	833, 836	
		Glycylglycylglycine	827, 848, 883	
		Glycyl-L-tyrosine	830	22
		Glycyl-L-leucine	835	10
		Glycyl-L-alanine	833	10
		Glycyl-L-asparagine	837	
Lysine γ-CH_2	2	L-Lysine	321	30
	2	Poly-L-lysine	315	35§
δ-CH_2	2	L-Lysine	375	25
β-CH_2	2	L-Lysine	412	25
δ-CH_2+β-CH_2	4	Poly-L-lysine	376	38§
ε-CH_2	2	L-Lysine	664	22
	2	Poly-L-lysine	660	22§
α-CH	1	L-Lysine	821	12
	1	Poly-L-lysine	947	22§

Appendix 5.1. (*continued*)

Hydrogen type	Equivalent hydrogens per residue	Compound	Resonance position, (Hz)†	Apparent half-width, (Hz)‡
Arginine γ-CH_2	2	L-Arginine	368	28
β-CH_2	2		412	24
δ-CH_2	2		709	14
α-CH	1		822	14
Serine β-CH_2	2	L-Serine	865	25
α-CH	1		840	17
Proline γ-CH_2	2	L-Proline	443	21
β-CH_2	1		456	25
β-CH_2	1		510	25
δ-CH_2	2		736	32
α-CH	1		905	23
Glutamic acid β-CH_2	2	L-Glutamic acid	454	20
γ-CH_2	2		512	20
α-CH	1		818	14
Glutamine β-CH_2	2	L-Glutamine	463	20
γ-CH_2	2		534	20
α-CH	1		820	12
Aspartic acid β-CH_3	2	L-Aspartic acid	595	55
α-CH	1		848	22
Asparagine β-CH_2	1	L-Asparagine	627	29
		Glycyl-L-asparagine	580	30
β-CH_2	1	L-Asparagine	641	29
		Glycyl-L-asparagine	615	30
α-CH	1	L-Asparagine	873	20
		Glycyl-L-asparagine	995	
Methionine CH_3	3	L-Methionine	467	10
β-CH_2	2		467	22
γ-CH_2	2		580	16
α-CH	1		842	12
Cysteine β-CH_2	2	L-Cysteine	671	12
α-CH	1		870	12
Histidine β-CH_2	2	L-Histidine	700	28
		L-Histidyl-L-leucine	702	17
		Glycyl-L-histidylglycine	690	
α-CH	1	L-Histidine	875	15
		L-Histidyl-L-leucine	ca. 895	
Imidazole C(4)H	1	L-Histidine	1558	10
		L-Histidyl-L-leucine	1571	
		Glycyl-L-histidylglycine	1558	
Imidazole C(2)H	1	L-Histidine	1725	10
		L-Histidyl-L-leucine	1773	
		Glycyl-L-histidylglycine	1758	
Tyrosine β-CH_2	1	Glycyl-L-tyrosine	628	27
		L-Tyrosyl-L-alanine	678	
β-CH_2	1	Glycyl-L-tyrosine	683	27
		L-Tyrosyl-L-alanine	694	
α-CH	1	Glycyl-L-tyrosine	972	
		L-Tyrosyl-L-alanine	906	
Aromatic *ortho* to OH	2	L-Tyrosine (80°)	1514	17
		Glycyl-L-tyrosine	1504	
		L-Tyrosyl-L-alanine	1513	

Appendix 5.1. (*continued*)

Hydrogen type	Equivalent hydrogens per residue	Compound	Resonance position, (Hz)†	Apparent half-width, (Hz)‡
Aromatic *meta* to OH	2	L-Tyrosine (80°)	1583	17
		Glycyl-L-tyrosine	1572	
		L-Tyrosyl-L-alanine	1578	
Phenylalanine β-CH_2	1	L-Phenylalanine	684	30
		Glycyl-L-phenylalanyl-L-alanine	ca. 651	
β-CH_2	1	L-Phenylalanine	720	30
		Glycylphenylalanine	ca. 702	
α-CH	1	L-Phenylalanine	876	
Aromatic	3	L-Phenylalanine	1627	
	2		1616	
	5	Glycylphenylalanine	1607	30
Tryptophan β-CH_2	2	L-Tryptophanyl-L-alanine	743	27
α-CH	1		932	15
Indole $C(2)H$	1		1610	10
Indole $C(5)H$	1		1583	15
Indole $C(6)H$	1		1602	15
Indole $C(4)H$	1		1690¶	18
Indole $C(7)H$	1		1658	18

† From internal DSS at 220 MHz.
‡ For half-width of all resonance components of 10 Hz.
§ Actual.
¶ Gerig, J. T. (1968). J. Am. Chem. Soc., **90**, 2681.

Appendix 5.2

Mathematical model for interacting groups in n.m.r. titration curves

Consider first the following equilibrium:

$$XH \rightleftharpoons X^- + H^+ \qquad K_A = [X][H^+]/[HX] \qquad (5.1)$$

If we assume that the chemical shift of X in the two environments is δ_{min}, when it exists as X^- and $\Delta_1 + \delta_{min}$ (see Fig. 5.26) for HX then the observed shift δ is given by

$$\delta = \frac{[HX]}{[HX]+[X]} \times (\Delta_1 + \delta_{min}) + \frac{[X]}{[HX]+[X]} \times \delta_{min} \qquad (5.2)$$

i.e. the observed shift is the weighted average of that in each environment. This obviously assumes fast exchange. From eqns. (5.1A) and (5.2A) we obtain:

$$\delta = \delta_{min} + \frac{\Delta_1 [H^+]/K_A}{1 + [H^+]/K_A}$$

which may be rewritten as

$$\delta = \delta_{min} + \frac{\Delta_1 \times 10^{(pH - pK_1)}}{1 + 10^{(pH - pK_1)}}.$$

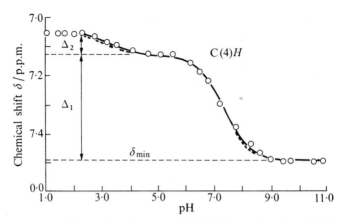

FIG. 5.26. Chemical shift data (relative to external TMS) as a function of pH (—○—) for the C(4) proton at 60 MHz of *N*-acetyl-L-histidine. The solid line is the fit for the sum of two equilibria. (From Shrager *et al.*, 1972 © 1972 Amer. Chem. Soc.)

If there is more than one equilibrium to consider, i.e.

$$HXYH \rightleftharpoons HXY^- + H^+$$
$$\Updownarrow$$
$$H^+ + XYH^-$$

then the titration curve can no longer be analysed in terms of a single pK_a value. If we assume that effects of the two equilibria may be treated in terms of a competition for protons, then by a simple extension of the above:

$$\delta = \delta_{min} + \frac{\Delta_1 \times 10^{(pH - pK_{a,1})}}{1 + 10^{(pH - pK_{a,1})}} + \frac{\Delta_2 \times 10^{(pH - pK_{a,2})}}{1 + 10^{(pH - pK_{a,2})}} + \cdots$$

Δ_1, Δ_2, and δ_{min} are defined as shown in Fig. 5.26. *In practice* this treatment will probably be good enough if the pK_a values are separated by ca. 1·5 units. When they are closer than this, fitting 'by eye' becomes difficult and a computer simulation of the curves is then needed to obtain the pK_a values (Shrager, Cohen, Heller, Sachs, and Schechter, 1972).

References

BRADBURY, E. M. and RATTLE, M. W. E. (1972). *Europ. J. Biochem.* **27**, 270.

BRADBURY, J. H. and KING, N. L. R. (1969). *Aust. J. Chem.* **22**, 1083.

BRADBURY, J. H. and CHAPMAN, B. N. (1972). *Biochem. biophys. Res. Commun.* **49**, 891

—— and KING, N. L. R. (1971). *Aust. J. Chem.* **24**, 1703.

—— —— (1972). *Aust. J. Chem.* **25**, 209.

CAMPBELL, I. D., DOBSON, C. M., WILLIAMS, R. J. P. and XAVIER, A. V. (1973). *J. mag. reson.*—in press and *Ann. N.Y. Acad. Sci.*—in press—(see page 372).

COHEN, J. S., GRIFFIN, J. H., and SCHECHTER, A. N. (1973). *J. Biol. Chem.* in press.

GLICKSON, J. D., McDONALD, C. C., and PHILLIPS, W. D. (1969). *Biochem. biophys. Res. Commun.* **35**, 492.

KING, N. L. R. and BRADBURY, J. H. (1971). *Nature (Lond.)* **229**, 404.

MARKLEY, J. L. (1969). *Ph.D. Thesis*, Harvard Univ., Cambridge, Mass.

McDONALD, C. C. and PHILLIPS, W. D. (1967). *J. Am. chem. Soc.* **89**, 6332.

—— —— (1969a). *J. Am. chem. Soc.* **91**, 1513.

—— —— (1969b). *Biochem. biophys. Res. Commun.* **35**, 43.

MEADOWS, D. H., MARKLEY, J. L., COHEN, J. S. and JARDETZKY, O. (1967). *Proc. Natn. Acad. Sci.* **58**, 1307.

——, JARDETZKY, O., EPAND, R. M., RÜTERJANS, H. H. and SCHERAGA, H. A. (1968). *Proc. Natn. Acad. Sci.* **60**, 766.

RÜTERJANS, H. H. and WITZEL, H. (1969). *Eur. J. Biochem.* **9**, 118.

SCHRAGER, R. I., COHEN, J. S., HELLER, S. R., SACHS, D. H., and SCHECHTER, A. N. (1972). *Biochemistry*, **11**, 541.

SCHECHTER, A. N., SACHS, D. H., HELLER, S. R., SHRAGER, R. I., and COHEN, J. S. (1972). *J. mol. Biol.* **71**, 39.

SCHULMAN, R. G., PEISACH, J., and WYLUDA, B. J. (1970). *J. molec. Biol.* **48**, 517.

WÜTHRICH, K. (1970). *Structure and Bonding*, **8**, 53.

——, SHULMAN, R. G., YAMANE, T., WYLUDA, B. J., HUGLI, T. E., and GURD, F. R. N. (1970). *J. biol. Chem.* **245**, 1947.

6

THE BINDING OF SMALL LIGANDS
TO MACROMOLECULES

6.1. Introduction

IN many instances the n.m.r. spectrum of a macromolecule may consist of a
broad envelope so that it is not possible to distinguish any individual resonance
lines. Alternatively, there may be insufficient concentration of the macromolecule to give a strong resonance absorption. The problem of studying
interactions between a small ligand or a macromolecule has then to be
approached by observation of any changes in the ligand spectrum caused by
the presence of the macromolecule. There are at least two environments for
the ligand, *free* and *bound* to the macromolecule (which may have several
ligand binding sites). Occasionally, it may be possible to detect the resonances
of bound ligand directly, but generally the concentration of the macromolecule, and thus that of the bound ligand, will be small and this detection
will be difficult. If, however, there is fast chemical exchange between the
bound and free environments, then the observed resonances will be a
weighted average of that in each environment and thus information on the
bound resonance signals can be obtained.

In the high-resolution spectrum of the ligand both relaxation times (or the
broadening) and chemical shifts may be measured, but the lack of any effect
on addition of a macromolecule does not, of itself, imply that binding has not
occurred. For example, if the amount of ligand bound is too small to detect,
then under conditions of slow exchange no effect on the bulk resonance will
be observed. Alternatively, conditions of fast exchange may not apply to all
the resonances (since they may all have different relaxation times and
chemical shifts in the bound site) and great care is then needed in interpreting
any differential effects in the spectrum.

Changes in the relaxation times of the ligand nuclei resulting from the
addition of the macromolecule may be difficult to interpret quantitatively, but
qualitatively may give some indication of the mobility of the particular
nucleus. In this respect the information available from measuring spin–lattice
relaxation times, T_1, may be more valuable than that from spin–spin relaxation times, T_2 (line-widths) for if there is a difference in chemical shifts ($\Delta\omega_M$)
between the bulk and bound site, T_2 will be affected more than T_1, i.e. the
conditions for slow and fast exchange under these conditions are defined
differently for T_1 and T_2 since the chemical shift difference will contribute to
T_2 but not T_1. (See p. 43.)

Conventionally, T_2 is measured from line-broadening experiments but the presence of a multiplet (from spin–spin coupling) may make this difficult. Even if conditions of fast exchange are valid, coupling constants of nuclei in the bound site may change from those of the bulk site and this can lead to an 'apparent broadening'. Further, in any spectroscopic technique in which a line-width is measured, the width may be dependent upon instrumental effects which may cause broadening. While some of these effects like magnetic inhomogeneity may sometimes be allowed for, there are cases (such as those involving 'noise decoupling' of protons or other nuclei) where this is not possible.

In this section we shall concentrate mainly on the type of information that can be obtained from shift measurements. With hydrogen nuclei the shifts caused by the macromolecule are often quite small. The shifts of other nuclei, e.g. ^{19}F, occur over a broader range and are much more sensitive to changes in environment, particularly when the interaction causing these changes is transmitted through chemical bonds to the nucleus. To this end, substituted ligands are often used and we shall give some examples of this in other chapters. Here we consider some examples that indicate the range of experiments that can be performed.

6.2. Observation of protein and ligand resonances

6.2.1. *Ribonuclease*

The addition of the inhibitor 3'-CMP (cytidine monophosphate) to RNase A at pH 5·5 gives downfield chemical shifts of the C(2) proton resonances of histidine-119, -12, and -48. The C(2)H and C(4)H resonances assigned to His-105 are however unaffected. The variation of the chemical shift with 3'-CMP concentration is shown in Figs. 6.1 and 6.2. The data also allow calculation of the 3'-CMP–enzyme dissociation constant (ca. 3 mM) at this pH value.

The titration curves of the histidine resonance in the absence and presence of 3'-CMP (in saturating amounts between pH 5 and 7) are shown in Fig. 6.3. Again we note that His-105 is unaffected while the resonance of His-48 is shifted downfield at low pH, but at pH $>5·5$ becomes too broad to detect. This behaviour may be indicative of an exchange process and is referred to later (see p. 118).

By a comparison of these pH titration curves with the curves obtained (by other techniques) for the pH dependence of the binding of 3'-CMP to the enzyme (Meadows and Jardetzky, 1968), it was concluded that the deprotonation, and consequent upfield shift of the C(2)H resonances, of histidine-12 and -119 only occurs when the enzyme is no longer saturated with 3'-CMP, i.e. the binding constant is pH dependent. Thus *in the inhibitor–enzyme complex both histidine-12 and -119 residues must be protonated.*

Similar experiments have been performed with various other inhibitors,

9

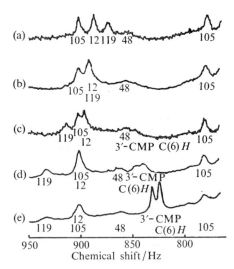

FIG. 6.1. Imidazole C(2)*H* and C(4)*H* region of the ¹H n.m.r. spectrum (100 MHz) of ribonuclease A [0·0065 M in 0·2 M NaCl (D₂O), pH (meter reading) 5·5, 32 °C] in the presence of increasing concentrations of cytidine 3′-monophosphate (3′-CMP). Peak 105 at 780 Hz is the imidazole C(4)*H* peak of His-105. Other peaks, labelled 105, 12, 119, and 48 are imidazole C(2)H peaks. (a) RNase alone, time-average of 25 scans; (b) +0·002 M 3′-CMP, 100 scans; (c) +0·005 M 3′-CMP, 39 scans; (d) +0·010 M 3′-CMP, 68 scans; (e) +0·030 M 3′-CMP, 100 scans. Chemical shift scale is in Hz downfield from hexamethyl-disiloxane. (From Meadows and Jardetzky, 1968.)

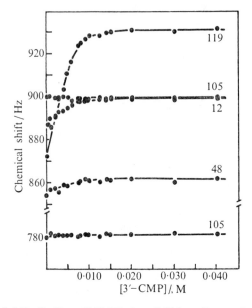

FIG. 6.2. Chemical shifts (in Hz at 100 MHz downfield from hexamethyldisiloxane) of the histidine C(2)*H* and (at 780 Hz) C(4)H peaks of ribonuclease A as a function of total added 3′-CMP concentration. Ribonuclease concentration 0·0065 M in 0·2 M NaCl (D₂O), pH (meter reading) 5·5, 32 °C. (From Meadows and Jardetzky, 1968.)

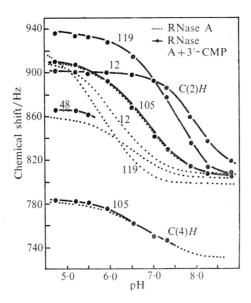

FIG. 6.3. Titration curves of the histidine residues of ribonuclease A in the presence (solid curve) and absence (broken curve) of 0·03 M 3′-CMP. The ordinate is the chemical shift of the C(2)H or C(4)H peaks, in Hz (at 100 MHz) downfield from hexamethyldisiloxane. (From Meadows and Jardetzky 1968.)

but we draw attention to 2′-CMP, 5′-CMP, and phosphate and the results of these experiments are listed in Table 6.1.

The results may be summarized as follows:

(1) 2′-CMP and 3′-CMP give *downfield* shifts of His-119 at low pH (fully protonated form) and *upfield* shifts of His-12. (His-105 is almost unaffected.)

(2) 5′-CMP has no effect on His-119 and produces an *upfield* shift of His-12 similar to that of 2′- and 3′-CMP.

(3) The pK_a value of His-12 is increased by 1·8 units irrespective of the inhibitor, but this is not the case for His-119 when the values are different.

It would seem that while His-119 is sensitive to the position of the phosphate group, (which has different orientations in the three inhibitors) His-12 is not and it is concluded that His-119 *forms a direct bond* to the phosphate group of 3′-CMP and 2′-CMP. On the other hand, His-12 shows an increase in pK_a which arises from being in the vicinity of the negatively-charged phosphate group (but has much less specific binding to it than has His-119).

One other change in the spectrum of RNase A can be observed when the inhibitors bind. An upfield shift of a peak in the main aromatic envelope

TABLE 6.1

Summary of chemical shift and pK$_a$ changes on binding of inhibitors to RNase† (from Meadows, Roberts, and Jardetzky, 1969)

| | RNase absorptions | | | | | | | | | | | Inhibitor absorptions at pH 5·5 | | |
| | His-12 | | | His-48 | His-105 | | | His-119 | | | Aromatic shift (Hz) | C(5)H (Hz) | C(6)H (Hz) | C(1')H (Hz) |
Inhibitor	pK$_a$	ΔpK$_a$‡	Δδ§ (Hz)	Δδ§ (Hz)	pK$_a$	ΔpK$_a$‡	Δδ§ (Hz)	pK	ΔpK	Δδ§ (Hz)				
3'-CMP	8·0	1·8	+10	−7	6·7	0	0	7·4	1·6	−20	+30	−6	−22	−3
2'-CMP	8·0	1·8	+8	−10	6·7	0	0	>8·0	>2·2	−25	Yes¶	−4	ca. −20	−12
5'-CMP	8·0	1·8	+10	−10	6·7	0	0	<7·0	<1·2	0	+30	−5	−17	−3
Phosphate	6·9	0·7	0	−6	6·7	0	0	6·6	0·8	+10	No			

† Downfield chemical-shift changes are given as negative numbers in Hz. In all cases inhibitor concentrations were sufficient to saturate the enzyme between pH 5 and pH 7.

‡ ΔpK = Change (increase) in pK$_a$ in the presence of inhibitors.

§ Δδ = change in chemical shift of the C(2)H peak of the fully-protonated histidine.

¶ Due to broadening of the peak, this shift could not always be followed over the whole concentration range.

corresponding to ca. five hydrogen nuclei occurs (possibly a phenylalanine residue) and this is thought to *implicate the binding of the pyrimidine ring* of the inhibitor since this shift is absent when phosphate alone binds.

Turning now to the observation of the inhibitor resonances, those of the pyrimidine ring C(5)H [doublet from splitting with C(6)H] and C(6)H and the

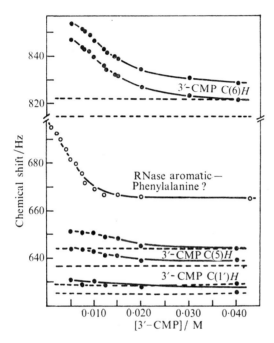

FIG. 6.4. Chemical shifts [in Hz at 100 MHz downfield from hexamethyldisiloxane (HMS)] of ribonuclease A aromatic peak and 3'-CMP pyrimidine C(6)H and C(5)H and ribose C(1')H peaks as a function of total added 3'-CMP concentration. Ribonuclease concentration, 0·0065 M in 0·2 M NaCl (D$_2$O); pH (meter reading), 5·5, 32 °C. Dashed lines indicate positions of peaks of 3'-CMP alone at equivalent concentrations. (From Meadows and Jardetzky 1968.)

C(1')H of the ribose ring are easily identified and monitored. These resonances show downfield chemical shifts on binding and those for 3'-CMP are shown in Fig. 6.4. The shifts for the C(5)H and C(6)H resonances are identical for all three inhibitors, thus again implicating the pyrimidine ring in binding (Fig. 6.4). The C(1')H resonance shows a much larger downfield shift when 2'-CMP binds to the enzyme than for the other inhibitors and thus the *conformation of the ribose moiety in this complex may be different.*

By combining the above conclusions from the n.m.r. data with the relative positions of the amino acids in the active site as determined from X-ray diffraction, (and of course assuming the conformation of RNase A in solution is unchanged) proposed structures for the inhibitor complexes were obtained such as that shown for 3'-CMP in Fig. 6.5. Meadows *et al.* (1969) conclude that if the pyrimidine ring of all the inhibitors is to bind in the same way and the phosphate group binds to His-105, the orientation about the glycosidic bond must be *syn* in 2'-CMP and this would account for the difference in the shift of the $C(1')H$ resonance of this inhibitor. The position of the phosphate group is fixed by two coulombic bonds to His-119 and Lys-41. X-ray data suggests that the cytosine ring is depicted in the *anti*-conformation and is fixed by the three hydrogen bonds shown, thus placing the 2'-OH group on the ribose ring very near to His-12. Phenylalanine-120 and the cytosine ring are in close contact.

Obviously there may have to be some refinements of this model since analysis of the His-C(2) titration data in presence of inhibitors indicate some asymmetry and 2'- and 3'-uridine monophosphates produce larger effects on His-12 than on His-119 (Fig. 6.6). In these cases, in contrast to the 2'-, 3'-, and 5'-CMP inhibitors, it is the pK_a of His-119 that appears insensitive to the position of the phosphate, while His-12 *is* sensitive. This is the reverse of point (2) on page 113 and would, on the same argument, mean that *His-12 forms a direct bond to the phosphate* group. Some care must therefore be taken in generalizing any effects observed in only one series of compounds. Nevertheless, as a first step, this analysis must rank as one of the major steps in the application of n.m.r. spectroscopy to biological systems.

Complex of 3'-CMP with RNase

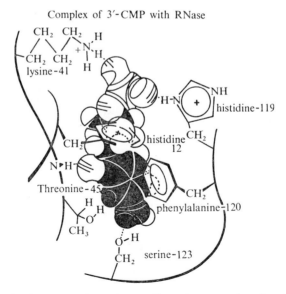

FIG. 6.5. Postulated structure of the 3'-CMP ribonuclease complex, viewed from the back of the active site cleft. (From Meadows *et al.* 1969.)

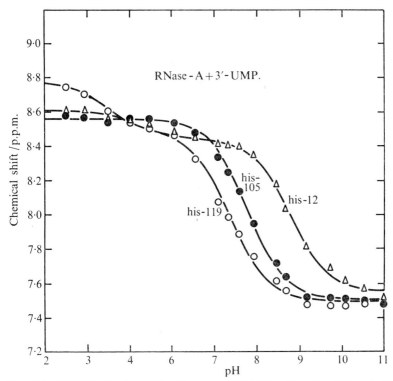

FIG. 6.6. 220 MHz titration of the ^1H n.m.r. of $C(2)H$ histidine-12, -105, and -119 in the presence of 3′-UMP (Chemical shifts downfield from external TMS) (From Griffin, Cohen, Schechter, Damodaran, Jones and Moffat, 1973.)

6.2.2. Exchange phenomena in RNase

We include two examples of the type of exchange behaviour that is often reported in enzyme systems:

(1) The addition of 2′-CMP to RNase at pH 5·5 results in a broadening of the His-119 resonance. However, when one molar equivalent of 2′-CMP is added, the peak narrows again, since the enzyme exists in the complexed form (Fig. 6.7). The broadening arises from the classical two-site exchange problem. The $C(2)H$ hydrogen may exist either in the enzyme–inhibitor complex or in the free enzyme. If the 2′-CMP (or even the hydrogen itself) exchanges between the two sites such that $\tau_M \Delta\omega_M \approx 1$ then broadening will occur. As the concentration of 2′-CMP is increased all the enzyme exists as the complex and there is effectively only one site which in this case has a similar relaxation time to that in the free enzyme—hence the broadening disappears.

(2) We have mentioned that the line-width and chemical shift of the His-48 $C(2)H$ hydrogen are altered on inhibitor binding despite the fact that the His-48 is 'buried' and removed from the active site. This

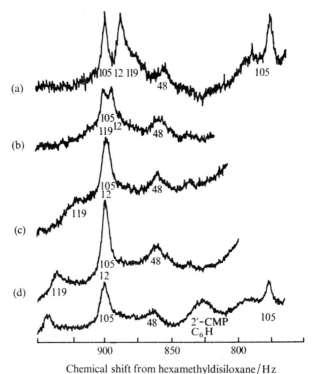

Chemical shift from hexamethyldisiloxane/Hz

FIG. 6.7. 100 MHz ^{1}H n.m.r. spectra of the histidine absorption region of 0·0065 M RNase with increasing amounts of 2′-CMP, pH 5·5. The C(2)H peaks of histidine-12, -48, -105, and -119, the C(4)H peak of histidine-105 (at approximately 870 Hz) and the broadened C(6)H doublet of 2′-CMP are indicated. Solutions are in 0·2 M NaCl(D$_2$O) at 32 °C. (a) RNase+0·001 M 2′-CMP; 38 CAT (computer of average transients) scans. (b) RNase+0·004 M 2′-CMP; 50 CAT scans. (c) RNase+0·006 M 2′-CMP; 55 CAT Scans. (d) RNase+0·008 M 2′-CMP; 100 CAT scans. (e) RNase+0·02 M 2′-CMP; 35 CAT scans. (From Meadows et al. 1969.)

observation could be rationalized by suggesting that there is a conformational equilibrium involving His-48 which is altered by inhibitor
binding. As previously, this would place the hydrogen in two sites
with different chemical shifts. Whenever such behaviour is observed,
additional experiments (which may involve measurements at different
temperatures and/or frequencies) are needed to confirm unambiguously
that the phenomenon actually results from chemical exchange. For
instance, an alternative explanation for (2) is that the broadening is produced by the C(2) hydrogen coming into close proximity with a hydrogen
of another group or even with a carboxyl group with consequent dipolar
broadening. This could result from the conformational change. The
observation of the His-48 C(4) hydrogens could help clarify this
(see p. 91).

6.2.3. *The binding of uridine 3'-monophosphate (3'-UMP) to ribonuclease A*

This is included to indicate the range of techniques that may be used in n.m.r. In this example the ^{31}P n.m.r. spectrum of the inhibitor was observed (Lee and Chan, 1971). As with ^{19}F, ^{31}P chemical shifts (in Hz) are larger than those of ^{1}H nuclei so that for a given chemical exchange rate τ_M, the possibility of being in the region $\tau_M \Delta\omega_M \approx 1$ is probably greater for ^{31}P nuclei than for ^{1}H. We noted above for RNase A, that, with several of the inhibitors, fast-exchange conditions may not always be valid in the presence of the enzyme. It is not too surprising, therefore, that the ^{31}P resonance of the inhibitor UMP exhibits characteristics of the intermediate exchange region. By suitable analysis of line-widths and chemical shifts as a function of the ratio of enzyme-inhibitor concentration, Lee and Chan (1971) obtained a rate constant for the chemical-exchange process ($k \approx 3200$ s^{-1}) and a value for the chemical shift of the ^{31}P in the enzyme complex (800 Hz upfield at 89 MHz).

Interestingly, ^{31}P n.m.r. studies of nucleotides showed that protonation of a primary phosphate shifts upfield by ca. 400 Hz at 89 MHz per protonation step. This would imply that 3'-CMP is bound in the enzyme simultaneously to two positively-charged side-chains of two amino acid residues, such as Lys-41 and His-119 as indicated in Fig. 6.5.

6.2.4. *Mechanism of ribonuclease A*

The mechanism proposed by Roberts, Dennis, Meadows, Cohen, and Jardetzky (1969) is shown in Fig. 6.8. It is based on X-ray data and the above n.m.r. conclusions and on a mechanism for phosphate ester hydrolysis which imposes certain geometrical restrictions on the substrate. The main restrictions are that the hydrolysis proceeds via a penta-covalent intermediate with the geometry of a trigonal bipyramid and that groups enter or leave the intermediate from apical positions. X-ray data suggest the spatial relationship between the two nucleotide binding sites as shown in the Fig. 6.8. By analogy with the inhibitor studies, the 2'-OH group of the substrate is close to His-12 and the phosphate group close to His-119. The His-12 residue accepts the proton from the 2'-OH activating the 2'-oxygen for attack on the phosphorus to form the penta-covalent intermediate. Since the attacking 2'-oxygen must be apical, geometry is constrained in such a way that the –OR group is apical,† and can thus leave directly resulting in formation of the cyclic phosphate.

The position of the RO$^-$ leaving group is such that it may receive a proton from His-119 (although the possibility that His-12 or water donates a proton cannot be ruled out). The pK_a value of His-119 is sensitive to the position of the phosphate group and the formation of the cyclic phosphate at opposite sides of the cleft would be expected to lower its pK_a value. This, combined

† On purely chemical grounds however, it is possible that, depending on the direction of the attack by the 2'-oxygen, the initial intermediate could have the –OR leaving group in an equatorial position. Pseudorotation would then have to occur to place –OR in an apical position (see Fig. 6.8II).

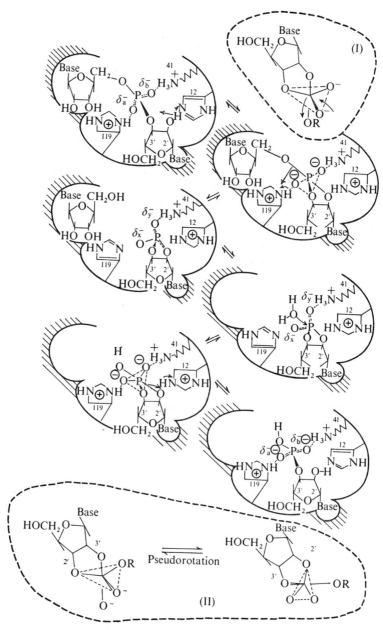

FIG. 6.8. Proposed mechanism of action of RNase. The relative positions of the substrate and the amino acid side-chains deduced from X-ray crystallography and n.m.r. spectroscopy are shown, as far as is possible in two dimensions. The shaded areas represent binding sites for the two nucleoside bases. The notation δ^- is used to indicate the uncertainty in the negative charge on the phosphate oxygens: $(\delta_a^- \approx \delta_b^- \approx \frac{1}{2})$ and $(0 < \delta_x^- < \frac{1}{2} < \delta_y^- < 1)$. In the last structure, the ionization states shown are those which result from the reaction; optimal binding of the product results when the phosphate group is doubly ionized and histidine-12 is protonated. (I) represents the initial intermediate in the linear mechanism. (II) represents the initial intermediate in the pseudo-rotation mechanism which would allow the same histidine to interact with the 2' oxygen and then with the RO$^-$ leaving group. (From Roberts *et al.* 1969.)

with the reduction in the negative charge density at the nearby phosphate oxygen, could result in facilitating the donation of a proton to RO⁻, or to a water molecule, with subsequent hydrogen bonding to a water molecule to initiate the second step. (Some movement of His-119 might be necessary to accommodate the water molecule.) In contrast to the pK_a of His-119, the value for His-12 is insensitive to the position of the phosphate and it is expected to remain protonated.

The hydrolysis of the cyclic phosphate is the reverse of the first step; His-119 donates the initial proton directly to the leaving group, and then His-12 acts as a proton donor to the 2'-oxygen.

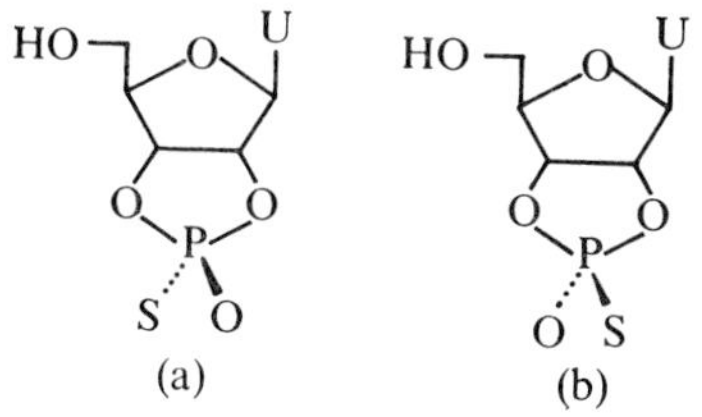

FIG. 6.9. The two diastereoisomers of uridine 2',3'-cyclic phosphorothioate. (From Usher *et al.*, 1972.)

The above mechanism is referred to as the *linear mechanism*. However a mechanism has been proposed by Hammes (1968) in which the same histidine residue interacts first with 2'-oxygen and then with the RO⁻ leaving group. On the basis of the binding configuration shown above this would require either a substantial movement of this histidine or a '*pseudo rotation*' of the intermediate to place the leaving group in an apical position (Fig. 6.8II). While the large movement necessary by the histidine seems unlikely from the results of X-ray diffraction, the possibility of pseudo-rotation cannot be ruled out, but it is considered unlikely on the basis of the geometry of substrate binding which is not compatible with having both the 2'-OH and OR⁻ leaving group close to the same histidine residue.

It should be emphasized that the conclusions on the mechanism are extrapolated from the case of the binding of the dianionic inhibitors only. Since the substrates are phosphodiesters which can only exist in the mono-anionic form, it could be that the di-anions induce conformational changes in the enzyme which affect the mode of binding of the inhibitor. That this is probably not so, is well illustrated by the following elegant experiments with the mono-anionic diester uridine 2',3'-cyclic phosphorothioate, a substrate intermediate which also provide a good example of the use of ³¹P n.m.r. spectroscopy in structural studies.

Usher, Erenrich, and Eckstein (1972) synthesized cyclic-Up(S)(often written Up̂(s)) which exists as two diastereoisomers (Fig. 6.9), one of which,

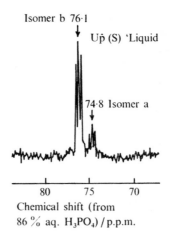

FIG. 6.10. ^{31}P n.m.r. spectrum of a mixture of the isomers of uridine 2′,3′-cyclic phosphoro-thioate, containing largely liquid isomer [isomer (b)]. (From Usher *et al.*, 1972.)

(a), could be obtained in crystalline form. The ^{31}P n.m.r. spectrum of a mixture is shown in Fig. 6.10. By taking advantage of the reversibility of the RNase reaction, it was possible to synthesize the intermediate in the enzyme reaction, starting from the crystalline isomer of cyclic Up̂(S) and cytidine. Cyclic Up̂(S) was then reformed by a non-enzymic reaction known to proceed via a *linear* mechanism. The reaction scheme is:

$$C + U\hat{p}(S) \xrightleftharpoons{\text{RNase A}} Up(S)C$$

$$Up(S)C \xrightarrow[\text{linear}]{\text{Base}} U\hat{p}(S) + C$$

where C = cytidine.

The nucleotide can approach the phosphorus atom from either of two directions (Fig. 6.11). If it attacks from the left, i.e. linearly, the 2′-oxygen assumes an apical position in the resulting trigonal bipyramidal complex and hence the hydrolysis can proceed, producing the same diastereoisomer of Up̂(S) as the starting material. On the other hand, if the attack is from the right (adjacent) then the 3′-oxygen is apical and pseudo-rotation must occur to

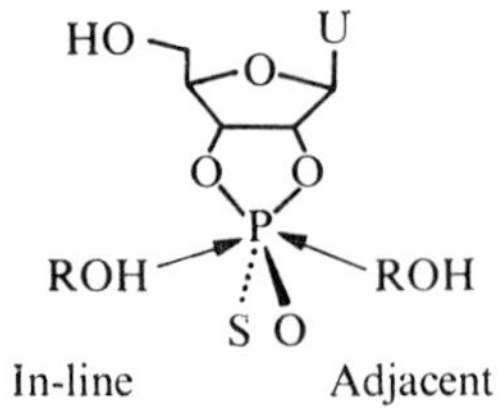

FIG. 6.11. The directions of attack for the in-line and adjacent mechanisms. ROH = Cytidine.

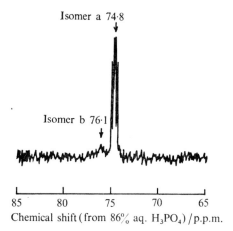

FIG. 6.12. ^{31}P n.m.r. spectrum of the final product of the reaction scheme as discussed in the text. (From Usher *et al.*, 1972.)

place the 2′-oxygen apical so that a different diastereoisomer of UpS is formed after hydrolysis. The ^{31}P n.m.r. spectrum of the product (Fig. 6.12) indicated that the original isomer had been reformed and thus the first step must involve a linear reaction. From the principle of microscopic reversibility the reverse reaction of the first step must be linear, and hence the pseudo-rotation mechanism is ruled out conclusively.

One disturbing point, however, emerges regarding the mechanism of protonation of the histidine as the first step. If the pH titration experiments are carried out for the *dinucleotide phosphate* which is thought to be a good substrate analogue and has a phosphonate bond (CH_2 group) to prevent hydrolysis, the titration curves (Fig. 6.13) show that at pH ca. 7 and above *no histidines are protonated*, in contrast to the results with the inhibitors above. The pH dependence of the catalytic action of RNase A thus poses a problem. Is the above mechanism wrong? Is this substrate analogue not a good model? Does only a small fraction of the histidine residue need to be protonated? Is the phosphate group protonated and not His-119? Is protonation not really the correct way to envisage the mechanism? These questions remain to be answered. Perhaps a thorough ^{31}P n.m.r. spectral study of the phosphate group itself may help to sort out some of these other possibilities.

The work with RNase does however provide a good model for illustrating the subtleties in interpretation and some of the experimental difficulties which are likely to be encountered in other systems. Even if the proposed mechanism and assignment of the histidines are not quite correct, a great deal of useful information has still been obtained. More importantly the work has pioneered the application of one aspect of n.m.r. spectroscopy to the studies of enzyme structure and function.

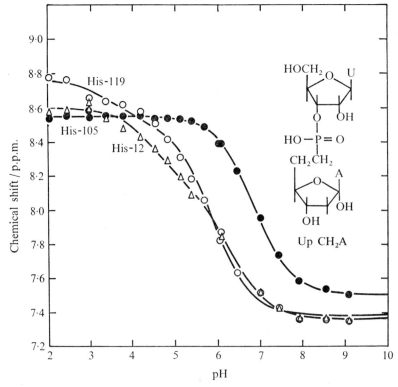

Fig. 6.13. pH Dependence of chemical shift (downfield from external TMS at 220 MHz) of His-12, -105, and -119 of RNase A in the presence of 3 M excess of UpCH₂A U = uracil, A = adenine. (From Griffin, Schechter and Cohen, 1973.)

6.3. The relative modes of binding of several inhibitors and substrates to lysozyme

6.3.1. *Monosaccharide inhibitors*

At 100 MHz the ¹H n.m.r. spectrum of a mutarotated solution of *N*-acetylglucosamine (NAG) consists of two sets of resonances for the anomeric hydrogen (corresponding to the α- and β-forms of NAG) but only one for the methyl group hydrogens in the acetamido side chain. Addition of lysozyme to NAG removes this degeneracy and two methyl resonances are observed, both of which are shifted upfield (Fig. 6.14). By examining methyl chemical shifts of the acetamido group for freshly prepared solutions of the pure α- and β-anomers of NAG, it was possible to assign the resonances to the α- and β-anomers. Figure 6.14 shows the changes in the spectrum for the methyl protons, starting with a solution of the α-anomer, as it undergoes muta-rotation to the equilibrium mixture of α- and β-anomers. If conditions of fast exchange apply to each resonance (i.e. $\tau_M \Delta\omega_M \ll 1$, where $\Delta\omega_M$ is the shift

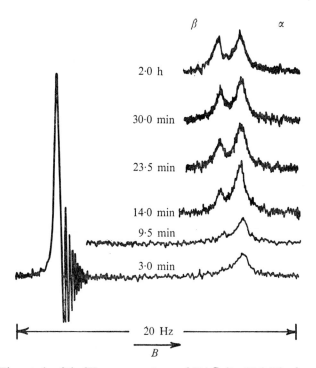

FIG. 6.14. Time study of the ^{1}H n.m.r. spectrum of NAG (5×10^{-2} M) after addition to lysozyme (3×10^{-3} M) in 0·1 M citrate in D_2O, pD 5·9. A solution of acetone (0·5 %) was used as an internal standard, and its resonance appears to lower field of the NAG resonances. (From Raftery, Dahlquist, Chan, Parsons, and Woolcott, 1968.)

of the enzyme-bound form of the enzyme from the unbound enzyme) the observed change in chemical shift will depend on the amount of each form of NAG bound and the chemical shift in the bound site. The observation of two separate resonances, which are shifted differently in the presence of lysozyme, indicates that either the two anomers have different binding constants to the enzyme or that the magnetic environment of the two methyl groups are different, or both. By competition studies with α-NAG (in which the methyl group was deuteriated and thus did not contribute to the observed signal), the upfield chemical shift of both resonances were decreased (because the fraction bound had decreased) and it was concluded that both the anomers bind at the same site. The problem of mutarotation is an added complication in measuring the binding of NAG to lysozyme by observation of the chemical shifts. From suitable plots of the chemical shift as a function of added NAG and using the procedures outlined in Appendix 6.1, values of K_I and $\Delta\omega_M$ for each anomer can be obtained. The values obtained are listed in Table 6.2.

The complication of mutarotation does not arise with α-Me- or β-Me-NAG, and in these cases the values of K_I and $\Delta\omega_M$ are obtained directly from

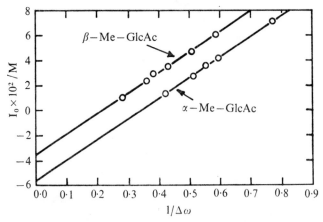

FIG. 6.15. The reciprocal of the observed chemical shift change of the acetamido methyl protons ($1/\Delta\omega$), as a function of total inhibitor concentrations (I_0) on addition of various concentrations of methyl N-acetyl-α-D-glucosamine or methyl N-acetyl-β-D-glucosamine to a constant concentration (0·003 M) of hen egg white lysozyme. (From Raftery *et al.* 1968.)

a plot of the chemical shift data as shown in Fig. 6.15. The results for the α-Me and β-Me-NAG (Table 6.2) suggest that the methyl groups of both compounds have similar magnetic environments on the enzyme and thus, presumably, bind with similar orientations. That they both bind in the same site as α-NAG (known from X-ray work to be site C) was again shown by competition studies with α-NAG (with a deuteriated methyl group) as described above. Thus all four of these inhibitors bind at the same sites on the enzyme, and three of them have almost identical values of $\Delta\omega_M$ for the acetamido methyl group. Since the only difference between α-NAG and β-NAG is the configuration of the hydrogen and hydroxyl at C(1), it has been suggested that these observed differences in K_I and $\Delta\omega_M$ between the two anomers could result from α-NAG forming a bond to the enzyme through the C(1) hydroxyl group.

TABLE 6.2

Data for binding of the anomers of NAG and Me-NAG to lysozyme

		$\Delta\omega_M$ (p.p.m.)	
Inhibitor	K_I (mM)	(a) NH·COMe	(b) OMe
β-NAG	33±2	0·51±0·03	
α-NAG	16±1	0·68±0·03	
β-Me-NAG	33±5	0·54±0·04	0·17±0·03
α-Me-NAG	52±4	0·55±0·02	0

Further information can be obtained by studying the pH dependence of the binding constant K_I and of $\Delta\omega_M$. The results for β-Me-NAG are shown in Fig. 6.16 (a and b). From Fig. 6.16a (a) a group on lysozyme with a $pK_a = 6\cdot1$ is implicated in the binding, while the pH dependence of $\Delta\omega_M$ implies groups with pK_a values of $4\cdot7\pm0\cdot1$ and $7\cdot0\pm0\cdot5$. The lower value comes from a group which may not be involved in the binding since it does not affect the

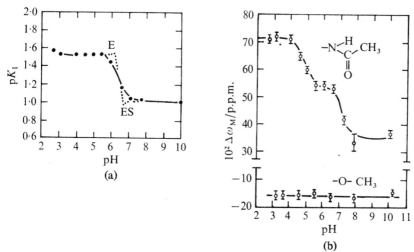

(a)

(b)

FIG. 6.16(a) The effect of pH on the dissociation constant, K_I, for β-Me NAG and lysozyme. To determine the pK_a values of ionizable groups on the enzyme and the enzyme-inhibitor complex, $-\log K_I(pK_I)$ was plotted against pH. (From Dahlquist and Raftery, 1968.) (b). The pH dependency of the chemical shifts, $\Delta\omega_M$ of the glycosidic methyl hydrogen and the acetamido methyl hydrogens of β-Me-NAG when bound to lysozyme. (From Dahlquist and Raftery, 1968.)

value of K_I (Fig. 6.16a) but the proximity of it could be such that it still may influence $\Delta\omega_M$ because it will affect the magnetic environment of the binding site. For example, the pK_a value of $4\cdot7$ suggest a carboxyl group which may cause a chemical shift because of electric field effects or because the magnetic field produced by it is anisotropic. Gross conformational changes can be ruled out since $\Delta\omega_M$ for the OCH_3 resonance shows no variation with pH. By inspection of the three-dimensional crystal structure of the inhibitor-lysozyme complex the group of $pK_a = 6\cdot1$ is assigned to Glu-35 (since its environment is non-polar and it would be expected to have an abnormally high pK_a), and the pK_a of $4\cdot7$ is assigned to Asp-103. It is worth stressing however that such analysis relies on knowledge of the crystal structure and in other enzymes where this is not so, the value of such studies will be limited.

6.3.2. Chitin oligosaccharides and their β-methyl glycosides

The acetamido methyl group resonances at 100 MHz of NAG, chitobiose, chitotriose, and chitotetraose are shown in Fig. 6.17. The resonance to

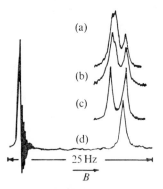

FIG. 6.17. ¹H N.m.r. spectra of the acetamido methyl groups of (a) chitotetraose, (b) chitotriose, (c) chitobiose, and (d) NAG. The sharp and intense resonance to lowest field is for an acetone internal standard. (From Raftery, Dahlquist, Parsons, and Wolcott, 1969.)

highest field corresponds to the methyl group at the reducing end of the saccharide, and the resonance at lowest field to the methyl group at the non-reducing end. Dahlquist and Raftery (1969) and Raftery *et al.* (1969) studied the changes in chemical shift and line broadenings of the various methyl resonances on addition of lysozyme to see if a general pattern would emerge. We briefly summarize their observations and conclusions.

The addition of lysozyme to a solution of *chitobiose* results in a large broadening of the resonance to high field and also a chemical shift (to high field) of this same resonance (Fig. 6.18). The extrapolated value of the line-width of the bound signal was estimated to be ca. 200 Hz, which is much larger than the value of ca. 20 Hz expected for methyl group residues in proteins (see p. 105). The most likely explanation is that of slow or intermediate chemical exchange (see p. 117) of the inhibitor between the free

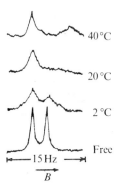

FIG. 6.18. ¹H n.m.r. spectra of the acetamido methyl group of chitobiose (5×10^{-2} M) free and in the presence of lysozyme (3×10^{-3} M) at various temperatures. (From Raftery *et al.*, 1969.)

TABLE 6.3

Chemical-shift data for inhibitors and substrates complexed with lysozyme at various pH and temperature values. (From Raftery et al., 1969.)

Compound	Temp. (°C)	pH	$CH_3-N_1\ddagger$	$\Delta\omega_M$ (p.p.m.)† $CH_3-N_2\ddagger$	$CH_3N_3\ddagger$	$OCH_2\S$
β-Methyl-NAG	31	4·9–5·4	0·54 ±0·04	—	—	0·17±0·03
	55	4 9–5·4	0·51 ±0·03	—	—	0·16±0·05
Chitobiose	45	4 9–5·4	0·57 ±0·04	0	—	—
Methyl-β-chitobiose	35	4 9–5·4	0·60 ±0·05	0	—	0·20±0·05
β-Methyl-NAG	31	9·7	0·36	—		0·16±0·02
Chitobiose	55	9·7	0·77 ±0·04			
Methyl-β-chitobioside	55	9·7	0·80 ±0·04	0		0·16±0·02
Chitotriose	65	9·7	0·61¶±0·12	0	0·08	
Methyl-β-chitotrioside	65	9·7	0·63¶	0	0·08	0·19

† The acetamido methyl groups are numbered 1, 2, 3 beginning at the reducing or glycosidic terminus of the inhibitor molecules.
‡ Values relative to acetone; all chemical shifts to higher field.
§ Values relative to methanol; all chemical shifts to lower field.
¶ Not in fast exchange limit.

and enzyme-bound environment. This is confirmed by changing the temperature, when the chemical shift is increased and the resonance narrows. In contrast to the results obtained for NAG, no distinction between the α- and β-anomers in the presence of lysozyme is apparent in the n.m.r. spectrum suggesting that the two anomers occupy magnetically equivalent positions in the enzyme, i.e. they occupy the same binding site.

The binding of *chitotriose* to lysozyme ($K_I = 6$ μM) shows little or no change in the methyl group resonances in the temperature range 10–30 °C. Increasing the temperature results in broadening of the resonance and an *upfield* chemical shift of the acetamido group at the reducing end of the molecule. However, the resonance corresponding to the central acetamido methyl group of the trisaccharide did not undergo any change in chemical shift but that of the acetamido methyl group at the non-reducing end displayed a slight *downfield* chemical shift.

The addition of lysozyme to *chitotetraose* results in a broadening of all the methyl resonances and a shift to higher field of the acetamido methyl group at the reducing end of the tetrasaccharide. Even at 65 °C Dahlquist and Raftery (1969) conclude that it is unlikely that fast-exchange conditions apply. (Higher temperatures lead to denaturation of the enzyme.)

Table 6.3 summarizes the results obtained with the oligosaccharides and their β-methyl derivatives; β-Me-NAG is included for comparison. The data suggest that the various subsites (labelled A, B, and C in Fig. 6.19) associated with the binding of chitobiose and chitotriose may be assigned specific values of $\Delta\omega_M$. The shifts of the glycosidic methyl resonances of the three saccharides

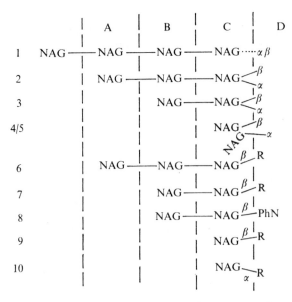

FIG. 6.19. Scheme for relative modes of association with lysozyme of various saccharide inhibitors and substrates. Where α- and β-anomeric forms are indicated on a single line, no information on relative binding modes was obtained. Where α- and β-forms are depicted separately (as with α-NAG and β-NAG), different binding modes were elucidated. Where α- and β-forms are shown on the same molecule on two levels, both anomeric forms bind identically. Methyl groups are denoted by R, nitrophenyl groups by PhN. (From Raftery *et al.* 1969.)

have almost the same chemical shifts and reinforce the scheme shown in Fig. 6.19. There are several important points to note from this example:

(1) The resonances that show measurable chemical shifts on addition of lysozyme are also those which are most broadened, but, in general, this is not necessarily the case.

(2) Temperature studies on the broadened resonances indicate that the broadenings arise from slow exchange. (The binding constants have been shown independently not to change significantly with temperature.) Where possible such studies should always be carried out.

(3) The lack of a shift or a broadening of a resonance, on addition of the enzyme does not imply that this residue is not interacting with the enzyme. It may be that several factors causing these effects act in opposite direction and thus cancel out or that the environment in the enzyme is not changed significantly from that of the bulk. Consideration of this point is particularly important in obtaining information from selective broadenings resulting from the binding of small molecules to macromolecules.

6.4. The conformation of the glucose ring of NAG-Glu-C$_6$H$_4$·NO$_2$† when bound at the active site of lysozyme

By monitoring changes in coupling constants in ligands when bound to macromolecules it may be possible to obtain some information of the geometry of the ligand on binding. Does, for example, the large macromolecule alter its shape by wrapping itself around the ligand or does the geometry of a relatively-rigid ligand remain unchanged? Problems such as these may often be solved using paramagnetic ion probes (see Chapters 9 and 10) but in some cases the measurement of coupling constants can help. For example, Sykes and Dolphin (1971) measured the changes in the coupling constant between the anomeric H(1) hydrogen of the glucose ring and the H(2) hydrogen in NAG–Glu–C$_6$H$_4$·NO$_2$ (Fig. 6.20) on binding to lysozyme. If there is fast chemical exchange of the substrate between the free and bound environments, then the observed coupling constant will be the weighted average of the coupling constants in each environment. The coupling constant $J_{H(1),H(2)}$ for NAG–Glu–C$_6$H$_4$·NO$_2$ free in solution is $7\cdot0\pm0\cdot1$ Hz and this corresponds to a dihedral angle ϕ of 180°. Any changes in ϕ when the inhibitor binds to the enzyme may be predicted by using the Karplus equation which Sykes and Dolphin (1971) wrote as

$$J_{H(1),H(2)}(\phi) = 7\cos^2\phi$$

i.e. B and C were taken to be zero. If ϕ changes to 120° then $J_{H(1),H(2)} = 1\cdot75$ Hz.

In the presence of the lysozyme the observed coupling constant is given by

$$J_{obs} = P_A J_A + P_M J_M$$

where the symbols A and M refer to free and bound and P is the mole fraction. If the substrate is bound with the glucose ring in the half-chair

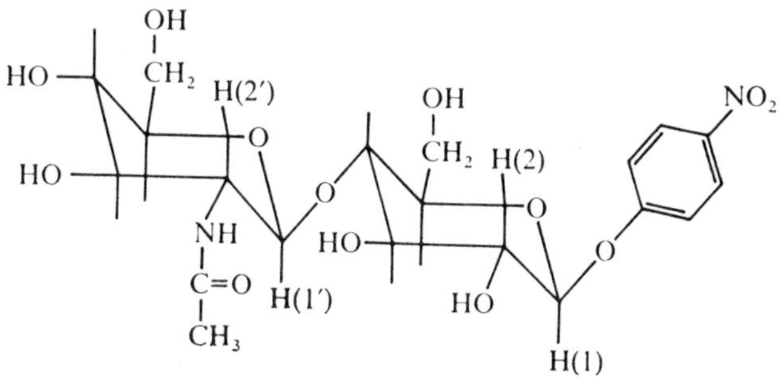

FIG. 6.20. *p*-Nitrophenyl-4-*O*-(2-deoxy-2-acetamido-β-D-glucopyranosyl)-β-D-glucopyranoside. (NAG-Glu-C$_6$H$_4$NO$_2$.)

† *p*-Nitrophenyl-4-*O*-(2-deoxy-2-acetamido-β-D-glucopyranosyl)-β-D-glucopyranoside.

TABLE 6.4

Observed and predicted coupling constants between
$H(1)$ *and* $H(2)$ *of* NAG–Glu–$C_6H_4\cdot NO_2$

[EI]/[I]†	$J_{H(1),H(2)}$ (observed) (H₂)	$J_{H(1),H(2)}$ (predicted)‡ (H₂)
0·0	7·0±0·1	7·0
0·1	6·7±0·2	6·5
0·14	7·2±0·2	6·3
0·27	7·2±0·2	5·9

† Calculated from the initial concentrations of enzyme and inhibitor on the basis of $K_D = 1·5 \times 10^{-2}$ M (Rand–Meir, Dahlquist and Raftery, 1969, © 1969 Amer. Chem. Soc.).

‡ $\phi = 120°$

conformation $\phi = 120°$ and then $J_M = 1·75$ Hz and J_{obs} should decrease as P_M increases. The results in Table 6.4 show that this is not the case and suggest that any changes between J_M and J_A are small, indicating, perhaps, that the glucose ring is not distorted when NAG–Glu–$C_6H_4\cdot NO_2$ is bound to lysozyme. However, as Sykes and Dolphin point out, one could, of course, argue that only a small fraction of the bound substrate is distorted. This does, however, illustrate the type of experiments that can be carried out. Results with a more tightly bound inhibitor should lead to more definitive results.

6.5. The binding of sulphonamides to bovine zinc carbonic anhydrase

The metallo-enzyme carbonic anhydrase has a molecular weight of 30 000, contains a single zinc atom, and catalyses the reaction:

$$CO_2 + H_2O \rightleftharpoons HCO_3^- + H^+$$

which has one of the largest turnover rates of any reaction so far studied. The enzyme also catalyses the hydrolysis of esters and aldehydes, but the rate constants here are more normal.

X-Ray crystallography has shown that in Human C-carbonic anhydrase, the zinc atom is co-ordinated to three imidazole groups, with a further two imidazole groups close by. The zinc can be reversibly replaced by Cd(II) and various divalent ions, of the first transition series e.g. Mn(II) and Co(II). In contrast to the other divalent ions, however, Co(II) substitution results in retention of similar catalytic and inhibitor binding properties and we shall discuss examples involving the Co(II) and Mn(II) enzymes in Chapter 10.

Sulphonamides are powerful inhibitors of carbonic anhydrase and X-ray crystallography has shown that sulphonamide inhibitors are bound in a cavity in the enzyme close to the zinc atom. All the strongest known inhibitors have an unsubstituted SO_2NH_2 group attached to an aromatic or heterocyclic residue. Some N'-substituted and aliphatic sulphonamides all inhibit carbonic anhydrase but bind less tightly to the enzyme. It has been suggested that the interaction of the aromatic group with a hydrophobic binding site assists the binding of the sulphonamide but the mechanism of inhibition remains a controversial point, even as to whether it is competitive or not. On the other hand, anions, such as F^-, Cl^-, CN^- are non-competitive inhibitors, and bind to the form of the enzyme which predominates at low pH.

Because of their powerful and important drug action the mode of binding of the sulphonamides has been the subject of intensive investigation and it is now believed that they bind in the anionic form. A mechanism proposed by Taylor, King, and Burgen (1970) is:

$$E + RSO_2NH_2 \rightleftharpoons E—RSO_2NH_2 \rightleftharpoons E—RSO_2NH^- + H^+.$$

The application of n.m.r. spectroscopy in studying such a mechanism will only be easily interpretable if the second stage is not, in fact, the controlling step. If it is, then it is quite possible that the mode of binding of $E + RSO_2NH_2$ may give little information on the mechanism. This, of course, is a limitation of any multi-step equilibrium process, in which binding is studied. Transient n.m.r. methods are then required to study the processes.

Here and in Chapter 10, we shall describe the effects of binding of various sulphonamide and anionic inhibitors to the enzyme. In the case of anionic binding, the above mechanism would not apply and one is then measuring the direct complex formation. In this Section, we shall consider the type of information that can be obtained from the measurement of relaxation times of the diamagnetic Zn-enzyme (Lanir and Navon, 1971). It is also worth pointing out that, in the zinc enzyme, relative high concentrations of enzyme are required for easily measurable n.m.r. effects to be observed, in sharp contrast to the cases where paramagnetic ion probes are added.

The inhibitors used in this study were sulphanilamide, N'-acetylsulphanilamide and toluene-p-sulphonamide. As a typical example of the changes caused in the line-widths of the various resonances, the spectrum of N'-acetylsulphanilamide is shown in Fig. 6.21, in the presence and absence of the enzyme. It is apparent that no chemical shift is observed on binding of the inhibitor to the enzyme and thus the terms in $\Delta\omega_M^2$ in eqn (2.32) are assumed to be small and the expression for $1/T_{2,E}$, the contribution to the inhibitor relaxation due to the presence of enzyme, is:

$$\frac{1}{T_{2,E}} = \frac{1}{T_{2,obs}} - \frac{1}{T_{2,A}} = \frac{P_M}{T_{2,M} + \tau_M}. \tag{6.1}$$

A typical plot of the relaxation rate versus temperature is shown in Fig. 6.22 and this must clearly correspond to the region of fast exchange (Region III in Fig. 2.19).

Before further interpretation, it is necessary to show that the broadening is not a result of any viscosity effects or any non-specific binding. This was

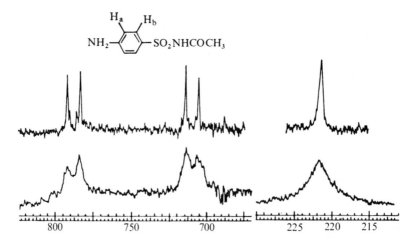

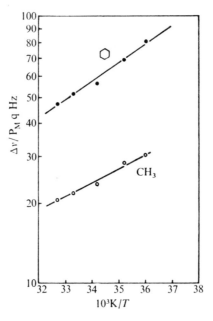

FIG. 6.21. 100 MHz spectra of N'-acetylsulphanilamide (0·02 M) in D_2O, pH 6·9, 27 °C. Shifts are expressed in Hz from external hexamethyldisiloxane. The upper trace is without enzyme. The lower one with $1·47 \times 10^{-3}$ M bovine carbonic anhydrase. (From Lanir and Navon, 1971. © 1971 Amer. Chem. Soc.)

FIG. 6.22. Temperature dependence of the specific broadening of toluene-p-sulphonamide, 0·015 M, pH 7·5. Enzyme concentration, $5·74 \times 10^{-4}$ M. (○) Methyl hydrogens; (●) phenyl hydrogens. (From Lanir and Navon, 1971. © 1971 Amer Chem. Soc.)

done by successively diluting a solution of the inhibitor–enzyme complex with D_2O, so that the ratio inhibitor:enzyme remained constant. No change was observed even when the concentration of the enzyme had been diluted tenfold. This control thus makes it unlikely that the changes in line-width results from a viscosity increase.

It is slightly more difficult to demonstrate that the effects arise from specific binding. An obvious way is to compare the binding constants from activity-inhibition experiments with those obtained in these experiments by (say) monitoring the changes in line-width. Details of this method are given in Appendix 6.2.

The next stage is to interpret the results. Since fast-exchange conditions obtain:

$$\frac{1}{P_M T_{2,E}} = \frac{1}{T_{2,M}}.$$

(6.2)

Three mechanisms for T_{2M} can be considered: (1) dipolar interaction between the hydrogen nuclei; (2) chemical-shift anisotropy; and (3) distribution of chemical shifts due to different binding modes of the inhibitor in which the bound and bulk inhibitor are rapidly exchanging.

Case (3) would actually correspond to Region II of Fig. 2.19 but, in view of the fact that no chemical shift of the inhibitor resonances are observed on binding, it is unlikely. The relaxation rates caused by fluctuating magnetic fields arising from chemical-shift anisotropy case (2), would be proportional to B_0^2. However the experimental magnetic-field dependence for the line broadening of N'-acetylsulphanilamide (Table 6.5; Appendix 6.3) shows that the resonance line-widths narrow at the higher field. Thus the relaxation mechanism must arise from dipole–dipole interactions.

As we have noted on p. 26, the expression for $1/T_{2,M}$ for *two* similar nuclei with $I = \frac{1}{2}$ is given by:

$$\frac{1}{T_{2,M}} = \frac{3}{20} \frac{\gamma_I^4 \hbar^2}{r^6} \left(3\tau_c + \frac{5\tau_c}{1+\omega_I^2\tau_c^2} + \frac{2\tau_c}{1+4\omega_I^2\tau_c^2} \right).$$

(6.3)

If the molecule contains more than two nuclei with $I = \frac{1}{2}$, the relaxation rates are given by the sum over all dipolar interactions. Thus, for example, if we consider a methyl group each hydrogen nucleus interacts with two other hydrogen nuclei and the right-hand-side of the above equation has to be multiplied by a factor of two.

There are two unknowns, τ_c and r in eqn (6.3) and τ_c can be evaluated as shown in Appendix 6.2. The following values are obtained:

$\tau_c = (1\cdot0\pm0\cdot5)\times10^{-8}$ s for the phenyl group and

$\tau_c = (2\cdot5\pm0\cdot5)\times10^{-9}$ s for the methyl group in N'-acetylsulphanilamide.

By combining this value of τ_c, together with the hydrogen–hydrogen distances of 2·48 Å in the phenyl group and 1·80 Å for the methyl groups, a value for $1/T_{2,M}$ can be calculated. This represents the *intra*molecular contribution. For the methyl group it accounts for ca. 86% of the value of $1/T_{2,M}$, whereas for the phenyl group it accounts for only ca. 10% of the observed value. The contribution due to hydrogens within the same inhibitor molecule is negligible because of their relatively large distance and the $1/r^6$ factor in eqn (6.3).

The conclusion must be that the phenyl hydrogens are broadened mainly by adjacent protein hydrogens. Since the specific broadening for the phenyl groups of all three inhibitors is about the same, it is reasonable to assume that the mode of binding of all three inhibitors is the same and that τ_c is the same for each $(1·0 \pm 0·5 \times 10^{-8}$ s). Interestingly, a value of the tumbling time of the molecule calculated using the Stokes–Einstein formula is $1·5 \times 10^{-8}$ s. The lack of independent rotation of the phenyl group indicates that this part of the inhibitor is strongly bound to the enzyme. The results thus conform to the X-ray data (Fridborg *et al.*, 1967) in which the benzene was shown to be in a narrow cavity of the protein.

Finally, since, in these experiments, the inequality $T_{2,M} \gg \tau_M$ is valid, for all the resonances studied, an upper limit for τ_M can be calculated as $\tau_M < 5 \times 10^{-3}$ s. The exchange rate may also be identified with the dissociation rate constant, k_{off}, in the relationship

$$E+I \underset{k_{off}}{\overset{k_{on}}{\rightleftharpoons}} EI$$

since $\tau_M = k_{off}^{-1}$; thus $k_{off} > 2 \times 10^2 \text{ s}^{-1}$. This value is several orders of magnitude larger than those derived for other inhibitors and possibly suggests that the n.m.r. technique is sensitive to one step in a complex mechanism which is different from steps that are measured by other methods. For example, the n.m.r. time scale may be sensitive to a primary fast-inhibitor association reaction that is followed by a slower conformational change of the enzyme–inhibitor complex.

Appendix 6.1.

Determination of binding constants from chemical shift measurements as in the lysozyme example

Consider the equilibrium between a macromolecule, E, and ligand or inhibitor, I

$$E+I \rightleftharpoons EI$$

where

$$K_I = [E][I]/[EI]. \tag{6.4}$$

If the binding of I to E results in a chemical shift of the I resonance $\Delta\omega$, then under conditions of fast exchange

$$\Delta\omega = P_M \Delta\omega_M, \tag{6.5}$$

where P_M is the fraction bound and $\Delta\omega_M$ the chemical shift of the fully-bound form with respect to its position in the unbound form. Since $P_M = [EI]/[I]_0$, where $[I]_0$ is the total concentration of I, then combining the above equations and assuming $[I]_0 \gg [EI]$ we obtain:

$$[I]_0 = \frac{\Delta\omega_M[E]_0}{\Delta\omega} - K_I. \tag{6.6}$$

(where E_0 is the total concentration of enzyme, and $[E] = [E]_0 - [EI]$).
Thus a plot of $[I]_0$ versus $1/\Delta\omega$ allows $\Delta\omega_M$ to be obtained from the slope and K_I from the intercept on the y-axis.

Equation (6.6) can be modified to take into account the effects of competition for the same site from a second inhibitor. Assuming $[I_2]_0 \gg [EI_2]$ then

$$[I]_0 = \frac{\Delta\omega_M[E]_0}{\Delta\omega} - K_I - \frac{K_I}{K_2}[I_2]_0, \tag{6.7}$$

where K_2 is the binding constant and $[I_2]_0$ the total concentration of the competing substrate. If $[I_2]_0$ becomes very large then, as expected, $\Delta\omega \to 0$.

In the special case such as the lysozyme example on p. 125 where there are two anomers α and β competing for the same site and they are in equilibrium with each other, then

$$[\alpha]_0 + [\beta]_0 = [I]_0$$

and

$$[\alpha]_0 = x[\beta]_0$$

where $[I]_0$ is the total concentration of substrate and $[\alpha]_0$ and $[\beta]_0$ are the total concentrations of each anomer which are present in the ratio $x:1$. Using these relationships, the relevant equation corresponding to eqn. (6.7) for the α form is

$$\frac{[I]_0}{(1+1/x)}\left[1 + \frac{K_\alpha}{K_\beta}(1/x)\right] = \frac{(\Delta\omega_M)_\alpha[E]_0}{(\Delta\omega)_\alpha} - K_\alpha) \tag{6.8}$$

and that for the β-form is:

$$\frac{[I]_0}{(1+x)}\left[1 + \frac{K_\beta}{K_\alpha}(x)\right] = \frac{(\Delta\omega_M)_\beta[E]_0}{(\Delta\omega)_\beta} - K_\beta. \tag{6.9}$$

The intercepts in the two cases are $-K_\alpha K_\beta(1+x)/K_\alpha + K_\beta \cdot x)$ and the ratio of the slopes is $K_\beta(\Delta\omega_M)_\alpha/K_\alpha(\Delta\omega_M)_\beta$. Thus the values of K and $\Delta\omega_M$ can be determined in each case.

In some cases, such as in the binding of the $\alpha + \beta$ anomers of N-acetylglucosamine to lysozyme, the intercepts are not sufficiently accurate to enable this type of procedure to be done. Differences in K and $\Delta\omega_M$ between the anomers have then to be shown by utilizing some extra (sometimes semi-empirical) information such as the extent of line broadening of the resonances arising from interaction with the enzyme. (Dalquest and Raftery 1968 b)

Appendix 6.2.

Determination of binding constants by line-width measurements using the example of sulphanilamide binding to carbonic anhydrase (Lanir and Navon, 1971)

The initial treatment is almost identical with that shown in Appendix 6.1. A slightly different form of the equation is used. For an excess of inhibitor $[I]_0 \gg [E]_0$, the concentration of enzyme–inhibitor complex, EI, is given by

$$[EI] = \frac{[E]_0[I]_0}{K_I + [I]_0},$$ (6.10)

where K_I is the dissociation constant of the complex and where it is assumed that there is only *one* binding site for the inhibitor. P_M, the fraction of the inhibitor bound to the enzyme is

$$P_M = \frac{[EI]}{[I]_0}$$ (6.11)

Combining eqns (6.2), (6.10), and (6.11) one obtains:

$$T_{2,E} = \frac{(T_{2M} + \tau_M)}{[E]_0}(K_I + [I]_0)$$ (6.12)

which can obviously be rearranged to a form similar to equation 6.6. Noting that $1/T_{2,E} = \pi \Delta \nu_E$ then K_I can be evaluated from a plot of $1/\Delta \nu_E$ versus $[I]_0$, provided that the concentrations of I are of the same order as K_1. In this example, from activiy inhibition measurements, $K_I \ll 1$ mM. Because of the tight binding in these cases it is difficult to carry our n.m.r. measurements at the low concentrations of inhibitors required. ($\ll 1$ mM) The binding constants are then best determined by competition experiments.

In the form of the equations used in this Appendix, the equation equivalent to (6.7) is

$$\frac{1}{\Delta \nu_E} = \pi \frac{(T_{2,M} + \tau_M)}{[E]_0}\left([I]_0 + K_I + \frac{K_I}{K_{I'}} \times \frac{[I']_0}{[I]_0}\right)$$ (6.13)

where $[I']_0$ is total concentration of the competing inhibitor, with dissociation constant K_I'. In this example conditions are such that $K_I \ll [I]_0$ thus

$$\frac{1}{\Delta \nu_E} = \pi(T_{2,M} + \tau_M)\frac{[I]_0}{[E]_0}\left(1 + \frac{K_I}{K_{I'}} \cdot \frac{[I']_0}{[I]_0}\right)$$ (6.14)

A plot of $1/\Delta \nu_E$ versus $[I']_0/[I]_0$ at constant $[I]_0/[E]_0$ yields an intercept of the x-axis of $-K_I'/K_I$. The ratio in the experiment here (Fig. 6.23) is in good agreement for the ratio of the dissociation constants found by activity inhibition measurements.

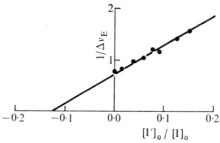

FIG. 6.23. Plot of sulphanilamide line-broadening in the presence of carbonic anhydrase as a function of the concentration ratio of *p*-toluene sulphonamide to sulphanilamide (From Lanir and Navon 1971.)

Appendix 6.3.

Determination of τ_c in diamagnetic systems

τ_c cannot be determined directly from the measured relaxation rates if the number and distance of interacting nuclei is unknown. There are essentially two methods for determining τ_c in diamagnetic systems: (1) Determination of the frequency dependence of $1/T_{1,M}$ or $1/T_{2,M}$, and (2) The ratio of $T_{1,E}/T_{2,E}$.†

Method 1. The theoretical ratios for $T_{2,M}$ at various frequencies is shown in Fig. 6.24. It is seen that the method is most sensitive for values of $\tau_c \approx 1/\omega_I$. At 60 and 100 MHz it is also apparent that for values of $\tau_c < 2 \times 10^{-10}$ s or $\tau_c > 2 \times 10^{-8}$, the ratio is insensitive to the value of τ_c. We also note that there are two τ_c values corresponding to each ratio, and thus *three* frequencies are needed to solve this ambiguity.

Method 2. Under conditions of rapid exchange the ratio of $T_{1,E}/T_{2,E}$ is equal to $T_{1,M}/T_{2,M}$. The theoretical ratios calculated at the frequencies, 60, 100, and 220 MHz are shown in Fig. 6.25. We recall that when $\tau_c > 1/\omega_I$, $1/T_{1,M}$ decreases with increasing τ_c unlike $1/T_{2,M}$. Thus the ratio $T_{1,M}/T_{2,M}$ increases dramatically when this value of τ_c is exceeded. It is also apparent that at 220 MHz, even for values of $\tau_c \approx 5 \times 10^{-9}$, the ratio $T_{1,M}/T_{2,M}$ is ca. 100. This arises because the condition $\tau_c > 1/\omega_I$ obviously occurs at lower τ_c values as ω_I increases.

EXAMPLE

Calculation of τ_c for various inhibitors bound to carbonic anhydrase. Table 6.5 lists the measured hydrogen relaxation times of various inhibitors bound to carbonic

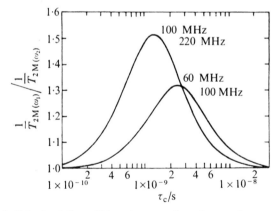

FIG. 6.24. Calculated line-width ratios at various frequencies as a function of the correlation time (From Navon and Lanir 1972.)

† For convenience the terms $\left(\dfrac{1}{T_{1,\text{obs}}} - \dfrac{1}{T_{1,A}}\right)$ and $\left(\dfrac{1}{T_{2,\text{obs}}} - \dfrac{1}{T_{2,A}}\right)$ in equations (2.31) (2.32) *et seq.* are abbreviated as $1/T_{1,E}$ and $1/T_{2,E}$.

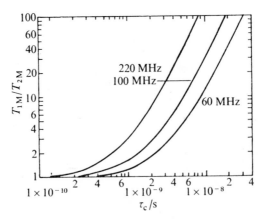

FIG. 6.25. Calculated T_1/T_2 ratios at various frequencies as a function of the correlation time. (From Navon and Lanir 1972.)

anhydrase and the calculated values of τ_c. The two methods give good agreement. By doing this at several temperatures and using $r = 1.80$ Å in eqn (6.3) the comparison between the theoretical and experimental results is seen in Fig. 6.26 to be quite good. The temperature dependence of $1/T_{2,E}$ and $1/T_{1,E}$ arises from that of τ_c and are opposite in sign as expected if $\tau_c > 1/\omega_I$ (see p. 26).

TABLE 6.5

Measured hydrogen relaxation times and calculated values of τ_c of various inhibitors bound to carbonic anhydrase

Inhibitor	Residue	$1/T_{2,M}$¶	$T_{1,E}/T_{2,E}$†	τ_c‡	$\dfrac{T_{2E}^{-1}(60 \text{ MHz})}{T_{2E}^{-1}(100 \text{ MHz})}$	τ_c§
Sulphanilamide	phenyl	225	88	1.7×10^{-8}	1·04	1.4×10^{-8}
Toluene-p-sulphon-	⎰phenyl	192	45	1.2×10^{-8}	1·04	1.4×10^{-8}
amide	⎱methyl	140	5·8	3.7×10^{-9}	1·06	1.1×10^{-9}
N'-acetylsulphanil-	⎰phenyl	201	32·3	1.0×10^{-8}	1·07	1.05×10^{-8}
amide	⎱methyl	72·5	3·72	2.62×10^{-9}	1·38	2.5×10^{-9}

† At a frequency of 100 MHz.
‡ Correlation time calculated using the $T_{1,E}/T_{2,E}$ ratio and Fig. 6.24.
§ Correlation time calculated using the ratio $T_{2,E}^{-1}$ at 60 and 100 MHz.
¶ From Lanir and Navon (1971).

The value of τ_c for the methyl groups differs considerably from that of the phenyl groups and it can be concluded that the methyl group has extra motions. This is considered in Appendix 6.4.

Appendix 6.4.

Motion of the methyl groups of the N'-acetylsulphanilamide and toluene-p-sulphonamide–carbonic anhydrase complexes (From Navon and Lanir, 1972)

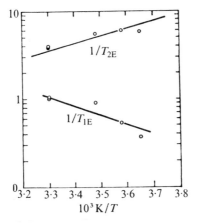

FIG. 6.26. Temperature dependence of relaxation rates of N'-acetylsulphanilamide methyl protons. The continuous lines are the calculated values. (From Navon and Lanir 1972.)

In Appendix 6.3, we noted that the correlation times of the methyl groups of the above inhibitors was different from that of the phenyl residues. It was concluded that the methyl groups exhibit extra motions. Navon and Lanir (1972) have considered this motion in terms of rotation of the methyl group about its symmetry axis with a correlation time τ_{rot} together with a random tumbling of the axis with a correlation time $\tau_{c,1}$. Using Woessner's (1962) general formulation the equations for the relaxation rates may be written as

$$\frac{1}{T_{1,M}} = \frac{3}{20}\frac{\gamma^4\hbar^2}{r^6}[J(\omega_1)+4J(2\omega_1)] \tag{6.15}$$

$$\frac{1}{T_{2,M}} = \frac{3}{20}\frac{\gamma^4\hbar^2}{r^6}[\tfrac{3}{2}J(0)+\tfrac{5}{2}J(\omega_0)+J(2\omega_0)] \tag{6.16}$$

where

$$J(\omega) = \tfrac{1}{4}(3\cos^2\Delta-1)^2\left(\frac{2\tau_{c,1}}{1+\omega^2\tau_{c,1}^2}+\tfrac{3}{4}\sin^22\Delta\frac{2\tau_{c,2}}{1+\omega^2\tau_{c,2}^2}+\tfrac{3}{4}\sin^4\Delta\frac{2\tau_{c,3}}{1+\omega^2\tau_{c,3}^2}\right)$$

$$\frac{1}{\tau_{c,2}} = \frac{1}{\tau_{c,1}}+\frac{1}{\tau_{rot}}, \quad \frac{1}{\tau_{c,3}} = \frac{1}{\tau_{c,1}}+\frac{1}{\tau_{rot}}.$$

The angle between the rotation axis and the radius vector connecting the two interacting hydrogens is represented by Δ.

In the case of a methyl group, which is randomly rotating about its symmetry axis, $\Delta = 90°$ and the expressions for the relaxation rates for each pair of hydrogens are

$$\frac{1}{T_{1,M}} = \frac{3}{20}\frac{\gamma^4\hbar^2}{r^6}[\tfrac{1}{4}f_1(\tau_{c,1})+\tfrac{3}{4}f_1(\tau_{c,3})], \tag{6.17}$$

$$\frac{1}{T_{2,M}} = \frac{3}{20}\frac{\gamma^4\hbar^2}{r^6}[\tfrac{1}{4}f_2(\tau_{c,1})+\tfrac{3}{4}f_2(\tau_{c,3})], \tag{6.18}$$

where
$$f_1(\tau_{c,i}) = \frac{2\tau_{c,i}}{1+\omega_1^2\tau_{c,i}^2} + \frac{8\tau_{c,i}}{1+4\omega_1^2\tau_{c,i}^2} \qquad i = 1,3, \tag{6.19}$$

$$f_2(\tau_{c,i}) = 3\tau_{c,i} + \frac{5\tau_{c,i}}{1+\omega_1^2\tau_{c,i}^2} + \frac{2\tau_{c,i}}{1+4\omega_1^2\tau_{c,i}^2} \qquad i = 1,3 \tag{6.20}$$

Navon and Lanir (1972) consider two limiting cases: (1) The rotation around the symmetry axis is slower than the isotropic tumbling of this axis, $\tau_{rot} \gg \tau_{c,1}$, when $\tau_{c,3} = \tau_{c,1}$ and eqns (6.17) and (6.18) are reduced to the usual equations for simple isotropic random motion. (2) A very fast rotation of the CH_3 group, $\tau_{rot} \ll \tau_{c,1}$. In this case, $\tau_{c,3} \simeq \frac{1}{4}\tau_{rot}$. For τ_{rot} short enough so that $f_1(\tau_{c,3}) \ll f_1(\tau_{c,1})$, the relaxation rates are reduced by a factor of 4, compared with their value for the usual case. In the two limiting cases the relaxation rates only depend on $\tau_{c,1}$ the isotropic correlation time.

If the value of $\tau_{c,1}$ is such that $(\omega_I\tau_{c,1})^2 \gg 1$, then $f_1(\tau_{c,1})$ will be small while $f_2(\tau_{c,2})$ is relatively large, because it contains the spectral density at zero frequency. Put another way, $1/T_{2,M}$ is then more sensitive to slow motions while $1/T_{1,M}$ is sensitive only to the fast motions. Thus in the presence of fast anisotropic rotation $f(\tau_{c,3})$ would contribute significantly to $1/T_{1,M}$ but not to $1/T_{2,M}$. In this case, the analysis of τ_c from the values of $1/T_{1,M}$ or $1/T_{2,M}$ would give values of $\tau_{c,3}$ and $\tau_{c,1}$ while the ratio $T_{1,M}/T_{2,M}$ would probably be in between. In N-acetylsulphanilamide, both methods give the same value of τ_c, which strongly suggests that the methyl group has an isotropic tumbling motion faster than that of the protein. On the other hand the value of the correlation times calculated for the methyl group of toluene-p-sulphonamide are different by the two methods. This could arise if the methyl group has a fast rotation about its symmetry axis, while this axis performs an isotropic tumbling motion with a correlation time ($\tau_{c,1} = 1 \cdot 1 \times 10^{-8}$ s, from the $T_{2,M}$ measurements) which is the same as that of the phenyl group and the whole protein molecule.

References

DAHLQUIST, F. W. and RAFTERY, M. A. (1968a). *Biochemistry* **7**, 3277.

—— (1968b). *Biochemistry* **7**, 3269.

—— (1969). *Biochemistry* **8**, 713.

FRIDBORG, K., KANNAN, K. K., LILJAS, A., LUNDIN, J., STRANDBERG, B., STRANDBERG, R., TILANDER, B., and WIRREN, G. (1967). *J. molec. Biol.* **25**, 505.

GRIFFIN, J. H., SCHECHTER, A. N., and COHEN, J. S. (1973). International Conference on Electron Spin Resonance and Nuclear Magnetic Resonance in Biology and Medicine and Fifth International Conference in Magnetic Resonance in Biological Systems, *Ann. New York, Acad. Sci.* in press.

GRIFFIN, J. H., COHEN, J. S., SCHECTER, A. N., DAMODARAN, N. P., JONES, G.H., MOFFATT, J. G., to be published.

HAMMES, G. G. (1968). *Accounts chem. Res.* **1**, 321.

LANIR, A. and NAVON, G. (1971). *Biochemistry* **10**, 1024.

LEE, G. C. Y. and CHAN, S. I. (1971). *Biochem. Biophys. Res. Comm.* **43**, 142.

MEADOWS, D. H. and JARDETZKY, O. (1968). *Proc. natn. Acad. Sci. U.S.* **61**, 406.

——, ROBERTS, G. C. K. and JARDETZKY, O. (1969). *J. molec. Biol.* **45**, 491.

NAVON, G. and LANIR, A. (1972). *J. magn. Resonan.* **8**, 144.

RAFTERY, M. A., DAHLQUIST, F. W., CHAN, S. I., and PARSONS, S. M. (1968). *J. Biol. Chem.* **243**, 4175.

——, PARSONS, S. M., and WOLCOTT, R. G. (1969). *Proc. natn. Acad. Sci. U.S.* **62**, 44.

RAND–MEIR, T., DAHLQUIST, F. W., and RAFTERY, M. A. (1969). *Biochemistry*, **8**, 4206.

ROBERTS, G. C. K., DENNIS, E. A., MEADOWS, D. H., COHEN, J. S., and JARDETZKY, O. (1969). *Proc. natn. Acad. Sci. U.S.* **62**, 1151.

SYKES, B. D. and DOLPHIN, D. (1971). *Nature (Lond.)* **233**, 421.

TAYLOR, P. W., KING, R. W. and BURGEN, A. S. V. (1970). *Biochemistry* **9**, 3894.

USHER, D. A., ERENRICH, E. S. and ECKSTEIN, F. (1972). *Proc. natn. Acad. Sci. U.S.* **69**, 115.

WOESSNER, D. E. (1962). *J. chem. Phys.* **36**, 1.

11

7

CARBON-13 STUDIES

7.1. Introduction

THE sensitivity of the ^{13}C nucleus is $1\cdot6\%$ that of ^1H for equal numbers of nuclei at the same magnetic field. However, its natural abundance is only $1\cdot1\%$ and so the relative sensitivity of ^{13}C is ca $1\cdot8\times10^{-4}$ that of ^1H. Thus to carry out ^{13}C magnetic resonance studies, signal enhancement is necessary. Several techniques have been used to achieve signal enhancement and these include: (1) Hydrogen noise decoupling, which removes all the ^1H–^{13}C spin–spin splittings. (This may also result in a nuclear Overhauser enhancement of the ^{13}C resonance which may have its integrated intensity increased up to a maximum of three times.) (2) The use of a computer of average transients (CAT) to add the signals from successive repetition of the same spectrum. (3) The Fourier transform technique (FT). (4) ^{13}C Isotopic enrichment. (See Chapter 14 for further details of (1), (2) and (3).)

The advantage of using FT over the conventional continuous wave technique is illustrated by the following experiments. Lauterbur (1970), measured the natural-abundance continuous wave spectrum (at 26 MHz) of a solution of 17 mM hen egg white lysozyme. The signal-to-noise ratio after 87 hours of accumulation (6260 scans of 50 s each) was ca. 5. By contrast using FT the signal-to-noise ratio at 15 MHz of a solution of 20 mM ribonuclease A was 35–40 obtained in 10 h (65 000 scans) (Allerhand, Cochran, and Doddrell, 1970). Thus, in effect, the method produces a 100-fold increase in sensitivity.

There can be little doubt that such a dramatic improvement will lead to more widespread use of ^{13}C n.m.r. spectra as a general method for analysis of problems in organic systems. The obvious advantages of ^{13}C spectroscopy are: (1) It is a non-perturbing 'probe' (2) The range of ^{13}C chemical shifts is greater than that for ^1H, and (3) The spectra are expected to be first-order, i.e. no ^{13}C–^{13}C spin–spin couplings will normally be detected because ^{13}C is only 1% abundant so that most neighbouring nuclei are non-magnetic ^{12}C. (This is obviously not the case for ^{13}C enrichment.)

The ^{13}C chemical shifts of most of the amino acids have been determined and to a good approximation the shift values observed in small peptides can be predicted on the basis of their amino acid composition. Figure 7.1 shows an example of the analysis of the spectra of three amino-terminal oligopeptides of ribonuclease. (RNaSe)

The use of ^{13}C enrichment to improve the observation of the resonance is well illustrated by reference to Figs. 7.1 and 7.3 in which the spectra of the peptide corresponding to the first fifteen residues of RNase are shown for two samples. In the lower spectrum the phenylaniline group at position 8

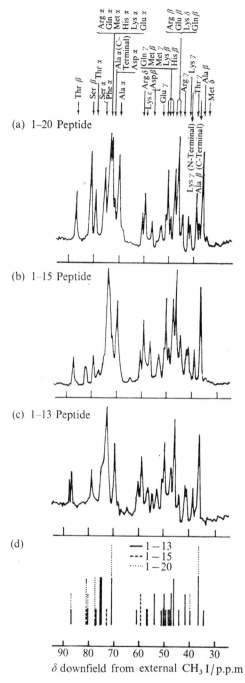

FIG. 7.1. Carbon-13 n.m.r. Fourier transform spectra at 25·1 MHz with proton noise decoupling of three N-terminal peptides of ribonuclease, showing the aliphatic carbon resonances. The assignments for the 1–20 peptide are shown above, and a stick diagram of the resonances of the amino acids are shown below, with normalized intensities. (From Freedman, Lyerla, Chaiken, and Cohen 1973.)

NH$_2$-Lys-Glu-Thr-Ala-Ala-Ala-Lys-Phe-Glu-Arg-Gln-His-Met-Asp-Ser-CO$_2$H

FIG. 7.2. Amino acid sequence of residues 1–15 of ribonuclease A.

is 15% enriched in ^{13}C. This enrichment technique not only increases the sensitivity of ^{13}C but, more importantly for biochemists, reduces the time required to obtain a good spectrum whilst providing a convenient 'reporter' group as a non-perturbing probe at a strategic site on the macromolecule. There are essentially two approaches that have been used for introducing a specifically-labelled ^{13}C amino acid residue into a large macromolecule: the biosynthetic approach the limitations of which we have discussed in Section 5.4.2 and semi-synthetic methods. In the semi-synthetic approach the macromolecule is selectively cleaved and the required peptides isolated and degraded (if necessary). An enriched amino acid is then incorporated into the appropriate peptide, and in turn all the peptides are then recombined to make the original macromolecule. Quite apart from the considerable experimental difficulties in doing this, the underlying assumption is that the tertiary structure of the macromolecule is determined by the primary sequence. While this is usually accepted as being so, it must be remembered that proteins formed by synthetic or semi-synthetic methods are not formed on the ribosome. Thus there may be differences in the tertiary structure, arising from differing environmental constraints in the two cases. In proteins

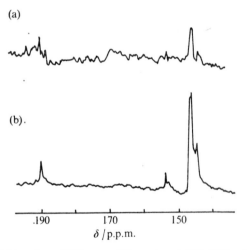

FIG. 7.3(a) Downfield region of ^{13}C n.m.r.-FT spectra at 25·1 MHz of (a) RNase-(1–15) peptide, natural abundance ^{13}C; 9 × 10^4 pulses (4 h.). pH 6·14 (b) [^{13}C-Phe-8]RNase-(1–15) peptide; 1·5 × 10^4 pulses (40 min) pH 5·88. Concentrations of these peptides were 4·5 × 10^{-2} M, acquisition time was 0·1 s. The following are the actual chemical shift values compared to those of Phe methyl ester (in parenthesis) at the same pH; C=O, 192·6 (191·6); Aromatic, 156·2 (155·0), 149·0 (149·2), 148·6 (148·8), 147·4 (147·0); C$_\alpha$, 75·1 (74·4); C$_\beta$, 57·2 (57·4). Chemical shift values are downfield from external ^{13}CH$_3$I (From Freedman. Cohen, and Chaiken, 1971.)

containing several subunits, the experimental problems may be even greater since it is, of course, necessary to separate the individual subunits as well. Nevertheless, as a first step, this enrichment approach should be quite successful with small peptides and small molecules. It may be that the future of this technique lies in studying model compounds and small molecules rather than the macromolecules themselves.

What then are the other difficulties in using ^{13}C for macromolecules? Obviously, a major limitation is the concentration available. It is difficult

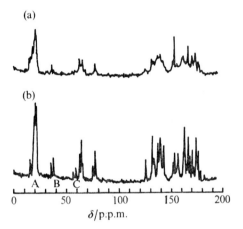

FIG. 7.4. Hydrogen-decoupled natural-abundance carbon-13 Fourier transform n.m.r. spectra of ribonuclease A at 45 °C and 15·08 MHz. Horizontal scale is in parts per million upfield from neat CS_2. (a) Normal spectrum of 0·017 M native protein at pH 4·12, resulting from 32 768 scans in 6 h. (b) normal spectrum of the random coil form of the protein 0·015 M at pH 1·46, from 31 284 scans in 12 h. (From Allerhand, Doddrell, Glushko, Cochran, Wenkert, Lawson, and Gurd, 1971.) © 1971 Amer. Chem. Soc.

to envisage any major developments in instrumental design which will lead to a dramatic improvement in sensitivity for ^{13}C nuclei. At present, concentrations of ca. 20 mM are required to observe the un-enriched (i.e. natural abundance) spectrum in a reasonable time (4 h). If spin–lattice relaxation times of the ^{13}C nuclei are to be measured, the time required may increase by an order of magnitude. Thus the stability of the macromolecule is an important factor too. Further, as with ^{1}H spectra of macromolecules, the presence of many ^{13}C resonances of the same type defy individual resolution. For while the range of ^{13}C shifts is about an order of magnitude greater than those of hydrogen, the shifts for a given type of ^{13}C are not very dependent upon neighbouring residues. For example, band A in Fig. 7.4 arises from 151 carbonyl carbons in RNase A. Band B in Fig. 7.4 can be assigned to a total of 10 carbon atoms: the aromatic C_6 carbons of the tyrosine residues (6) and the tertiary carbon atoms of the arginine residues (4). The peaks in region

C can be assigned to the imidazole C(2) carbons of the four histidine residues and the quaternary carbon atoms of the three phenylalanine residues, and so on for the other peaks. As expected too the difference in chemical shifts between the random coil and native forms is not very dramatic (Fig. 7.4).

We indicated above that relaxation-time measurements on macromolecules may be too time consuming to make their determination generally worthwhile. This means that most biochemical experiments will involve monitoring changes in chemical shifts, as a function of added ligands, pH, etc. The successful observation of change in the shift will depend on the ratio of this shift to the line-width. An estimate of the line-width can be obtained as follows.

We shall deal only with the case when the ^{13}C nuclei are decoupled from the ^1H nuclei. (When this is not so the theory is more complex.) We shall assume that the relaxation times of carbons bonded to hydrogen, can be interpreted in terms of a dipole–dipole interaction with the directly-bonded hydrogens. The modulation of this interaction arises from rotation of the C—H fragment. If we assume that the rotational motion is isotropic the equations for $1/T_1$ and $1/T_2$ are given by:

$$\frac{1}{T_{1,2}} = \frac{N\gamma_C^2\gamma_H^2\hbar^2}{10r_{C-H}^6} f_{1,2}(\tau) \tag{7.1}$$

where

$$f_1(\tau) = \left(\frac{\tau_c}{1+(\omega_C-\omega_H)^2\tau_c^2} + \frac{3\tau_c}{1+\omega_C^2\tau_c^2} + \frac{6\tau_c}{1+(\omega_C+\omega_H)^2\tau_c^2}\right)$$

and

$$f_2(\tau) = \left(2\tau_c + \frac{0\cdot5\tau_c}{1+(\omega_C-\omega_H)^2\tau_c^2} + \frac{1\cdot5\tau_c}{1+\omega_C^2\tau_c^2} + \frac{3\tau_c}{1+\omega_H^2\tau_c^2} + \frac{3\tau_c}{1+(\omega_C+\omega_H)^2\tau_c^2}\right)$$

where N is the number of hydrogens attached to the carbon atom; τ_c is the correlation time; r_{C-H} is the C—H distance; ω_C and ω_H are the ^{13}C and ^1H resonance frequencies (in rad s^{-1}); γ_C and γ_H are the gyromagnetic ratios of ^{13}C and ^1H. The equations are formally similar to those for paramagnetic ions derived by Solomon (1955). Assuming a C—H bond distance of 1·1 Å, a value of $\tau_c = 3\times10^{-9}$ s (as would be expected for the rotation of a macromolecule such as lysozyme, mol wt $\approx$ 14 000), and $N = 2$, then with $\omega_C = 2\pi\times22\cdot628\times10^6$ rad s^{-1} and $\omega_H = 2\pi\times90\times10^6$ rad s^{-1};

$$\frac{1}{T_1} \approx 50 \text{ s}^{-1} \quad \text{and} \quad \frac{1}{T_2} \approx 60 \text{ s}^{-1}$$

Thus since $1/T_2 = \pi\Delta\nu$, the line-width $\Delta\nu \approx 18$ Hz. The T_1/T_2 ratio >1 arises from the condition $(\omega_{c,H}\tau_c) \approx 1$. As the macromolecule becomes larger, τ_c will increase, $1/T_1$ will eventually decrease, but $1/T_2$ and hence the line-width will increase (cf. Fig. 2.12). Increasing the magnetic field (i.e. ω_C) further increases the T_1/T_2 ratio.

- There are additional mechanisms that may well contribute to $1/T_2$. For example, if the attached hydrogen atom is relaxing very rapidly,† faster than the value of the ^{13}C–^{1}H coupling constant, J, the hyperfine structure is 'washed out' (noise decoupling then has no effect on this ^{13}C) and a single resonance line is observed. This implies that the local magnetic field produced by the hydrogen and 'seen' by the ^{13}C nucleus, fluctuates at a rate that is rapid compared with the frequency of the splitting that is produced in the absence of any fluctuations. Under these conditions this fluctuating field becomes a relaxation mechanism. The relaxation effects of this spin–spin or scalar interaction may be formally treated by addition of a second term to the above equations viz.

$$\frac{1}{T_{1,\text{scalar}}} = \frac{J^2}{2}\left(\frac{\tau_{1,\text{H}}}{1+(\omega_C-\omega_H)^2\tau_{1,\text{H}}^2}\right). \tag{7.2}$$

$$\frac{1}{T_{2,\text{scalar}}} = \frac{J^2}{4}\left(\tau_{1,\text{H}}+\frac{\tau_{1,\text{H}}}{1+(\omega_C-\omega_H)^2\tau_{1,\text{H}}^2}\right). \tag{7.3}$$

A similar theory is applicable if the ^{13}C is attached to a quadrupole nucleus, e.g. ^{14}N, which relaxes very rapidly 'washing out' the ^{13}C–^{14}N spin–spin coupling.

Thus, in macromolecules we do not expect the ^{13}C line-widths for carbons bonded to hydrogen to be significantly different from the hydrogen line-widths. In contrast, the line-widths of quaternary carbons should be quite narrow.

Turning now to the observation of changes in chemical shift, little work on biochemical systems has been done apart from the titrations of several amino acids and peptides. However, here too there are some unexpected effects.

7.2. The anomalous pH dependence of some ^{13}C shifts in amino acids

On an inductive model, protonating a particular group attached to a carbon atom draws electronic charge from the neighbouring atom, which in turn draws electrons from carbons or hydrogens attached to it, e.g.

$$-\overset{\displaystyle |}{\underset{\displaystyle |}{C}}-\overset{\displaystyle \overset{\textstyle H}{|}}{\underset{\displaystyle |}{C}}-\overset{+}{N}H_3$$

† with a relaxation time $\tau_{1\text{H}}$

Thus the hydrogens and carbons involved in this are deshielded and would be expected to show downfield chemical shifts. Indeed while this is the case for hydrogens, the carbon-13 nuclei shift *upfield*. Similarly upfield chemical shifts of hydrogen and carbon-13 resonances would be predicted upon ionization of a carboxyl group. Again *downfield* chemical shifts are generally observed for ^{13}C resonances. These observations have been explained by Horsley and Sternlicht (1968) in terms of a 'through space' electric-field effect and a 'through-bond' inductive effect which falls off more rapidly. The two effects are opposed, and, in general, this gives smaller shifts in the α-carbon resonances than in those of the β-carbons. Of course other effects, such as hybridization changes around the carbon atoms, may contribute. Theoretically the chemical shift should be correlated with the electron density at a particular carbon atom, but it is difficult at present to estimate these quantitatively for any but the simplest systems.

The above points are well illustrated for the pH titration of the ^{13}C resonances of L-histidine, Figs. 7.5 and 7.6.† The carboxyl carbon resonance shifts very little in the pH range 1–4, despite the ionization of the carboxyl

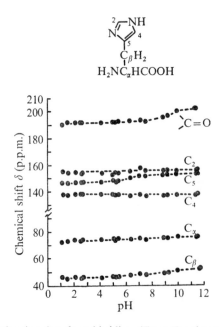

FIG. 7.5. Titration data for L-histidine. (From Freedman *et al.*, 1971.)

† The shifts of L-histidine are *downfield* relative to $^{13}CH_3I$ used as a reference standard here, (CH_3I is approximately 216 p.p.m. upfield from CS_2). Thus increasing δ values in the titration curves means that the resonance is shifting further downfield.

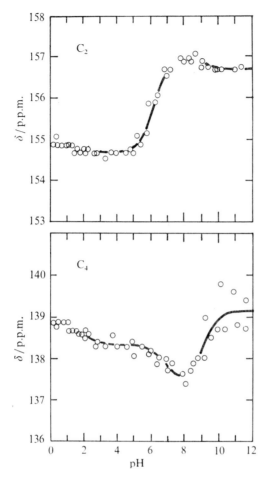

FIG. 7.6. The shifts of the C(2) and C(4) nuclei shown in more detail. (From Freedman *et al.* 1973.)

group. This probably, in part, results from a near cancellation of the above two effects. However in the higher pH range, both the carboxyl and β-carbons show changes reflecting the titration of the amino terminal groups ($pK_a = 9\cdot5$). We also note that although most of the shifts are in general downfield there are definite trends towards upfield shifts, e.g. in parts of the titration curves of the C(4) and C(2) resonances. This probably indicates that there is near cancellation of the inductive and through-space effects. The total titration curves of these resonances shifts only ca. 2 p.p.m. for these resonances. Interestingly, titration of imidazole group ($pK_a = 6\cdot5$) affects the quaternary C(5) resonance far more than those of the C(2) and

C(4). (The shift of the *hydrogens* attached to C(2) and C(4) change by ca. 1·0 and 0·5 p.p.m. respectively on titration of this group.) Again this may be explained on the basis of the largest changes in electron density occurring at C(5) (Quirt, Lyerla, Peat, Cohen, Reynolds and Freedman, 1973).

In a pH titration of a peptide, we see much the same behaviour. The peptide, L-valyl-L-leucyl-L-seryl-L-glutamylglycine comprises the first five residues of the amino terminus of sperm whale myoglobin. The pH titration is shown in Fig. 7.7 (Gurd, Lawson, Cochran, and Wenkert, 1971). The largest changes occur on titration of the α-amino group and are most easily observed for the carbonyl carbon of the amino-terminal (valyl) residue. As above, the effect of this titration is smaller for the vicinal α-carbon than for the β-carbon, and is negligible for the other carbon atoms. Titration in the carboxyl range results in larger changes in the chemical shifts of the side-chain group of the glutamyl residue than in the terminal carboxyl group. The pH shifts of the δ-, γ-, and β-glutamyl carbon atoms imply a pK_a of 4·3 for the side-chain carboxyl,

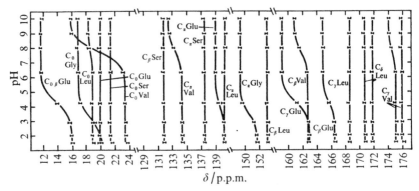

FIG. 7.7. Dependence on pH of chemical shifts of individual carbon atoms relative to CS_2 for L-valyl-L-leucyl-L-seryl-L-glutamylglycine. (From Gurd *et al.* 1971.)

while those of the carboxyl and α-carbon in the terminal residue correspond to a group with a $pK_a = 3·4$.

In dealing with macromolecules where many of the ^{13}C resonances overlap, care will therefore be needed in interpreting the effects of pH. The use of enriched samples may remove some of the ambiguity and generally the largest pH shifts are expected on titration of amino groups. In the titration of the imidazole group in histidine the quaternary carbon is most affected but this resonance has the lowest intensity, because of the absence of any $^{13}C-^{1}H$ Overhauser effect. On the other hand, the lack of any directly coupled hydrogens means that the resonance will be reasonably narrow, and the changes in shifts easily monitored.

7.3. Some tentative conclusions on the use of ^{13}C

The advent of Fourier transform techniques will probably revolutionize the use of ^{13}C spectroscopy in the field of organic chemistry when dealing with relatively small molecules. In macromolecules, the main drawback at present is one of sensitivity and here enrichment offers the major hope. Although ^{13}C shifts occur over a much wider range than ^{1}H chemical shifts, ^{13}C shifts are dominated by the paramagnetic term σ_p (see p. 28) and thus unlike ^{1}H chemical shifts are relatively insensitive to their environments. For example, ^{13}C shifts in the observed spectrum of lysozyme (Lauterbur, 1970) are well predicted by the shifts of the constituent amino acids. In macromolecules the ^{13}C resonances will, in general, be about the same width as those of hydrogens, and the problems of assignment are much the same. The interpretation of chemical shifts is slightly more complicated, and in cases where the crystal structure is unknown, it will be difficult to interpret quantitatively any changes in chemical shift as a result of ligand binding. The enrichment of a particular residue may still provide valuable information on structure–function relationships in macromolecules, but in such cases one is using the n.m.r. technique as a monoparametric one.

In macromolecules the values of the correlation time, τ_c, are long, and although outside the scope of this section it is worth noting that the Overhauser enhancement decreases as τ_c increases. Further, it also decreases with increasing magnetic field (Dwek, Richards, and Taylor, 1968) and thus a balance has to be obtained between higher magnetic fields (which increase the signal-to-noise ratio) and the consequent decrease of the Overhauser effect. The majority of biochemical applications with macromolecules will probably therefore involve monitoring the ^{13}C resonances of *ligands*. (In this respect the work on ^{13}CO binding to myoglobin and haemoglobin looks promising (Conti and Paci, 1971).) Under conditions of rapid exchange of ligands between a site on a macromolecule and the bulk solution, the observed resonance will be a weighted average of those in the two environments. The larger the effects of the 'bound' site the higher the concentrations of the ligand that can be used, again alleviating, somewhat, the sensitivity problem.

The use of ^{13}C n.m.r. as an analytical method has recently been demonstrated elegantly by Brown, Katz, and Shemin (1972) in a study of the origin of the methyl group on the C(1) of corrin ring A in vitamin B_{12}. By the biosynthesis of B_{12} from 5-[^{13}C]-δ-aminolaevulate they were able to show the absence of a ^{13}C signal from the C-1 methyl group which indicated that this methyl group does not have the expected origin in C(5) of δ-aminolaevulate acid. In a problem such as this, n.m.r. can offer a faster and more convenient method of ascertaining the position of an isotopically-labelled nucleus than techniques which involve chemical breakdown.

Appendix 7.1.

^{13}C *Shifts of Amino Acids at 25 °C relative to* CS_2†

Amino acid (Partial structures)	C atom	δ
Alanine C_β | N—C_α—CO	C_0 C_α C_β	16·6 142·1 176·1
Arginine C_γ—N—C_ε—N | C_γ | C_β | N—C_α—CO	C_0 C_α C_β C_γ C_δ C_ε	18·2 138·3 164·9 168·5 151·9 35·9
Asparagine $C_\gamma ONH_2$ | C_β | N—C_α—CO	C_0 C_α C_β $C_{\gamma,0}$	(16·3) (140·9) (157·6) (17·8)
Aspartic acid $C_\gamma O_2 H$ | C_β | N—C_α—CO	C_0 C_α C_β $C_{\gamma,0}$	17·9 140·2 156·1 14·6
Cysteine C_β | N—C_α—CO	C_0 C_α C_β	20·0 136·5 165·5
Glutamine $C_\delta ONH_2$ | C_γ | C_β | N—C_α—CO	C_0 C_α C_β C_γ $C_{\delta,0}$	18·6 138·2 166·2 161·6 15·0
Glutamic acid $C_\delta O_2 H$ | C_γ | C_β | N—C_α—CO	C_0 C_α C_β C_γ $C_{\delta,0}$	17·8 137·7 165·3 158·9 11·1
Glycine N—C_α—C_0	C_0 C_α	19·9 150·9 ·

Appendix 7.1 (*Continued*)

Amino acid (Partial structures)	C atom	δ
Histidine	C_0	18·5
	C_α	137·6
	C_β	164·4
	C_2	56·3
	C_4	75·2
	C_5	61·2
Isoleucine	C_0	18·2
	C_α	132·5
	C_β	156·3
	C_γ	167·7
	C_δ	181·1
	C_ε	177·5
Leucine	C_0	16·8
	C_α	138·7
	C_β	152·4
	C_γ	168·0
	C_δ	170·2
	C_ε	171·3
Lysine	C_0	18·0
	C_α	138·1
	C_β	162·7
	C_γ	171·0
	C_δ	166·2
	C_ε	153·4
Methionine	C_0	18·1
	C_α	138·3
	C_β	162·4
	C_γ	163·3
	C_δ	178·2
Phenylalanine	C_0	18·3
	C_α	136·1
	C_β	155·9
	C_1	56·7
	$C_{2,6}$	
	$C_{3,5}$	
	C_4	64·3

Appendix 7.1   *(Continued)*

Amino acid (Partial structures)	C atom	δ
Proline C_δ—C_γ ; N ; C_α—C_β ; C_0	C_0	18·8
	C_α	131·8
	C_β	163·7
	C_γ	169·0
	C_δ	146·9
Serine C_β ; N—C_α—CO	C_0	20·3
	C_α	136·0
	C_β	132·1
Threonine C_γ ; C_β ; N—C_α—CO	C_0	19·4
	C_α	131·9
	C_β	126·3
	C_γ	172·9
Tryptophan‡ (indole ring, 1N—H) ; C_β ; N—C_α—CO	C_0	20·1
	C_α	139·6
	C_β	166·0
	C_2	55·5
	C_3	65·3
	C_4	66·3
	C_5	69·7
	C_6	72·3
	C_7	73·5
	C_8	80·0
	C_9	85·6
Tyrosine OH (phenol ring) ; C_β ; N—C_α—CO	C_0	20·3
	C_α	137·6
	C_β	157·1
	C_1	66·1
	$C_{2,6}$	60·7
	$C_{3,5}$	75·6
	C_4	36·6
Valine C_γ C_δ ; C_β ; N—C_α—CO	C_0	18·1
	C_α	131·8
	C_β	163·1
	C_γ	174·2
	C_δ	175·8

† This table was compiled from data obtained at pH = 6·8 by Drs M. H. Freedman, J. Lyerla, Jr., and J. S. Cohen, but see also Horsley, Sternlicht, and Cohen (1970), Allerhand *et al.* (1970), and Freedman *et al.* (1971).
‡ At pH = 1·0.

References

ALLERHAND, A., COCHRAN, D. W., and DODDRELL, D. (1970). *Proc. natn. Acad. Sci.* (*U.S.*) **67**, 1093.

ALLERHAND, A., DODDRELL, D., GLUSHKO, V., COCHRAN, D. W., WENKERT, E., LAWSON, P. J., and GURD, F. R. N. (1971). *J. Am. chem. Soc.* **93**, 544.

BROWN, C. E., KATZ, J. J., and SHEMIN, D. (1972). *Proc. natn. Acad. Sci.* (*U.S.*), **69**, 2585.

CONTI, F. and PACI, M. (1971). *FEBS Letts.* **17**, 149.

DWEK, R. A., RICHARDS, R. E. and TAYLOR, D. (1968). *A. Rev. N.M.R.* **2**, 394.

FREEDMAN, M. H., COHEN, J. S., and CHAIKEN, I. M. (1971), *Biochem. biophys. Res. Commun.* **42**, 1148.

——, LYERLA, JR. J. R., CHAIKEN, I. M. and COHEN, J. S. (1973). *Eur. J. Biochem.* **32**, 215.

GURD, F. R. N., LAWSON, P. J., COCHRAN, D. W., and WENKERT, E. (1971). *J. biol. Chem.* **246**, 3725.

HORSLEY, W. J. and STERNLICHT, H. (1968). *J. Am. chem. Soc.* **90**, 3738.

——, ——, and COHEN, J. S. (1970). *J. Am. chem. Soc.* **92**, 680.

LAUTERBUR, P. C. (1970). *Appl. Spectrosc.* **24**, 450.

QUIRT, A., LYERLA, J. R. JR., PEAT, I. R., COHEN, J. S., REYNOLDS W. F., and FREEDMAN, M. H. (1973). to be published.

SOLOMON, I. (1955). *Phys. Rev.* **99**, 559.

8

THE USE OF FLUORINE-19 AS A DETECTING SHIFT PROBE

8.1. Introduction

IN general, fluorine coupling constants and chemical shifts are at least an order of magnitude larger than those of the corresponding hydrogen analogues. At a given magnetic field, the sensitivity of fluorine nuclei is second only to that of hydrogen. These properties together with the small size (roughly that of an OH group) and low reactivity of a fluorine atom (whose introduction into a molecule should therefore cause only small perturbations) provide the ingredients of an excellent detecting probe.

In proteins, two approaches using fluorine probes are most commonly used. The first involves introduction of a fluorine moiety covalently bound to the protein, and observations are then made on the ^{19}F resonance, as various other perturbations (e.g. ligands, changing pH) are introduced. In principle, this is a very similar study to the observation of hydrogen resonances of a protein in which all but a few selected hydrogens have been replaced by deuterium atoms. The advantage of using a fluorine probe in such cases arises from the potentially larger changes that may occur when the inhibitor binds, and the relative ease of selective fluorination compared with gross deuteriation. Of course such procedures are limited to proteins where chemical modifications are possible.

The second approach is to incorporate the fluorine atom or moiety at specific sites in the ligand molecule. The chemical shift and line-width of the fluorine resonances are expected to alter when the ligand binds to the protein since the environment of the fluorine nucleus is different. In some cases the bound resonance signal itself may be observed, as illustrated in the example below for the binding of the trifluoroacetylated analogue of chitotriose to lysozyme. If however, there is a significant amount of free ligand, then, in principle, two resonances corresponding to each environment will be observed. If chemical exchange between the environments is important then we have the usual exchange behaviour. Information on the bound ligand is then most easily obtained under conditions of rapid chemical exchange.

We now give some examples illustrating the range of fluorine as a probe.

8.2. Fluorine as a covalent probe

8.2.1. *Ribonuclease A*

As we have already mentioned in chapter 5, the enzyme ribonuclease A (RNase A) can be cleaved by subtilisin at the peptide bond between residues-20 and -21 to give the ribonuclease S-peptide, RNase S (residues 1–20) and ribonuclease S-protein (residues 21–124). The two components, together, have the same enzymic activity as the parent RNase A. A ¹⁹F probe was covalently introduced by trifluoroacetylating lysine residues-1 and -7 of the S-peptide with ethyl thiol trifluoroacetate, and then recombining the S-peptide with the S-protein. The ¹⁹F-labelled enzyme showed no change in activity.

Ethyl thiol trifluoroacetate normally acylates specifically at the ε-amino group but Huestis and Raftery (1971) used a large excess of the acylating reagent so that some doubly-acylated Lys-1 as well as the singly-acylated residue resulted. The ¹⁹F spectrum of the labelled S-peptide is shown in Fig. 8.1. The next stage is, of course, the assignment of the ¹⁹F resonances

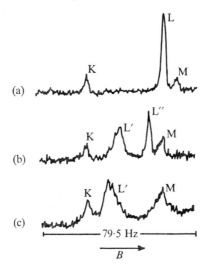

FIG. 8.1. ¹⁹F N.m.r. spectra of trifluoroacetylated RNase S; (a) trifluoroacetylated RNase S-peptide; (b) trifluoroacetylated RNase S-peptide associated with RNase S-protein; (c) trifluoroacetylpeptide–protein complex, after removal of Lys-1 from the peptide (from Huestis and Raftery, 1971. © 1971 Amer. Chem. Soc.).

and the usual procedure is a comparison with resonances from model compounds. In this case these were ε-trifluoroacetyl-lysine and α,ε-bis-(trifluoroacetyl)lysine (Fig. 8.2). Peak L (Fig. 8.1) is thus assigned to the ε-acylated groups of Lys-7 and the singly acylated Lys-1. Peaks K and M are attributed to the trifluoroacetyl groups on the α- and ε-amino groups of the doubly-acylated Lys-1.

Addition of a molar equivalent of RNase S-protein caused the following changes in the spectrum: the resonance L splits into two (L′ and L″) indicating that the ε-trifluoroacetyl amino groups Lys-1 and Lys-7 experience significantly different environments, the resonance K does not change, and the

12

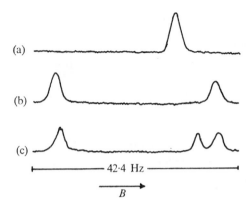

FIG. 8.2. ^{19}F N.m.r. spectra of trifluoroacetylated lysine: (a) ε-trifluoroacetyl lysine; (b) α,ε-bis(trifluoroacetyl)lysine; (c) a mixture of ε-mono- and α,ε-bis-(trifluoroacetyl)-lysine (from Huestis and Raftery, 1971. © 1971 Amer. Chem. Soc.)

ε-trifluoroacetyl group (M) of bis-modified Lys-1 shifts downfield. Specific assignment of L″ and L′ to the ε-trifluoroacetyl (ε-TFA) groups of Lys-1 and Lys-7 was achieved by an Edman degradation which results in removal of the ε-TFA amino terminal group of Lys-1 from the S-peptide. Upon recombination with the S-protein resonance, L″ is absent (Fig. 8.1). The assignment also agrees qualitatively with predictions from X-ray data which suggest that Lys-1 is exposed to the solvent near the surface of the protein and consequently would experience a smaller chemical shift (0·07 p.p.m) whilst Lys-7 is inside close to the active site and its environment in RNase S is quite different (shift = 0·22 p.p.m.). The above shifts on combination of the modified S-peptide and S-protein, indicate that the ^{19}F probe is sensitive to changes in its environment. However, a detailed interpretation of the shifts is only possible from a comparison of the crystal structures of the individual moieties and with that of RNase S.

The addition of inhibitors results in further changes (see Fig. 8.3) the main effect being a shift of the TFA-ε-Lys-7 resonance (L′) to higher field and in some cases a splitting of this resonance into a major and minor peak. The minor peak is approximately equal in magnitude to that observed for the resonances K and M (of TFA-α-Lys-1) and it was suggested that this peak arises from slightly different modes of association of TFA-α-Lys-1 in the modified S-peptide with the S-protein. If this is so, then it may be possible, by extrinsic perturbations, to obtain conditions of rapid chemical exchange between these two environments and thus derive information on the modes of association of the S-peptide and S-protein.

Judged by the magnitude of the shift, the effect of the inhibitors is to cause a change in the environment of Lys-7 which places it in a similar environment to Lys-1. Examination of the molecular model indicates that if, as a result of

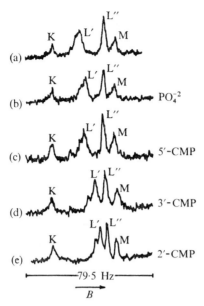

FIG. 8.3. ^{19}F N.m.r. spectra of trifluoroacetylated RNase S (4×10^{-3} M) in the presence of various inhibitors: (a) free RNase S; (b) $+10^{-2}$ M phosphate; (c) $+4 \times 10^{-3}$ M 5′-CMP; (d) $+4 \times 10^{-3}$ M 3′-CMP; (e) $+4 \times 10^{-3}$ M 2′-CMP (from Huestis and Raftery, 1971. © 1971 Amer. Chem. Soc.)

inhibitor binding His-12 is drawn closer to His-119 in the active site, Lys-7 would be pulled away from the active site and become more exposed to the solvent. Some information on the cause of these shifts can be obtained from qualitative and quantitative consideration of the effects of other inhibitors and pH. For example, phosphate causes a similar shift as 5′-CMP and thus ring-current effects of the bound inhibitor are unlikely to be the major cause. With 3′-CMP the shift of the TFA-ε-Lys-7 resonance is pH dependent (Fig. 8.4) and implies a group with a $pK_a = 7.4$. Reference to Table 6.1 suggests that this group is most probably His-119. His-12 has a pK_a value of 8.0, yet within the experimental error no change in the chemical shift of L′ is observed in the pH range 8–9 with 3′-CMP present. The charge on His-12 has no effect, therefore, on the observed resonance, and yet the X-ray data indicate that His-12 and His-119 are almost equidistant from Lys-7. Thus it seems unlikely that only His-119 would exert an electric-field effect (with consequent chemical shift) on Lys-7. One tentative explanation along the lines suggested by an examination of the molecular model is that the inhibitor binding results in a change in the conformation of His-12, bringing it closer to His-119 in the complexed form, and the shift then results from interaction with both these residues. When His-119 is deprotonated ($pK_a \approx 7.4$), His-12 moves back towards its position in the native enzyme. As the distance between them increases, and His-119 ceases to be strongly bound to the

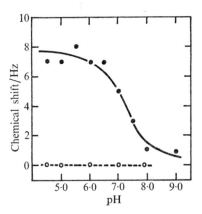

Fɪɢ. 8.4. Chemical shift of ε-trifluoroacetyl Lys-7 (peak L') on binding of 3'-CMP as a function of pH: (●) position of peak L' in free RNase S: (○) position of peak L' in RNase S saturated with 3'-CMP. The solid line is the theoretical titration curve of an ionizable group of pK_a 7·25. (From Huestis and Raftery, 1971. © 1971 Amer. Chem. Soc.)

phosphate (as is the case with 3'-CMP above pH 5·6), the molecular model shows that Lys-7 can be carried as close as 6 Å to the guanidino group of Arg-39, thus changing its environment and producing a further chemical shift.

8.2.2 *Conformational effects in selectively trifluoroacetonylated haemoglobin A*

The use of fluorine as a probe at a specific site in haemoglobin (Hb) (Huestis and Raftery, 1972) affords in principle an easier experimental method than the observation of hydrogen resonances, for monitoring conformational changes, since changes in chemical shift will probably be larger than those of the hydrogen nuclei. By reaction with 1-bromo-3-trifluoroacetone, a 3,3,3-trifluoroacetonyl group (CF_3COCH_2, here designated TFA) was covalently and specifically attached to the SH group of cysteine-β-93, thus placing the probe close to the α_1, β_2-region of intersubunit contact. The binding of oxygen to the modified haemoglobin was essentially the same as that to native haemoglobin.

In Fig. 8.5, the chemical shifts of oxy- and deoxy-TFA-Hb are shown, together with those of various forms bound to ligands. The small differences in chemical shifts caused by the ligands suggest similar, but not identical, environments for the ¹⁹F probe in each case. Apart from the chemical shift of deoxy-TFA-Hb, the chemical shifts were independent of pH. The pH dependence of the shift of deoxy-TFA-Hb implies a group with a pK_a value of 7·4. As usual assignment of this pK_a to a specific group is only possible if the structures of oxy- and deoxy-Hb are known. The crystallographic results of Perutz (1970) suggest that structural changes on ligation should involve

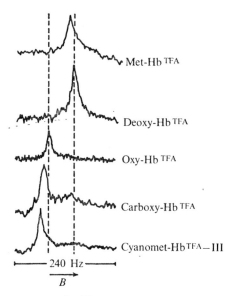

FIG. 8.5. ¹⁹F N.m.r. spectra of trifluoroacetonylated haemoglobin with and without various ligands. (From Huestis and Raftery, 1972. © 1972 Amer. Chem. Soc.)

the position of the carboxyl-terminal amino acid residues of the β-chain. In deoxy-TFA-Hb, the terminal histidine is doubly bonded by salt bridges; an interchain bond from its α-carboxyl group to Lys-α-40 and an intrachain bond from its imidazolium side chain to Asp-β-94. Ligand binding results in breaking both of these salt bridges, and expulsion of the carboxyl terminus of the β-chain so that it can move freely on the outside of the β-subunit. Without invoking very large conformational changes, possible groups with this pK_a value are the β-histidine-146, -143, or -92 (next to Cys-β-93). His-β-92 is the proximal haeme-linked histidine and from other studies on met-Hb has been assigned a pK_a value of 5·1–5·3, consistent with forming a bond to the iron which would be expected to lower its pK_a value. 2,3-Diphosphoglycerate (DPG) is thought to bind to His-β-143 and does not change the pK_a value (7·4) of the deoxy TFA–Hb complex (Fig. 8.6). Thus, this pK_a value is assigned to His-β-146. This argument also implies that the probe will be sensitive primarily to conformational changes within the β-chain.

Huestis and Raftery (1972) used this probe to show that, contrary to previous evidence (dye binding, crystal form), met-Fe^{3+}-Hb is different from oxy-Fe^{2+}-Hb, since the curve for the pH dependence of the chemical shift of the ¹⁹F probe was different (Fig. 8.7). (The X-ray structure does not yield precise information on this point at present since the oxy-Hb crystals used contained some met-Hb (Perutz, 1970)). Assignment to a specific

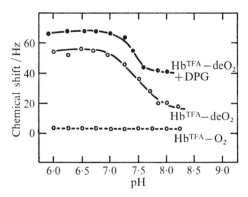

FIG. 8.6. ^{19}F Chemical shift, as a function of pH, of trifluoroacetonylated haemoglobin (Hb(TFA)): oxygenated (Hb(TFA)-O$_2$), in the presence or absence of 2,3-diphosphogly-cerate (DPG) deoxygenated (Hb(TFA)-deO$_2$), and deoxygenated in the presence of DPG (Hb(TFA)-deO$_2$). (From Huestis and Raftery, 1972. © Amer. Chem. Soc.)

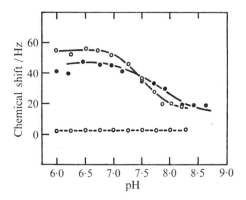

FIG. 8.7. As for legend to Fig. 8.6, except that ● represents met-Fe^{3+}-TFA-Hb (demon-strating that the latter differs from oxy-Fe^{2+}-TFA-Hb). (From Huestis and Raftery 1972. © 1972 Amer. Chem. Soc.)

residue of the group with a pK_a of 7·8 (Fig. 8.7) is slightly more complex and because of perturbations introduced by the ligand there are various possibilities based on structural considerations (such as histidines-97, -143 and -146).

8.3. Fluorosubstrates as probes

8.3.1 *The binding to lysozyme of several fluorinated inhibitors*

An example in which 'slow-exchange conditions' obtain is provided by the binding of the trifluoroacetylated analogue of chitotriose to lysozyme (Millet and Raftery, 1972a). The signal of a one-to-one mole ratio of enzyme to tri-saccharide, each at 3 mM is shown in Fig. 8.8 at three pH values (at these

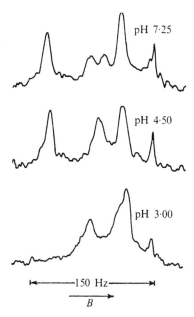

pH 7·25

pH 4·50

pH 3·00

|——————150 Hz——————|

B

FIG. 8.8. ^{19}F n.m.r. spectra of solutions of 3 mM trifluoroacetylated chitotriose and 3 mM lysozyme at 10 °C at three pH values. The resonance to highest field is due to the trifluoroacetate anion (from Millet and Raftery 1972a).

concentrations, it can be calculated from the binding constants that the concentration of the complex is close to 100%). The sharp peak at highest field is due to the trifluoroacetate anion, which was used as an internal reference. Assignments were made on the basis of experiments at higher temperatures and, at pH 4 under conditions of fast exchange with excess trisaccharide present, and also by comparison with the spectra of the pure trisaccharide. In the spectra at pH 4·50, the bound inhibitor resonances are assigned as follows: the resonance at lowest field corresponds to the trifluoroacetyl (F_3Ac) group at the non-reducing end of the trisaccharide; the one next to high field corresponds to the F_3Ac group at the reducing end of the trisaccharide (this resonance separates into α- and β-anomeric forms at pH $> 7·25$); the other resonance corresponds to the middle F_3Ac group of the trisaccharide.

The effects of pH on the position of the resonance is illustrated in Fig. 8.9, from which it can be observed that the shift of the central F_3Ac group of the trisaccharide $(\Delta\omega_M)_2$ is almost independent of pH. The pH dependence of the shifts is shown in detail in Fig. 8.9, in which $(\Delta\omega_M)_3$ represents the shift of the non-reducing F_3Ac end group and $(\Delta\omega_M)_{1\alpha}$ that of the reducing F_3Ac end group of the α-anomeric form of the trisaccharide. The pH dependence of $(\Delta\omega_M)_3$ implies a group with a pK_a of 3·2. This is identified as

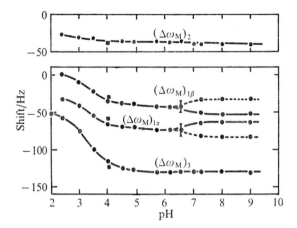

FIG. 8.9. Chemical shift data for the trifluoroacetyl group ¹⁹F resonance for the pertri-fluoroacetylchitotriose lysozyme complex (3 mM) over a range of pH at 10 °C. The squares represent the chemical shifts measured under conditions of fast exchange at pH 4·0 and 65 °C (from Millet and Raftery 1972a)

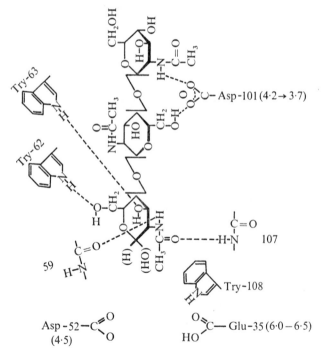

FIG. 8.10. Representation of hydrogen-bonded interactions of chitotriose and lysozyme as determined by crystallographic methods. (From Blake *et al.* 1967.)

Asp-101 in the top of the lysozyme-binding cleft, since previous experimental work suggested that the pK_a of this group in the enzyme (4.2) changes to 3·6 upon binding of chitotriose and, further, a group of pK_a 4·2 in the enzyme changes to 3·2 on binding F$_3$Ac-chitotriose.

The interactions of chitotriose with specific groups on the enzyme, as determined by X-ray analysis (Blake, Johnson, Mair, North, Phillips, and Sarma, 1967) are shown in Fig. 8.10. If the fluorinated analogue binds similarly, the formation of a hydrogen bond between Asp-101 and the F$_3$Ac group would give a change in the ^{19}F chemical shift of this group. From the model, it is apparent that the F$_3$Ac group at the reducing end of the trisaccharide is the furthest F$_3$Ac group from Asp-101, and yet, unlike the central F$_3$Ac group, it experiences an obvious chemical shift. The origin of the shift is unlikely to be electronic, as is that for the F$_3$Ac group at the non-reducing end. If however the formation of the Asp-101 hydrogen bond causes a conformational change in the complex such that the position of the reducing end of the trifluoroacetamido group is shifted with respect to the aromatic ring of Try-108, then the differences in ring-current effects could account for the observed shift.

In contrast to the above the binding of fluorinated monosaccharide inhibitors to lysozyme is an example where fast-exchange conditions are valid (i.e. increasing the temperature does not alter the shift significantly, see Fig. 2.15). For example, Fig. 8.11 shows the shifts of the α- and β-anomers of N-fluoroacetylglucosamine (FNAG) on binding to lysozyme (under conditions where the concentration of both anomers is saturating). It is apparent that the α-anomer is shifted more than that of the β. If conditions of fast exchange are valid then the chemical shift of the inhibitor on the enzyme, $\Delta\omega_M$, is obtained from the relationship:

$$\Delta\omega = P_M \Delta\omega_M$$

where $\Delta\omega$ is the observed chemical shift measured with respect to that of the pure inhibitor and P_M is the fraction of inhibitor bound to the enzyme (assuming one-binding site per enzyme molecule—i.e. $q = 1$ in eqn (2.31)). It is possible to arrange conditions so that all the enzyme is bound, or alternatively, as discussed in Chapter 6 (eqn 6.6) to evaluate $\Delta\omega_M$ by means of the relationship:

$$[I]_0 = \frac{\Delta\omega_M [E]_0}{\Delta\omega} - K_I$$

where $[I]_0$ is the total concentration of inhibitor and $[E]_0$ is the enzyme concentration. It must be stressed that this equation assumes that fast chemical exchange occurs between the bulk and bound inhibitor.

The values of $\Delta\omega_M$ for several fluoromonosaccharides are listed in Table 8.1.

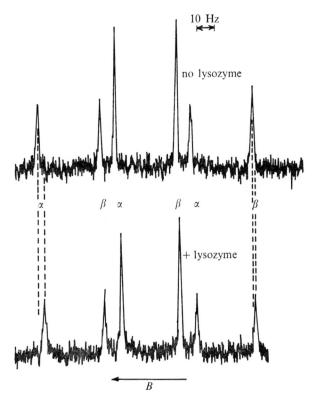

FIG. 8.11. ¹⁹F shifts for α- and β-FNAG in the presence and absence of lysozyme. Conditions: lysozyme (3 mM), FNAG (200 mM), imidazole buffer, 99·8% ²H₂O, pH (nominal) 5·0, 26 °C. (From Dwek, Kent, and Xavier, 1971.)

TABLE 8.1

Values of the chemical shift ($\Delta\omega_M$) of some sugar lysozyme complexes

	$(\Delta\omega_M)$†‡
α-FNAG	−326
β-FNAG	−166
α-Me-FNAG	−140
β-Me-FNAG	−140

† Measured at 84·60 MHz at 22 °C.
‡ Negative values indicate shifts to low field.

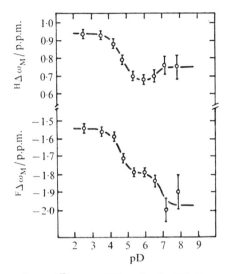

FIG. 8.12. pD dependence of $^H\Delta\omega_M$ and $^F\Delta\omega_M$ for the CH_2F group in β-Me-FNAG bound to lysozyme at 35 °C. Solutions contained 0·01 M phosphate, 0·01 M citrate, 0·01 M NaCl. (From Millet and Raftery 1972b.)

and we note that the value for α-FNAG is different from those for β-FNAG, and either α- or β-Me-FNAG. This is consistent with the suggestion from work on the protonated analogues (see p. 126) that the orientation of α-NAG in subsite C is different from that of the other inhibitors.

It is also interesting to compare the pH variation of the protons and fluorine chemical shifts of the CH_2F group in β-Me FNAG (Fig. 8.12). (Millet and Raftery, 1972b). The titration of both sets of shifts implies groups of pD = 4·5 and 6·5 which are ascribed to ionization of Asp-52 and Glu-35. On the other hand the pH dependence of the binding constant only implies a

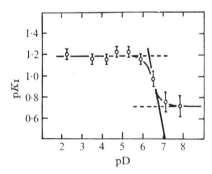

FIG. 8.13. pK versus pD for the β-Me-FNAG-lysozyme complex at 35 °C. Solutions contained 0·01 M phosphate, 0·01 M citrate, 0·01 M NaCl. (From Millet and Raftery, 1972b.)

group of pD = 6.5, due to the ionization of Glu-35 (see Fig. 8.13). These results can be interpreted in terms of the X-ray structure of β-NAG lysozyme, since β-Me-FNAG has been shown to bind similarly. For example, at pH = 5·5, the hydrogen chemical shift of the fully-bound form $^H\Delta\omega_M$ agrees reasonably well with a ring-current shift due to Try-108, calculated on the basis of the X-ray data. This ring-current shift must be the same for ^{19}F nuclei. The observed difference between $^H\Delta\omega_M$ and $^F\Delta\omega_M$ reflects other contributions to the ^{19}F shifts. The change in chemical shifts of both nuclei on the ionization of Asp-52 is approximately the same (in p.p.m.) One interpretation of this is on the basis of a conformational change, in which the ionization of Asp-52 leads to the formation of a hydrogen bond to the β-NH$_2$ group of Asn-59, which 'moves' this residue slightly. Asn-59 is known from the crystal structure to be bonded through its α-amide NH to the carbonyl group of the acetamido group of β-Me-FNAG. Thus movement of Asn-59, can cause movement of the acetamido group to a region of lower positive ring-current field from Try-108. By similar arguments to those used in myoglobin and deoxymyoglobin, (see p. 53) the movement of this group can be estimated as ca. 1 Å.

The use of F$_3$NAG as the fluorosubstrate has the advantage that the spin–spin coupling is between protons and fluorine nuclei is absent with concomitant increase in signal height. Addition of lysozyme causes the ^{19}F resonance of the α-form to broaden and shift *upfield* ($\Delta\omega_M \approx$ 1 p.p.m.) while the β-resonance is not broadened and shifts very slightly *downfield* ($\Delta\omega_M \approx$ 0·3 p.p.m.) The interpretation of such line broadening must be attempted with great caution, for we have already observed (p. 130) that the lack of broadening or chemical shift does not necessarily indicate that binding does not occur.

The interpretation of chemical shifts from a knowledge of the crystal structure is possible if the mode of binding of the inhibitor is known or assumed. For example, reference to the lysozyme model with β-F$_3$NAG positioned in site C according to the co-ordinates of β-NAG suggests that the shift can be explained on the basis of ring-current effects, the downfield shift is probably the result of tryptophan-108 on the CF$_3$-group since this residue is known to tilt when the inhibitor binds (see Fig. 8.14) (L. N. Johnson, personal communication). The upfield chemical shift of the α-anomer arises from interaction of the CF$_3$-group and Try-63.

8.4. Some conclusions regarding the use of ^{19}F probes in proteins

In view of the large shifts normally observed for fluorine nuclei, the magnitude of these shifts (in the examples when fluorine is covalently bound) is rather small, although for the inhibitors, usually larger than those of the corresponding hydrogen analogues. In general, when the origin of the shifts is a consequence of through-space interactions (e.g. electrostatic or ring current effects) the chemical shifts will be relatively small. The

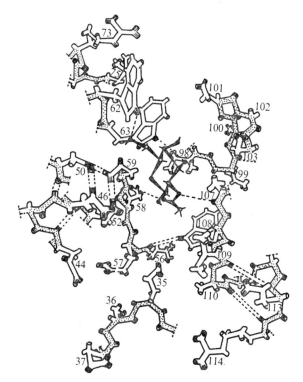

FIG. 8.14. Model of the relevant part of the lysozyme/β-NAG complex based on the X-ray crystallographic work. The Gd(III) binds between the residues Glu-35 and Asp-52. (From Butchard, Dwek, Kent, Williams, and Xavier, 1972.)

contribution (in p.p.m.) of neighbouring-group effects to the chemical shift of a particular nucleus in a molecule, is really independent of that nucleus, provided the screening terms σ_P and σ_D of the nucleus itself are unaltered. Thus through-space effects should be much the same magnitude for fluorine and hydrogen nuclei.

The ¹⁹F spectrum of pure FNAG (Fig. 8.11) illustrates the sensitivity of fluorine to interactions which are transmitted through bonds, viz. the position of the OH group on the anomeric carbon atom five bonds away. It will be recalled that with the hydrogen analogue NAG, no distinction between the methyl resonances of the two anomers was observed in the absence of the enzyme. The extra sensitivity of fluorine is also apparent with coupling constants too, as illustrated in Fig. 8.15 for the single fluorine atom in 5-fluoro-L-iodose, although most cases are not as extreme as this. The extensive coupling with hydrogens raises the point of careful designing of fluorine probes, that have no detectable coupling with hydrogens, e.g. $COCF_3$ groups, or the use of suitable n.m.r. instrumentation to 'decouple' the hydrogens and thus give a singlet fluorine resonance.

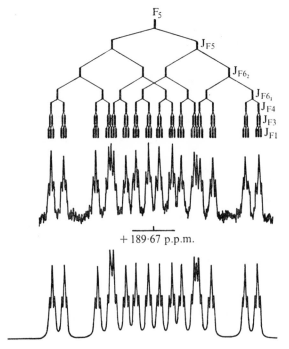

FIG. 8.15. Normal ^{19}F n.m.r. spectrum of 3,6-anhydro-5-deoxy-5-fluoro-1,2-O-isopropyli-dene-α-L-idofuranose in benzene-d_6 solution (upper trace). Simulated spectrum (lower trace). (From Hall and Steiner, 1970.)

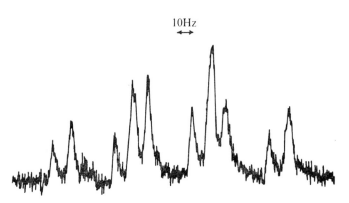

FIG. 8.16. ^{19}F N.m.r. spectum of N-fluoroacetylglucosamine at 84 MHz (25 °C, pD = 5·0, imidazole buffer, 25 % D_2O). (From Dwek *et al.*, 1971.)

The sensitivity of fluorine is further illustrated by the spectrum of FNAG in 25% D_2O (Fig. 8.16). Analysis of the spectrum indicates that there are now four sets of triplets, corresponding to a hydrogen or deuterium attached to the nitrogen atom. The isotope shift between the NH or ND forms is 11 Hz. The resonances are also broadened because of exchange between the hydrogen and the deuterium atoms. The exchange rate between the two forms can be increased on heating, until at 80 °C only two sets of resonances are observed; their chemical shifts are at the mean of the two positions of those for the NH or ND compound.

References

BLAKE, C. C. F., JOHNSON, L. N., MAIR, G. A., NORTH, A. C. T., PHILLIPS, D. C., and SARMA, V. R. (1967). *Proc. Roy. Soc. (Lond.)* B **167**, 378.

BUTCHARD, C. G., DWEK, R. A., KENT, P. W., WILLIAMS, R. J. P., and XAVIER, A. V. (1972). *Eur. J. Biochem.* **27**, 548.

DWEK, R. A., KENT, P. W., and XAVIER, A. V. (1971). *Eur. J. Biochem.* **23**, 343.

HALL, L. D. and STEINER, P. R. (1970). *Can. J. Chem.* **48**, 451.

HUESTIS, W. H. and RAFTERY, M. A. (1971). *Biochemistry*, **10**, 1181.

—— and RAFTERY, M. A. (1972). *Biochemistry*, **11**, 1648.

MILLET, F. and RAFTERY, M. A. (1972a). *Biochemistry*, **11**, 1639.

——, ——. (1972b). *Biochem. biophys. Res. Commun.* **47**, 625.

PERUTZ, M. (1970). *Nature (Lond.)* **228**, 726.

9

RELAXATION IN PARAMAGNETIC SYSTEMS

9.1. General introduction

THE possibilities of obtaining structural and kinetic information in bio-chemical systems from studying nuclear relaxation effects caused by para-magnetic probes was first realised in 1962 (Eisinger, Shulman, and Szymanski, 1962). In that year, it was noted that the binding of several transition metals to DNA gave a large enhancement in the hydrogen relaxation rates in the solvent, water. This phenomenon was termed proton relaxation enhancement (PRE). In the same year the first PRE results on Mn(II)–protein and Mn(II)–protein-substrate complexes were reported (Cohn and Leigh, 1962).

In the next few years, most of the effort was devoted to studies of the water relaxation rates, in various enzyme systems containing the para-magnetic ion Mn(II), generally added as a probe for Mg(II), (since their chemistries are very similar). On the basis of PRE results it has been possible to obtain binding parameters and even to propose various schemes of enzyme mechanism (Mildvan and Cohn, 1970; Birkett, Dwek, Radda, Richards, and Salmon, 1971).

Enzyme regulation is apparently associated with ligand-induced con-formational changes, which in some cases may be detected by relaxation enhancement of the water hydrogens around a paramagnetic probe site. However, this technique gives no information on the nature and extent of these conformational changes or even the relationship between the regulatory and catalytic sites. To do this it is necessary to use several probes, and to observe the ligand nuclei themselves. Under suitable conditions, the relax-ation rates of these nuclei, in the presence of a paramagnetic probe, may be related to their distance from this probe. Clearly, a paramagnetic ion such as Mn(II) would be one such a probe.

In 1965, the first description of another type of probe appeared, that of the nitroxide spin label (Stone, Buckman, Nordio, and McConnell, 1965). These labels could be attached covalently to specific groups on macromolecules and the effects of ligands on the label studied by e.s.r. spectroscopy. Further, as in the case of a paramagnetic metal ion, the effect of this probe on the relaxation rates of the ligand nuclei is related to the distance between them.

To map out the ligand completely, the distance between two such probes must be determined and this can be done by analysis of the effect of the paramagnetic metal ion on the e.s.r. spectrum of the spin label probe (Leigh, 1970). Similar procedures are of course applicable if pseudo-contact shifts (see Chapter 4) are measured instead of relaxation rates.

In this section we start with a qualitative discussion of how relaxation rates of diamagnetic nuclei alter in the presence of a paramagnetic ion; we then give the relevant equations. The importance of these equations is such that a section outlining the assumptions and difficulties in their application is included and the reader new to relaxation methods might omit this on the initial reading.

The possibilities of obtaining kinetic data from relaxation measurements, although mentioned briefly in diamagnetic systems is then discussed. Unlike the case for diamagnetic systems, measurements of relaxation rates can be more easily interpreted in terms of molecular motion and some examples are included to give perspective to the problems involved. The following sections are then concerned with the main uses of proton relaxation enhancement of solvent water and the use of spin labels as probes. In general, the examples chosen throughout are the more straightforward ones, and an attempt has been made to indicate the main assumptions, and to discuss some of the theoretical points that have yet to be developed adequately.

9.2. Some qualitative aspects of relaxation in paramagnetic systems

The nuclear relaxation times in a diamagnetic liquid are often reduced markedly when very small concentrations of paramagnetic solutes are added. The presence of dissolved oxygen, for example, in benzene, reduces the hydrogen relaxation times by a factor of ca. 5. The reason for this is that the magnetic moments of unpaired electrons are ca. 10^3 times greater than the nuclear magnetic moments, so that the local fields generated by them are much greater. In general, it is the fluctuation of these local fields which leads to relaxation and the larger fields resulting from the presence of a paramagnetic species give more efficient nuclear spin relaxation.

In a dilute solution of a paramagnetic solute, the nuclear relaxation is often entirely dominated by pair-wise interactions between an unpaired electron spin, S, and the spin of the nucleus, I. The strong local fields produced by the electron can be coupled to the nuclei by simple *dipole–dipole interaction*, or sometimes by a *scalar* or *hyperfine* coupling transmitted through a chemical bond (which may be transient), by the same mechanism which is responsible for the hyperfine structure of e.s.r. spectra or the spin–spin multiplets in n.m.r. spectra.

The Zeeman energy levels for a nuclear magnetic dipole, I, and an electron magnetic dipole, S, in a magnetic field are shown in Fig. 9.1, where the levels are labelled according to their value of the spin quantum numbers I_z, $S_z = \pm\frac{1}{2}$. The lowest electron magnetic level is labelled − (minus) because its magnetic moment has the opposite sign to that of the hydrogen (and most other nuclei). The random magnetic-field fluctuations can induce nuclear spin transitions involving changes in the spin quantum number,

13

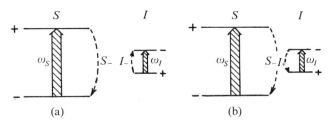

FIG. 9.1. Zeeman energy levels for electron spin, S, and nuclear spin, I. (a) relaxation transition S_-I_-. (b) relaxation transition S_-I_+.

I_z, of $-\frac{1}{2}$ to $+\frac{1}{2}$ and of $+\frac{1}{2}$ to $-\frac{1}{2}$ conveniently denoted I_+ and I_-, respectively. These transitions involve the nuclear Larmor frequency ω_I. The strong pair-wise interaction between the electron and nuclear magnetic dipoles can also induce coupled two-spin transitions, such as S_+I_- (S_-I_+) in which the electron and nucleus make simultaneous transitions. The probability of any of the transitions shown in Fig. 9.1 occurring depends on the nature of the interaction between the electron and the nucleus. Since simultaneous nuclear–electron spin transitions can occur, the nuclear relaxation processes will depend not only on the intensity of fluctuations at ω_I (i.e. $J(\omega_I)$) but also on the intensity of fluctuations at $\omega_S \pm \omega_I$, which characterize the double-spin transitions. Since $\omega_S \gg \omega_I$, the intensity of fluctuations at $\omega_S \pm \omega_I$, i.e. $J(\omega_S \pm \omega_I)$, which are important in relaxation processes, can be approximated by those at $J(\omega_S)$.

As we mentioned in Chapter 2, any interaction that affects T_1 must also affect T_2. However scalar interactions can shorten T_2 by a mechanism which does not affect T_1. One definition of T_2 involved a measure of the time taken for the nuclear magnetization in the xy-plane to decay by the loss of phase of the precessing nuclei. This phase loss may be induced by any static component of a local field which causes a spread in the precession frequencies of the nuclei. The scalar interaction produces such a component that does not vary as the complex rotates. The longer the component is applied, the more the nuclei will lose phase, so the effect will be proportional to τ_e, the lifetime of the scalar interaction.

In the light of the above discussion we now consider the relevant equations for the relaxation of a nucleus bound near or at the paramagnetic centre.

9.3. Relaxation of nuclei bound to, or near, a paramagnetic centre

In this section we shall consider the equations for the relaxation rates of a nucleus bound to, or near, a paramagnetic centre, and the validity of these equations in the light of several complications.

9.3.1 *The Solomon–Bloembergen equations* (From Solomon, (1955); Bloembergen, (1957))

The relaxation times T_1 and T_2 of nuclei bound near a paramagnetic site, are usually well represented by the Solomon–Bloembergen equations. Assuming $\omega_S \gg \omega_I$, these are

$$\frac{1}{T_{1,M}} = \frac{2}{15} \frac{\gamma_I^2 g^2 S(S+1)\beta^2}{r^6}\left(\frac{3\tau_c}{1+\omega_I^2\tau_c^2}+\frac{7\tau_c}{1+\omega_S^2\tau_c^2}\right)+$$

$$+\tfrac{2}{3}S(S+1)\left(\frac{A}{\hbar}\right)^2\left(\frac{\tau_e}{1+\omega_S^2\tau_e^2}\right) \quad (9.1)$$

and

$$\frac{1}{T_{2,M}} = \frac{1}{15} \frac{\gamma_I^2 g^2 S(S+1)\beta^2}{r^6}\left(4\tau_c+\frac{3\tau_c}{1+\omega_I^2\tau_c^2}+\frac{13\tau_c}{1+\omega_S^2\tau_c^2}\right)+$$

$$+\tfrac{1}{3}S(S+1)\left(\frac{A}{\hbar}\right)^2\left(\frac{\tau_e}{1+\omega_S^2\tau_e^2}+\tau_e\right) \quad (9.2)$$

The first terms in both equations arise from dipole–dipole interaction between the electron spin, S, and the nuclear spin, I, which is characterized by a correlation time τ_c which modulates this interaction. The second terms arise from modulation of the scalar interaction (often called isotropic nuclear–electron spin exchange interaction) which is characterized by a correlation time τ_e. ω_S and ω_I are the electronic and nuclear Larmor precession frequencies, γ_I is the magnetogyric ratio, β is the Bohr magneton, S is the total electron spin, r is the distance between the nucleus and the paramagnetic ion, and $A/\hbar$ is the electron–nuclear hyperfine coupling constant in Hz. In many cases, the magnetic moment is given by $\mu = gB\sqrt{[S(S+1)]}$ and this may be substituted into the above equations. In other cases this formula is not strictly true but it is permissible to replace this quantity in the equations by the *effective magnetic moment*, μ_{eff}.

9.3.2. *Correlation times*

The correlation times in the above equations are defined as:

$$\frac{1}{\tau_c} = \frac{1}{\tau_S}+\frac{1}{\tau_M}+\frac{1}{\tau_R} \quad (9.3)$$

$$\frac{1}{\tau_e} = \frac{1}{\tau_S}+\frac{1}{\tau_M} \quad (9.4)$$

where τ_M is the life-time of a nucleus in the bound site, τ_R is the rotational correlation time of the bound paramagnetic ion, and τ_S is the electron–spin relaxation time.†

9.3.2.1. *Temperature dependence of correlation times.* The temperature dependence of τ_M is assumed to be given by the Eyring relationship [eqn (9.5)]:

$$\frac{1}{\tau_M} = \frac{kT}{h}\left[\exp\left(-\frac{\Delta H^{\ddagger}}{RT} + \frac{\Delta S^{\ddagger}}{R}\right)\right] \tag{9.5}$$

where k is the Boltzmann constant, T the temperature in K, $\Delta H^{\ddagger}$ and $\Delta S^{\ddagger}$ are the enthalpy and entropy of activation, respectively, for the first-order reaction of chemical exchange. τ_R is given by:

$$\tau_R = \tau_R^{\circ} \exp(E_R/RT), \tag{9.6}$$

where E_R is an activation energy for the rotational motion.

The temperature dependence of τ_S may be quite complex and we therefore examine this next.

9.3.3. *The value of τ_S and its temperature dependence.*

At this stage we assume that the electron–spin lattice relaxation time is given by eqn (9.7) from Bloembergen and Morgan (1961):

$$\frac{1}{\tau_S} = B\left[\frac{\tau_v}{1+\omega_S^2\tau_v^2} + \frac{4\tau_v}{1+4\omega_S^2\tau_v^2}\right] \tag{9.7}$$

where τ_v is a correlation time, which is related to the rate at which the zero-field splitting is modulated by impact of the solvent molecules on the aquocomplex, and B is a constant containing the value of the resultant electronic spin and the zero-field splitting parameters. This theory assumes that $\tau_v < \tau_S$, and also that the electron spin–lattice relaxation time is much larger than the electron spin–spin relaxation time.

Equation (9.7) predicts that $1/\tau_S$ will be frequency dependent. This can be illustrated by plotting the value of

$$f(\tau_v) = \frac{\tau_v}{1+\omega_S^2\tau_v^2} + \frac{4\tau_v}{1+4\omega_S^2\tau_v^2}$$

as a function of τ_v at different frequencies, as shown in Fig. 9.2. *The temperature dependence of $1/\tau_S$ arises from that of τ_v which is given by:*

$$\tau_v = \tau_v^{\circ}\exp(E_v/RT) \tag{9.8}$$

where E_v is the activation energy for the motion characterized by τ_v.

A change in temperature is equivalent to a change in τ_v, and reference to Fig. 9.2 shows that as τ_v is decreased, the change in $1/\tau_S$ will be *positive* if

† Strictly, τ_S has at least two possible values, the longitudinal and transverse relaxation times. This is discussed later.

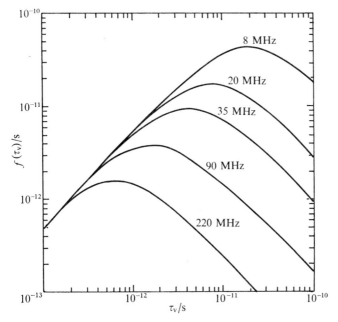

FIG. 9.2. Variation of $f(\tau_V)$ with τ_V as a function of frequency.

$$f(\tau_V) = \frac{\tau_V}{1+\omega_S^2\tau_V^2} + \frac{4\tau_V}{1+4\omega_S^2\tau_V^2}$$

$\omega_S^2\tau_V^2 \ll 1$, but *negative* if $\omega_S^2\tau_V^2 \gg 1$. Thus the activation energy may 'appear' to change at different frequencies depending on the value of $\omega_S^2\tau_V^2$ (e.g. it could be $\ll 1$ at one frequency but $\gg 1$, at a higher one). For this reason many workers often write

$$\tau_S = \tau_S^{\circ}\exp(\pm E_s/RT) \tag{9.9}$$

where the $\pm$ sign allows for a negative or positive temperature dependence.

9.3.4. *Some assumptions in the Solomon–Bloembergen equations*

In general the applications of the Solomon–Bloembergen equations as stated, have several assumptions.

(1) There is only *one* electron–spin relaxation time to consider, i.e. the e.s.r. spectrum would consist of a single Lorentzian line with equal electron spin–spin and spin–lattice relaxation times.

(2) There is no splitting of the electron energy levels in the absence of a magnetic field, i.e. no zero-field splitting (ZFS). (The equations however are valid if the ZFS $\ll \omega_S$).

(3) There is no splitting of the electron–spin resonance line by interaction with the nuclear spin of the paramagnetic ion itself. (However, as in (2) the equations are still valid if this splitting $\ll \omega_S$.)

(4) The tumbling motion is isotropic.

(5) The electron g-value is isotropic.

(6) The correlation functions for the different types of motions are exponential.

We shall deal with each of these in turn.

9.3.4.1(a). *The nature of* τ_S. So far we have assumed that there is only one electronic relaxation time, τ_S, but strictly speaking, both the spin–lattice $\tau_{1,S}$ and spin–spin $\tau_{2,S}$ relaxation times of the electrons have to be considered. As we have mentioned, and just as in the case for the nuclear relaxation times, the values of $\tau_{1,S}$ and $\tau_{2,S}$ may be field (or frequency) dependent. Further they may also depend on more than one single correlation time, i.e. several physical processes could be responsible for the electronic relaxation. For example, processes could include rotational diffusion or symmetry distortions of the complex by impact of solvent molecules. This latter mechanism is the one most frequently suggested as dominating electronic relaxation in paramagnetic metal ions. Explicitly the electronic relaxation occurs by modulation of the zero field splitting interactions, the time dependence arising from the modulation caused by the collisions of solvent molecules with the paramagnetic species.

In general $\tau_{1,S}$ is very short ca. 10^{-8} s or less and thus its determination is very difficult experimentally. On the other hand $\tau_{2,S}$ may often be estimated from the line-width in the e.s.r. spectrum. However, only in the simplest case, when the paramagnetic species has a spin quantum number, $S = \frac{1}{2}$, and the spectrum consists of a single line is the measurement relatively easy. In such a case, $\tau_{2,S}$ is calculated from the line-width, and $\tau_{1,S}$ is calculated from a suitable theory, e.g. eqn (9.7), if the value of B can be estimated from an analysis of the line-width. Otherwise a lower limit can be obtained since $\tau_{1,S} \geqslant \tau_{2,S}$.

The possibility of hyperfine structure in the e.s.r. spectrum arising from interaction of the electron spin with the nuclear spin of the paramagnetic species also exists. For instance the spectra of nitroxide radicals consists of three lines corresponding to the three ^{14}N nuclear spin quantum states. In general, the lines are of unequal width, the differences arising from anisotropies in the hyperfine interactions and in the g-values. The line-widths are given by an expression of the type (Stone *et al.*, 1965):

$$\frac{1}{\tau_{2,S}} = A + Bm_I + Cm_I^2 \tag{9.10}$$

where A, B, and C are constants and m_I is the nuclear spin quantum state. There are thus three values of $\tau_{2,S}$ to consider.

In the case when the paramagnetic species has a spin quantum number $S > \frac{1}{2}$, the interaction of the resultant spin with the magnetic field means that there are several energy levels to consider and the e.s.r. spectrum consists of a multiplicity of lines representing different transitions. Thus it is no longer meaningful to speak of a single $\tau_{2,S}$ or single transition probability $\tau_{1,S}$. For example, in the case of Mn(II), $S = \frac{5}{2}$, the e.s.r. spectrum consists of six lines of slightly different intensities and line-widths arising from hyperfine coupling with the Mn(II) nuclear spin. However, because of coupling with the zero-field splitting in this case, each line is split into five components which overlap leading to an apparent broadening of the six lines. Thus identification of the transitions corresponding to each line does not help much in obtaining values of $\tau_{2,S}$. It has to be done by a comparison of theoretical and observed spectrum.

For Mn(II), it can be shown that for values of $\omega_S\tau_v < 1$ there will only be one spin–lattice and one spin–spin relaxation to consider and further that $\tau_{1,S} = \tau_{2,S}$ (Rubinstein, Baram, and Luz, 1971; Luckhurst, 1972). When $\omega_S\tau_v > 1$, it turns out that there are three spin–spin and three spin–lattice relaxation times to consider, the relative weights of which vary with ω_S and τ_v. However *one* spin–lattice relaxation time has a far greater weight than the other two at almost every frequency. (The case $\omega_S\tau_v > 1$ is probably encountered most in Mn(II)–macromolecular complexes.) On the other hand, all three electronic spin–spin relaxation times have to be considered at most frequencies and this may affect the interpretation of the n.m.r results in systems (such as that of the pure aquo-ion) in which $\tau_{2,S}$ may be important [see eqns (9.13) and (9.14)].

In contrast, in Gd(III) ($S = \frac{7}{2}$), where the e.s.r. spectrum of an aqueous solution should consist of seven lines arising from interaction with the zero-field splitting, it can be shown numerically (Reuben, 1971b) that only one of these lines, identified as the $-\frac{1}{2}$ to $+\frac{1}{2}$ transition contributes significantly to the observed spectrum (the other transitions contributing $<6\%$ at 9·14 GHz, for values of $\tau_v < 10^{-11}$ s, where τ_v is the relevant correlation for the modulation of the zero-field splitting). The values of $\tau_{2,S}$, for Gd(III) in two different magnetic fields have been measured (Reuben, 1971b). Using the theory of Hudson and Lewis (1970), $\tau_{2,S}$ can be related to the zero-field splitting tensor B and τ_v. This allows the value of B to be obtained and thus $\tau_{1,S}$ from eqn (9.7).

We must however also consider the possibility that $\tau_{1,S}$ and $\tau_{2,S}$ may be sensitive to different motions. When 'non-white' spectrum conditions apply, i.e. $\omega_S\tau_v \gg 1$, then as with nuclear relaxation, $\tau_{2,S}$ will be sensitive to 'slow' motions, while $\tau_{1,S}$ is sensitive to the 'fast' motions. One could envisage equations of the type

$$\frac{1}{\tau_{1,S}} = B^0\left(\frac{\tau_R}{1+\omega_S^2\tau_R^2}+\frac{4\tau_R}{1+\omega_S^2\tau_R^2}\right)+B^1\left(\frac{\tau_v}{1+\omega_S^2\tau_v^2}+\frac{4\tau_v}{1+\omega_S^2\tau_v^2}\right)+\dots \quad (9.11)$$

and

$$\frac{1}{\tau_{2,S}} = B^0\left(a\tau_R + \frac{b\tau_R}{1+\omega_S^2\tau_R^2} + \ldots\right) + B^1\left(a\tau_v + \frac{b\tau_v}{1+\omega_S^2\tau_v^2}\right) + \ldots \quad (9.12)$$

where a and b are constants and B^0 and B^1 represent different zero field splittings. For example, B^0 may be the *static* zero-field splitting which may be modulated by, say, rotation, characterised by a correlation time τ_R. On the other hand, B^1 might be a *transient* zero-field splitting arising from impact of solvent molecules or asymmetric ligand vibrations. Suppose B^0 is large and B^1 relatively small. If τ_v is very long, then $\omega_S^2\tau_v^2 \gg 1$ and only the terms in B^1 contribute to $1/\tau_{1,S}$. Conversely, terms in $B^0\tau_v$ will make a dominant contribution to $1/\tau_{2,S}$. If this effect is important, an analysis of the e.s.r. spectrum could yield information on B^0, but an analysis of the nuclear relaxation rates gives information on B^1. The type of approach used by Reuben (1971b) could well be invalid in macromolecular systems where only the terms in $\tau_{1,S}$ are involved. It is in such situations where the metal ion resides in non-symmetric environment and has a large static zero field splitting (τ_R is very long) that this behaviour is to be expected.

It is relevant to point out some of the analyses of the e.s.r. spectra of Mn(II) complexes in aqueous solutions (Reed and Cohn, 1970, 1972, Cohn, *et al.* 1971, Reed, Leigh, and Pearson 1971). We have seen that the e.s.r. spectrum of an aqueous solution of Mn(II) consists of six hyperfine lines (each line consisting of a superposition of five components because of interaction with the zero-field splitting (ZFS)). Figure 9.3 shows a theoretical comparison of a single hyperfine line of Mn(II) in an isotropic environment (ZFS = 0) and

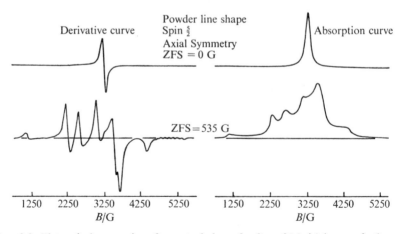

FIG. 9.3. Theoretical comparison for a *single hyperfine line* of Mn(II) in a perfectly symmetrical or isotropically averaged environment (ZFS = 0) and in an asymmetric environment with a uniaxial distortion (ZFS = 535 G). (From Cohn *et al.* 1971.)

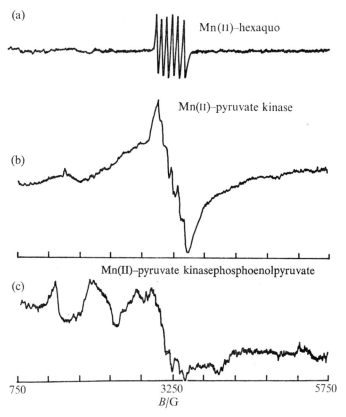

FIG. 9.4. X-Band e.s.r. spectra for solutions of Mn(II)-hexaquo complex and Mn(II) complexes in the pyruvate kinase reaction. (a) 0·2 mM MnCl$_2$; (b) 200 mg ml^{-1} pyruvate kinase, 1 mM MnCl$_2$; and (c) 200 mg ml^{-1} pyruvate kinase, 1 mM MnCl$_2$, 2mM phosphoenol-pyruvate (From Cohn, Leigh, and Reed 1971.)

in an asymmetric environment with an axial distortion of the ZFS of 535 G. Figure 9.4 shows the e.s.r. spectra of $[Mn(H_2O)]_6^{2+}$, Mn–pyruvate kinase (M–PK), and Mn–pyruvate kinase–phosphoenolpyruvate (M–PK–P-Py). Comparisons in Fig. 9.4 shows that the substitution of protein ligands for water in the Mn(II) first co-ordination sphere produces considerable asymmetry in the ZFS of Mn(II) in the M–PK–P-Py complex and less in the Mn–PK complex. Relaxation measurements on the hydrogen nuclei of the solvent water have suggested that at least three water molecules are displaced in the Mn–PK complex (Navon, 1971; Reuben and Cohn, 1970) (see pp. 214–219). Perhaps the greater asymmetry in the ternary complex arises because even more water molecules are displaced. In favourable cases therefore, it is possible that the change in the e.s.r. spectrum of Mn(II) on binding can be used to detect changes in the Mn(II) environment on addition

of ligands etc. Reed and Ray (1971) have indeed suggested a classification
in Mn(II)–phosphoglucomutase–ligand complexes based on the magnitude
of the ZFS.

All the above suggests that the e.s.r. spectrum of Mn(II) when bound to a
macromolecule will usually be more complicated than that of the corresponding 'free' metal ion. Use of this fact is often made in monitoring the binding of
Mn(II) to macromolecules (see p. 254). On the other hand, with the other
major relaxation probe that we shall encounter, Gd(III), it may be relatively
easy to observe the e.s.r. signal due to the bound form. Such is the case for
the Gd(III)–BSA (Reuben, 1971b) and Gd(III)–lysozyme (Dwek, Morallee,
Nieboer, Richards, Williams, and Xaivier, 1971) complexes and this may
have its origin in a higher symmetry of the Gd(III) ion in the complex (thus
reducing the zero-field splitting). For example, the fully-hydrated Gd(III)
ion may be nine- or ten-fold co-ordinated but in the complexes the co-
ordination number could be eight (or six).

9.3.4.1(b). *Modified form of the Solomon–Bloembergen equations to incorpo-
rate the different correlation times* $\tau_{1,S}$ *and* $\tau_{2,S}$. Assuming that there is only
one $\tau_{1,S}$ and one $\tau_{2,S}$, the Solomon–Bloembergen equations become (Connick
and Fiat, 1966; Reuben, Reed, and Cohn, 1970):

$$\frac{1}{T_{1,M}} = \frac{2}{15}\frac{\gamma_I^2 g^2 S(S+1)\beta^2}{r^6}\left(\frac{3\tau_{c,1}}{1+\omega_I^2\tau_{c,1}^2}+\frac{7\tau_{c,2}}{1+\omega_S^2\tau_{c,2}^2}\right)+\frac{2}{3}\left(\frac{A}{\hbar}\right)^2 S(S+1)\left(\frac{\tau_{e,2}}{1+\omega_S^2\tau_{e,2}^2}\right)$$

and   (9.13)

$$\frac{1}{T_{2,M}} = \frac{1}{15}\frac{\gamma_I^2 g^2 S(S+1)\beta^2}{r^6}\left(4\tau_{c,1}+\frac{3\tau_{c,1}}{1+\omega_I^2\tau_{c,1}^2}+\frac{13\tau_{c,2}}{1+\omega_S^2\tau_{c,2}^2}\right)+$$

$$+\frac{1}{3}\left(\frac{A}{\hbar}\right)^2 S(S+1)\left(\tau_{e,1}+\frac{\tau_{e,2}}{1+\omega_S^2\tau_{e,2}^2}\right)\quad(9.14)$$

where, now,

$$\frac{1}{\tau_{c,1}}=\frac{1}{\tau_R}+\frac{1}{\tau_{e,1}}\;;\qquad\frac{1}{\tau_{e,1}}=\frac{1}{\tau_{1,S}}+\frac{1}{\tau_M}$$

and

$$\frac{1}{\tau_{c,2}}=\frac{1}{\tau_R}+\frac{1}{\tau_{e,2}}\;;\qquad\frac{1}{\tau_{e,2}}=\frac{1}{\tau_{2,S}}+\frac{1}{\tau_M}$$

The value of $\tau_{1,S}$ may be obtained from the Bloembergen–Morgan equation
[eqn (9.7)] if B, the zero-field splitting parameter, is known. $\tau_{2,S}$ may be
obtained either from comparison with the theoretical spectrum or, in some
cases, such as Gd(III), by direct measurement. If there is more than one
$\tau_{1,S}$ or $\tau_{2,S}$ eqns (9.13) and (9.16) can be modified by substituting expressions

of the form (Karger and Pfeifer, 1968; Rubinstein *et al.*, 1971):

$$\sum_k P_k \frac{(\tau_{c,2})_k}{(1+\omega_S^2\tau_c^2)^2}, \text{ etc.}$$

i.e. the sum over all k transitions for $\tau_{c,2}$ and a corresponding expression for $\tau_{c,1}$. P_k is the relative intensity of the kth transition, and the correlation times now involve $(\tau_{1,S})_k$ and $(\tau_{2,S})_k$.

9.3.4.1(c). *The nature of τ_S when the paramagnetic ion is bound to a macromolecule.* In the majority of cases, it turns out that the correlation times will be such that the product $\omega_S\tau_{c,2} \gg 1$ and thus the terms containing $\tau_{c,2}$ in eqns (9.13) and (9.14) become small and may be neglected; This removes the need to know $\tau_{2,S}$. The complication of a value for $\tau_{1,S}$ still remains, and no adequate theory exists at present for the magnitude of electronic spin–lattice relaxation times when the metal ion is bound to a macromolecule. Most of the detailed studies carried out in macromolecular systems have involved Mn(II) and for this case it has usually been assumed that the Bloembergen–Morgan equation [eqn (9.7)] is valid. However, in that equation, τ_v is the correlation time for symmetry distortions of the cubic symmetry of the aquo-complex by impact with the solvent molecules. For ions bound to macromolecules, the effectiveness of such a process may be reduced since the location of the binding site may be such that the approach of solvent molecules is restricted. Other processes for the symmetry distortion have then to be considered, such as ion–ligand vibrations coupled to the slower vibrations and other motions of the whole polypeptide chain. Such processes, however, may involve more than a single correlation time adding a further complication. We have discussed some difficulties which may be involved in the study of macromolecular systems, however in most of the biochemical problems discussed later, which involve Mn(II), only one value of $\tau_{1,S}$ is actually considered. In the detailed molecular-motion studies (Chapter 10), the addition of more than one $\tau_{1,S}$ would involve a further parameter. The theoretical fit of the experimental points would be even more ambiguous and difficult than at present. In the distance studies (see Chapter 10) errors in τ_c (because of $\tau_{1,S}$) are also not too serious, as r depends approximately on the sixth root of τ_c.

9.3.4.2. *The magnitude of the zero field splitting (ZFS).* The Solomon–Bloembergen equation may be used as long as the ZFS $\ll \omega_S$, for in the derivation of these equations the ZFS is neglected. In practice, for Gd(III) and Mn(II), the most commonly used probes, the ZFS $\approx 10^3$ MHz, so that in magnetic fields for which $\omega_S \gg 10^3$ MHz this complication can be largely ignored. In practice for hydrogen nuclei this means nuclear frequencies >3 MHz. For example, for the hydrogen nuclei of the solvent water in solutions of Mn(II)–t-RNA complexes (Danchin and Guéron, 1970), the value of the relaxation times are no longer satisfactorily predicted by the Solomon–Bloembergen equations at 3 and 8 MHz. However there are some cases, e.g. Ni(II) where the magnitude of the ZFS is $>10^4$ MHz such that even in magnetic fields of ca 14 000 G (^{1}H n.m.r. frequency ca. 60 MHz), it is comparable with ω_S, bringing the validity of the Solomon–Bloembergen equations into question.

9.3.4.3. *The nuclear spin–electron spin interaction in a paramagnetic ion* (Pfeifer, Michel, Sames, and Sprinz, 1966). If a paramagnetic ion has a nuclear spin, its e.s.r. spectrum may exhibit hyperfine splitting. If the nuclear spin is I, there will be $2I+1$ equally-spaced and equally-intense lines. When this hyperfine interaction is ca. ω_S, then, as in Section 9.3.4.2, the Solomon–Bloembergen equations may be invalid. In the case of Mn(II), this hyperfine interaction is ca. 150 MHz and so again at low magnetic fields we may expect significant deviations from the Solomon–Bloembergen equation. In fact such behaviour has been observed for the hydrogen relaxation rates of water molecules in aqueous solutions of Mn(II) at 90 °C and at 1 MHz ($\omega_S = 660$ MHz). It turns out that a correction to the basic equations is necessary when $\tau_M < \tau_S$ occurs for this system at high temperatures. At room temperature $\tau_M > \tau_S$ and the results can be interpreted using the normal equations. In Mn(II)–macromolecular systems, such as the Mn(II)–t-RNA system both the effects discussed in Section 9.3.4.2 and 9.3.4.3 will probably occur at low frequencies i.e. ca. 3 MHz.

9.3.4.4. *Anisotropic rotation* (Woessner, 1962). The equations for the relaxation times $1/T_{1,M}$ and $1/T_{2,M}$ assume that the paramagnetic ion undergoes isotropic rotation. If, however, the rotation is anisotropic, then the overall rotational correlation time, τ_R, may be an involved function of the re-orientational times of the different axes. When a paramagnetic ion is bound via only *one* ligand to a macromolecule, the rotation of the paramagnetic ion may be no longer isotropic. If the paramagnetic ion is bound via *two* or more ligands it is then held rigidly to the macromolecule. Its rotational motion is either that of the whole macromolecule or that part containing the paramagnetic ion which may have its own segmental motion.

Consider the case when the paramagnetic ion is bound via only one ligand to a macromolecule. This is illustrated in Fig. 9.5, for a metal which binds via an oxygen. We have several correlation times to consider depending on which nuclei are considered here, unlike the usual case for $1/\tau_o$ in which the fastest motion dominates. For anisotropic rotational motion it is not, in general, the shortest rotational correlation time which dominates the relaxation rates. Strictly, three correlation times have to be defined for anisotropic rotation and if these are all shorter than ω_I it can be shown that it is the longest correlation time that is important. The other faster motions

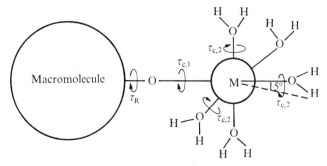

FIG. 9.5. Schematic representation of binding of metal ion to a macromolecule *via* one ligand. Various rotational correlation times are indicated.

will reduce the intensity of the dipolar interaction. For example, if the nuclei under consideration in Fig. 9.5 are on the macromolecule, the faster rotational motion, designated $\tau_{c,1}$ in Fig. 9.5, will lead to a partial averaging of the dipolar interaction between these nuclei and the paramagnetic metal ion. To a good approximation the reduction in dipolar coupling is given by:

$$R = \left(\frac{3\cos^2\!\Delta - 1}{2}\right)^2 \tag{9.15}$$

where Δ is the angle between the rotation axes and the distance vector of the nuclei in question. In the case considered, the nuclei could be either on the macromolecule or those of a small molecule which binds to it.

If the nuclei concerned are those of the bound water molecules themselves the situation is somewhat more complex. Here, τ_R may still be the relevant correlation time in the relaxation rate equations, but the dipolar interaction is partially averaged out by the two faster rotational times, $\tau_{c,1}$ and $\tau_{c,2}$. The rotational motion characterized by $\tau_{c,2}$ leads to a reduction of 20% in the dipolar interaction (i.e. $\Delta \approx 15°$, see Fig. 9.5) for each of the water molecules, while the motion characterized by $\tau_{c,1}$ leads to a reduction of 75% for four water molecules ($\Delta \approx 90°$). The fifth water molecule experiences no reduction since $\Delta = 0$. The total reduction in dipolar interaction is then $\frac{17}{25}$ of that from both effects. If the metal ion is bound to more than one site on the macromolecule, then only the first of these effects has to be considered.

The case of the bound water molecules merits further attention since the majority of relaxation studies have been on the hydrogen nuclei of solvent water. In some enzyme systems it is possible that the metal ion could bind in a hydrophobic region (Valee and Williams, 1968). If this were the case, there would be some structuring of the water around the metal ion such that for this case the correlation time $\tau_{c,2}$ would approach τ_R, leading to a much smaller reduction in the dipolar interaction (possibly invalidating the treatment described here).

The reduction caused by the motion characterised by $\tau_{c,1}$ is only important as long as the correlation times for the motions leading to this reduction are shorter than those of the other correlation times, such as τ_S and τ_M which could contribute to τ_c and τ_e. This reduction in dipolar coupling may be most simply treated by multiplying the dipolar parts of eqns (9.1) and (9.2) by a factor $(1-R)$. No doubt even more complex theories could be applied.

9.3.4.5(a). *Anisotropic g-values.*   The orbital angular momentum associated with the electrons of an atom gives rise to local fields. In a crystal field the orbital motion of the electrons is quenched to an extent which is determined by the ligand (electrostatic) field. The g value reflects the resultant field that the unpaired electrons now feel and so must be a directional measure of the magnetic properties of the entire complex. If the electric field produced at the electron by the ligands is symmetrical, then the value of g is isotropic. If the field is anisotropic it is necessary to introduce different components of g, e.g. for an axial field $g_{\parallel}$ and $g_{\perp}$ which are measured with respect to the direction of the applied magnetic field. The number of unpaired electrons on the metal ion, the orbitals they occupy, and the number and arrangement of the ligands are all factors that are important in determining the nature of g. However not only is the symmetry of the environment important in determining g, but also the relative energies of different states.

For the first-row transition metal ions, it seems that if the complex has a singlet ground state (no net orbital angular momentum) the value of g is isotropic. Conversely, if the ground state has a net orbital angular momentum, or if there are several excited states near in energy to the ground state, such that a mixture of energy states may be obtained, the g-value will be anisotropic. In the transition metal ions, it is probably the ligand field which is important in determining the value of g, and spin orbit coupling is relatively unimportant.

9.3.4.5. *Modified form of the Solomon–Bloembergen equations for the case of an axially symmetric g-value.*   For the anisotropic case, the basic eqns (9.1) and (9.2) can be modified. If we assume axial symmetry for g with values of $g_{\parallel}$ and $g_{\perp}$, as the values along each axis, and if we consider only the *dipolar contribution* to the relaxation rate, then the relevant relaxation equations in the limit of short correlation times (white spectrum $\omega\tau_c \ll 1$) are (Sternlicht, 1965; Karger and Pfeifer, 1968):

$$\frac{1}{T_{2,M}} = \frac{1}{T_{1,M}} = \frac{2}{3}\frac{\gamma^2\beta^2 S(S+1)}{r^6} \tfrac{1}{3}[(g_{\parallel}^2 + 2g_{\perp}^2) + g_{\parallel}^2 \cos 2\chi + g_{\perp}^2 \sin^2\chi]\tau_c \quad (9.16)$$

where χ is the angle between the main symmetry axis and the electron and nuclear dipole distance vector.

When $\omega\tau_c \gg 1$ the eqn (9.16) becomes more complex, e.g.

$$\frac{1}{T_{2M}} = \frac{1}{15}\frac{\gamma_I^2\beta^2 S(S+1)}{r^6}f(\tau_c)$$

$$f(\tau_c) = \tfrac{1}{3}(12g_\parallel^2 + 8g_\perp^2 - 8g_\parallel g_\perp) + 6(g_\parallel - g_\perp)^2\sin^4\chi +$$

$$+ [(9g_\parallel - 5g_\perp)(g_\perp - g_\parallel)\sin^2\chi]\tau_c +$$

$$+ \tfrac{1}{3}(24g_\parallel^2 + 11g_\perp^2 + 4g_\parallel g_\perp) - 3(g_\parallel - g_\perp)^2\sin^4\chi +$$

$$+ [(3g_\parallel + 10g_\perp)(g_\perp - g_\parallel)\sin^2\chi]\tau_c/[1+(\omega_S\tau_c)^2] +$$

$$+ \tfrac{1}{3}(4g_\parallel^2 + g_\perp^2 + 4g_\parallel g_\perp) - 3(g_\parallel - g_\perp)^2\sin^4\chi -$$

$$- [(2g_\parallel - 5g_\perp)(g_\perp - g_\parallel)\sin^2\chi]\tau_c/[1+(\omega_I\tau_c)^2]$$

We note that when $g_\parallel = g_\perp$, the equation reduces to the usual Solomon–Bloembergen form. For further details the reader is referred to the above references. Even if $g_\parallel$ and $g_\perp$ are known, χ now provides one further parameter, and then it is often difficult to know precisely what is the principal symmetry axis.

One interesting aside is to consider this complication for an approximately *octahedral complex* (Karger and Pfeifer, 1968) for which $g_\parallel \neq g_\perp$. At any given time two ligands will be in the direction of the g-symmetry axis and four will be in the equatorial plane. If the rotation of the metal axes is rapid with respect to the examined nucleus, then for values of the ratio of $g_\perp/g_\parallel$ between 0·4 and 4, it has been shown that the error introduced by only considering an isotropic g-value is $<10\%$. So in simple inorganic complexes, e.g. Ni(MeCN)$_6$ or Co(MeCN)$_6$ which have anisotropic g-values but tumble rapidly, this complication may be ignored.

Of the transition metal ions, Mn(II), is most commonly used at present in macromolecular studies. Mn(II) has a half-filled shell ($3d^5$) and has an isotropic g-value. Its chemistry has many similarities to that of the 'A' metals and thus Mn(II) tends to co-ordinate preferentially with oxygen ligands which produce little or no perturbation in its symmetrical ground state. This does not rule out the possibility that once Mn(II) had co-ordinated to an oxygen atom, it may then co-ordinate to a suitably-orientated nitrogen or sulphur which may probably make the g-value anisotropic. In the rare earth ions, to a first approximation all the g values are isotropic (Bleaney, 1972) and so this complication may be absent. However spin–orbit coupling is very significant in the lanthanides and the value of J should be substituted for S in the basic equations (see p. 65).

9.3.4.6. *Exponential correlation functions.* The different types of motion characterized by τ_M, τ_S, and τ_R may have different correlation functions, which may lead to slightly modified forms of the basic equations. By assuming, however, that all the correlation functions have the same (exponential) form, this complication is effectively ignored.

9.4. Outer sphere relaxation

In most experiments involving measurements of nuclear relaxation times in the presence of paramagnetic ions, only a small fraction of the nuclei will be bound at, or near, the paramagnetic centre. It is necessary to consider the contribution to the observed relaxation times of the effects of the paramagnetic centre on the remaining (bulk) nuclei.

While the scalar or hyperfine interaction is expected to be important only for nuclei directly bonded to the paramagnetic ion, we have at this stage to include the possibility of dipole–dipole interaction between the paramagnetic ion and the bulk nuclei. A general mathematical treatment is unfortunately not straightforward and the exact approach will depend on the system and the nature of the correlation times involved.

It is convenient to consider initially two limiting cases to indicate the approaches that can be used in calculating the outer sphere contribution to the relaxation rates: (1) The dipolar interaction is modulated by the relative translational (diffusion) motions of the paramagnetic ion complex and the ligands in the bulk solution. The modulation is characterized by a correlation time τ_D. (2) The dipolar interaction is modulated by motions characterized by τ_S.

The equations turn out to be slightly different in the two cases because the form of the correlation functions used in calculating the relaxation times are not the same. While for motions characterized by τ_S, the correlation function is exponential, it is more complex for translational diffusion.

9.4.1. Translational diffusion

A quantitative treatment (Abragam, 1961; Hubbard, 1966) leads to eqns (9.17) and (9.18) (for the 'outer sphere' relaxation times ($T''_{1,A}$ and $T''_{2,A}$)

$$\frac{1}{T''_{1,A}} = \frac{8\pi}{225} \frac{N_S \gamma_I^2 \gamma_S^2 \hbar^2}{\mathscr{D}d} S(S+1)[7f_t(\omega_S\tau_D) + 3f_t(\omega_I\tau_D)] \tag{9.17}$$

and

$$\frac{1}{T''_{2,A}} = \frac{8\pi}{225} \frac{N_S \gamma_I^2 \gamma_S^2 \hbar^2}{\mathscr{D}d} S(S+1)[\tfrac{13}{2}f_t(\omega_S\tau_D) + \tfrac{3}{2}f_t(\omega_I\tau_D) + 2] \tag{9.18}$$

where N_S is the number of spins per ml of solution; $f_t(\omega\tau_D) = \tfrac{15}{2}I(u)$; $u = |\omega\tau_D|^{\frac{1}{2}}$, $I(u) = u^{-5}\{u^2 - 2 + e^{-u}[(u^2-2)\sin u + (u^2+4u+2)\cos u]\}$. τ_D, the diffusional correlation time is defined as $\tau_D = d^2/\mathscr{D}$ where d is the distance of closest approach between a ligand molecule in the bulk solution (radius a_1), and the paramagnetic ion complex (radius a_2). $\mathscr{D} = \tfrac{1}{2}(\mathscr{D}_I + \mathscr{D}_S)$ and

$\mathscr{D}_I$ and $\mathscr{D}_S$ are the diffusion coefficients of the ligand molecule and of the paramagnetic ion complex respectively.

If we assume that the relative motions are correctly described by the diffusion of rigid spheres in a medium of viscosity η, then:

$$\mathscr{D}_I = \frac{kT}{6\pi a_1 \eta} \quad \text{and} \quad \mathscr{D}_S = \frac{kT}{6\pi a_2 \eta}. \tag{9.19}$$

From these relationships and from $d = a_1 + a_2$ one obtains expressions like

$$\frac{1}{T''_{1,A}} = \frac{32\pi^2\eta}{45kT}\frac{a_1 a_2}{(a_1+a_2)^2} N_S\gamma_I^2\gamma_S^2\hbar^2 S(S+1)[7f_t(\omega_S\tau_D)+3f_t(\omega_I\tau_D)]. \tag{9.20}$$

If we consider the example of an aqueous solution of $(100\ \mu\text{M})$ Mn(II) then at 60 MHz, for $\tau_D \approx 10^{-10}$ s, $f_t(\omega_S\tau_D) \approx 0.1$ and $f_t(\omega_I\tau_D) = 1$. In water at 20 °C, the viscosity is $\eta = 0\cdot001$ poise and assuming a molecular radius of $1\cdot5$ Å for the water molecule and of ca. 3 Å for the $\text{Mn(OH}_2)_6$ complex, then $1/T''_{1,A} \approx 0\cdot2$ s^{-1}. Most relaxation data are 'normalized' to unit paramagnetic ion concentration, when reported; in this case the normalized value would be $\approx 2\times10^3$ s^{-1}. This may be compared with an experimental value of the spin–lattice relaxation time (see Fig. 9.11) of ca. 3×10^4. Thus in this case outer sphere relaxation is responsible for ca. 6% of the observed relaxation. For the spin–spin relaxation rate of aqueous Mn(II) solutions scalar inter-action is very important in the first co-ordination sphere and the outer sphere relaxation is consequently less important.

9.4.2. τ_S important

This is the case commonly found with aqueous solutions of Ni(II), Co(II) and the majority of the lanthanide ions. The value of $1/T''_{1,A}$ may be estimated by averaging the dipolar contribution over the volume between the sphere of closest approach, radius d, and infinity (Luz and Meiboom, 1964). Thus:

$$\frac{1}{T''_{1,A}} = \tfrac{2}{15}N_S\gamma_I^2 g^2 S(S+1)\beta^2\left(\frac{3\tau_c}{1+\omega_I^2\tau_c^2}+\frac{7\tau_c}{1+\omega_S^2\tau_c^2}\right)\int_d^\infty\frac{4\pi r^2\,dr}{r^6} \tag{9.21}$$

and

$$\frac{1}{T''_{2,A}} = \tfrac{1}{15}N_S\gamma_I^2 g^2 S(S+1)\beta^2\left(4\tau_c+\frac{3\tau_c}{1+\omega_I^2\tau_c^2}+\frac{13\tau_c}{1+\omega_S^2\tau_c^2}\right)\int_d^\infty\frac{4\pi r^2\,dr}{r^6} \tag{9.22}$$

From which expressions like eqn (9.23) are obtained

$$\frac{1}{T''_{1,A}} = \tfrac{8}{45}N_S\gamma_I^2 g^2 S(S+1)\left(\frac{3\tau_c}{1+\omega_I^2\tau_c^2}+\frac{7\tau_c}{1+\omega_S^2\tau_c^2}\right)\frac{\pi}{d^3} \tag{9.23}$$

9.4.3. Other cases

In many systems, it may not be possible to use the above equations since $\tau_S \approx \tau_D$. This may occur for instance for some paramagnetic ions bound

14

to macromolecules. Again the generalized treatment may be rather complex and we merely point out to the reader to use care in interpreting any such effects.

For completion we also mention the case when ligands exchange with those in the bulk from a second co-ordination sphere (as well as the first). This is a rather unusual 'outer sphere' problem and implies a stable second co-ordination sphere. If the residence time in such a sphere is τ'_M then if $\tau'_M > \tau_R$ or τ_S the calculation is formally the same as that for the nuclei in the first co-ordination sphere. The number of ligands in the second co-ordination sphere and their distance from the paramagnetic centre must, of course, be known.

9.5. Chemical exchange in paramagnetic systems

The treatment is very similar to that already discussed for diamagnetic systems. We consider initially that the nuclei may exist in either the bulk solvent or bound to the paramagnetic ion. These sites are denoted A and M respectively. Again we shall confine the discussion to the situation in which only a small fraction of nuclei are bound. We shall consider the relaxation rates initially in each site and then how the observed relaxation rates alter when chemical exchange, with the appropriate time scale, occurs between these two sites.

9.5.1. *Bound site; no exchange*

The relaxation rates of the bound nuclei are given by the Solomon–Bloembergen equations [eqns (9.1) and (9.2)] (q.v.).

9.5.2. *Unbound site; no exchange*

The relaxation rates of the unbound nuclei, $1/T_{1,A}$ and $1/T_{2,A}$, can, in the absence of chemical exchange, be considered to arise from two contributions: (a) the relaxation rates in the absence of the paramagnetic ion, $1/T_{1,A}$ and $1/T_{2,A}$, and (b) the relaxation rates arising from dipolar interaction with the solvated paramagnetic ion, $1/T''_{1,A}$ and $1/T''_{2,A}$:

$$\frac{1}{T_{1,A}} = \frac{1}{T_{1,A}} + \frac{1}{T''_{1,A}} \tag{9.24}$$

The forms of $1/T''_{1,A}$ and $1/T''_{2,A}$ have already been discussed.

9.5.3. *The equations for the relaxation rates in the unbound site in the presence of chemical exchange.*

Having considered the relaxation rates in the two sites separately, we now consider what happens if chemical exchange between the two sites occurs. Although both the relaxation rates in the A and M sites may change, we shall confine our attention to the A (bulk) site since this is much easier to measure experimentally because of its much larger concentration.

If the observed relaxation rates in the presence of chemical exchange are $1/T_{1,\text{obs.}}$ and $1/T_{2,\text{obs.}}$, the contribution of the paramagnetic ion to these rates, $1/T_{1,\text{P}}$, may be shown to be (Luz and Meiboom, 1964; Swift and Connick, 1962):

$$\frac{1}{P_\text{M}q}\frac{1}{T_{1,\text{P}}} = \frac{1}{P_\text{M}q}\left(\frac{1}{T_{1,\text{obs.}}}-\frac{1}{T_{1,\text{A}}}\right) = \frac{1}{P_\text{M}qT''_{1,\text{A}}}+\frac{1}{(T_{1,\text{M}}+\tau_\text{M})} \tag{9.25}$$

$$\frac{1}{P_\text{M}q}\frac{1}{T_{2,\text{P}}} = \frac{1}{P_\text{M}q}\left(\frac{1}{T_{2,\text{obs.}}}-\frac{1}{T_{2,\text{A}}}\right) = \frac{1}{P_\text{M}qT''_{2,\text{A}}}+\frac{1}{\tau_\text{M}}\left[\frac{\dfrac{1}{T_{2,\text{M}}}\left(\dfrac{1}{T_{2,\text{M}}}+\dfrac{1}{\tau_\text{M}}\right)+\Delta\omega_\text{M}^2}{\left(\dfrac{1}{T_{2,\text{M}}}+\dfrac{1}{\tau_\text{M}}\right)^2+\Delta\omega_\text{M}^2}\right] \tag{9.26}$$

We note immediately that the equations are essentially identical to those for the diamagnetic case, but an outer-sphere term is included here. As before, these equations assume that the fraction of bound nuclei $P_\text{M}q \ll 1$, where P_M is the mole fraction of ligand nuclei bound to the paramagnetic ion in the solution and q is the co-ordination number. $\Delta\omega_\text{M}$ is the chemical shift between the two sites.

9.5.3.2. *The value of* $\Delta\omega_\text{M}$. This shift may have its origin in either contact (hyperfine) or pseudo-contact (dipolar) interactions, or possibly both. The forms of $\Delta\omega_\text{M}$ for these cases have been discussed in some detail on pp. 54–59. We note, however, that its temperature dependence varies usually with $1/T$ but with $1/T^2$ in certain cases.

9.5.3.3. *The value of* $\Delta\omega$. As a result of chemical exchange, the resonance position of the unbound bulk nuclei may be shifted with respect to a solution in which there is only the A environment. The observed shift, $\Delta\omega$ is given by:

$$\Delta\omega = \frac{P_\text{M}q\,\Delta\omega_\text{M}}{(1+\tau_\text{M}/T_{2,\text{M}})^2+\tau_\text{M}^2\,\Delta\omega_\text{M}^2}. \tag{9.27}$$

The temperature dependence of $\Delta\omega$ arises from that of $\Delta\omega_\text{M}$ and $1/\tau_\text{M}$. If we assume, for simplicity, that the temperature dependence of $\Delta\omega_\text{M}$ is proportional to $1/T$ and that, τ_M is given by the usual Eyring relationship, a graph of $\Delta\omega$ versus $1/T$ will have the form shown in Fig. 9.6 (cf. Fig. 2.18). If it is possible to observe the shifts over a large enough temperature range, values of τ_M, and, thus, $\Delta H^\ddagger$ and $\Delta S^\ddagger$ can be estimated.

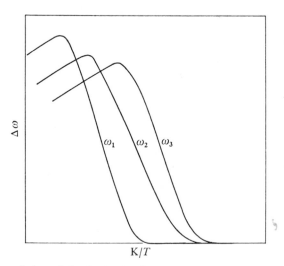

FIG. 9.6. The variation of the 'bulk' chemical shift with exchange rate at different frequencies ($\omega_1 > \omega_2 > \omega_3$).

9.5.3.4. *Temperature dependence of observed relaxation rates.* The temperature dependence of these equations arises from the temperature depend-
ence of $\Delta\omega_M$, τ_M, $T_{1,M}$ and $T_{2,M}$. That of $T_{1,M}$ or $T_{2,M}$ depends on τ_M, τ_S and
τ_R. As for the diamagnetic case we consider pictorially the temperature depend-
ence of $1/T_{1,P}$ and $1/T_{2,P}$ in eqns (9.25) and (9.26). This is shown in Fig. 9.7.
Four distinct regions are apparent for $1/T_{2P}$ and three for $1/T_{1,P}$. The
theoretical high-resolution spectrum is superimposed to enable some physical
insight into the graphs to be obtained. Basically, the only difference from the
diamagnetic case is the inclusion of an outer-sphere contribution, Region '0'.
From eqns (9.1), (9.2), (9.25), and (9.26), it can be seen that for 'short'
correlation times or when the values of $(\omega_s\tau_c)$ and $(\omega_s\tau_e) \ll 1$, $1/T_{1,P}$ and
$1/T_{2,P}$ in this region will be equal. For the other regions we again consider
some limiting forms of eqns (9.25) and (9.26). Initially, we start with $1/T_{2,P}$.
   Region I is the 'slow-exchange' region with the condition such that
$\Delta\omega_M^2 \gg 1/T_{2,M}^2$, $1/\tau_M$, when eqn (9.26) becomes

$$\frac{1}{P_M q T_{2,P}} = \frac{1}{P_M q T_{2,A}''} + \frac{1}{\tau_M} \tag{9.28}$$

i.e. the relaxation rate is governed by the rate of chemical exchange of
molecules between the bulk and bound site.
   Equation (9.28) may also be obtained if the relaxation in the first co-
ordination sphere of the paramagnetic ion is very rapid, i.e. $1/T_{2,M}^2 \gg 1/\tau_M^2$,
$\Delta\omega_M^2$ or $1/T_{2,M}^2 \approx \Delta\omega_M^2 \gg 1/\tau_M^2$.

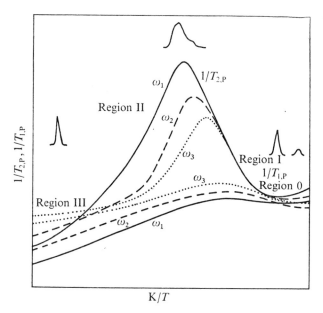

FIG. 9.7. Schematic representation of the frequency ($\omega_3 > \omega_2 > \omega_1$) and temperature dependence of the relaxation rates for a system undergoing chemical exchange. The high-resolution spectrum is superimposed. To make the diagram general the relaxation rates in Regions 0 and III are shown as being frequency dependent though this may not always be the case.

After subtraction of the outer sphere contribution, (obtained from an appropriate extrapolation of Region '0') the value of $1/\tau_M$ may thus be obtained. By plotting

$$\log_{10} \frac{1}{P_M q} \left(\frac{1}{T_{2,P}} - \frac{1}{T_{2,A}''} \right) \text{ versus } K/T,$$

$\Delta H^{\ddagger}$ may be obtained from the slope ($\Delta H^{\ddagger}/4\cdot6$). It is important to note that $1/\tau_M$ is *independent of the frequency* at which the relaxation rates are measured. This is often important experimentally in confirming that the slow-exchange region obtains.

In Region II the high-resolution spectrum would show initially the two separate resonances coalesced into one broad signal. Increase of temperature then leads to 'exchange narrowing'. In this region $1/T_{2,P}$ will be determined by the change in precessional frequency (chemical shift) and, in the limit, we may write:

$$\frac{1}{\tau_M^2} \gg \Delta \omega_M^2 \gg \frac{1}{T_{2,M} \tau_M} \tag{9.29}$$

from which we obtain

$$\frac{1}{P_M q T_{2,P}} = \frac{1}{P_M q T_{2,A}''} + \tau_M \Delta\omega_M^2.$$

(9.30)

and since $1/T_{2,P}$ depends on $\Delta\omega_M$, it will vary with the magnetic field, i.e. $1/T_{2,P}$ will be dependent on the frequency at which the measurements are made. Note that the lower the frequency, the lower the temperature at which Region II is reached.

Region III is that of 'fast exchange' where now the measured relaxation rates are essentially the weighted average of the two environments, i.e. $1/T_{2,M}\tau_M \gg 1/T_{2,M}^2$, $\Delta\omega_M^2$, leading to

$$\frac{1}{P_M q T_{2,P}} - \frac{1}{P_M q T_{2,A}''} = \frac{1}{T_{2,M}}$$

(9.31)

Turning now to the temperature dependence of $1/T_{1,P}$ this is represented by the dotted curves in Fig. 9.7. In Region I, $1/T_{1,M} \gg 1/\tau_M$ and the graph of $1/T_{1,P}$ versus (K/T) follows initially that of $1/T_{2,P}$ since eqn (9.25) then becomes

$$\frac{1}{P_M q T_{1,p}} = \frac{1}{P_M q T_{1,A}} + \frac{1}{\tau_M}$$

(9.32)

When $1/\tau_M \gg 1/T_{1,M}$, the relevant equation for $1/T_{1,P}$ is

$$\frac{1}{P_M q T_{1,P}} = \frac{1}{P_M q T_{1,A}''} + \frac{1}{T_{1,M}}$$

(9.33)

i.e. fast-exchange conditions apply.

As before (Section 2.17.4) we note that there is no region for $1/T_{1,P}$ corresponding to Region II for $1/T_{2,P}$. Thus, there are differences in the behaviour of the two relaxation times when chemical exchange is important.

In Region III (fast-exchange region) we have represented the temperature dependence of $1/T_{1,P}$ and $1/T_{2,P}$ and consequently those of $1/T_{1,M}$ and $1/T_{2,M}$, as decreasing with increasing temperature. The temperature variation depends on the magnitude and the nature of the correlation time. (Strictly speaking this also applies to the outer-sphere region, Region '0'.)

In most systems in which the above type of chemical exchange behaviour has been observed, only a small section of Region III is observed. To the biochemist, however, this region is probably most attractive, since analysis of the relaxation rates gives information on the values of distances between the ligand nuclei under consideration and the paramagnetic centre. The paramagnetic centre may be intrinsic to the system or added as a probe. As we shall see, by using several paramagnetic probes on a macromolecule, it is possible in principle to 'map out' the orientation of ligand molecules. In order to apply eqns (9.31) and (9.33), we shall need to know the binding constant (to calculate $P_M q$), and the value of τ_c to be used in the Solomon–Bloembergen equations for $1/T_{1,M}$ and $1/T_{2,M}$. We return to these points later. At present it is however relevant to discuss the variation of $1/T_{1,M}$ and $1/T_{2,M}$ with τ_c.

9.6. Variation of $1/T_{1,M}$ and $1/T_{2,M}$ with correlation time τ_c and frequency

The equations for $1/T_{1,M}$ and $1/T_{2,M}$ indicate that, at fixed ω_S and ω_I, the values of $1/T_{1,M}$ and $1/T_{2,M}$ will depend on that of τ_c. Conversely, for a fixed τ_c, the values of $1/T_{1,M}$ and $1/T_{2,M}$ will depend on those of ω_S and ω_I. We have noted that the value of τ_c is determined by those of τ_R, τ_M, and τ_S. Of these τ_S will be dependent on ω_S [eqn (9.7)], when the condition $\omega_S \tau_v \gtrsim 1$ is true. To describe the variation of the relaxation rates with τ_c and frequency, it is convenient to consider the cases when: (1) τ_c is frequency independent, i.e. τ_R, τ_M, or τ_S (when $\omega_S \tau_v \ll 1$) and (2) τ_c is frequency dependent, i.e. τ_S ($\omega_S \tau_v \approx 1$).

9.6.1. Variation of $1/T_{1,M}$ and $1/T_{2,M}$ with τ_c

We start by considering the dipolar contributions to the relaxation rates. Since for a given system:

$$\frac{1}{T_{1,2}} = D_{1,2} f_{1,2}(\tau_c)$$

it is sufficient to plot $1/T_{1,2}$ versus $f_{1,2}(\tau_c)$ to illustrate the necessary features (1 and 2 refer to the respective relaxation rates). Inspection of Figs. 9.8 and 9.9, indicates that at 'short τ_c' (or when $\omega \tau_c \ll 1$):

$$\frac{1}{T_{1,M}} = \frac{1}{T_{2,M}}.$$

However as τ_c increases so that $\omega_S \tau_c \simeq 1$ this equality is no longer true. At longer τ_c values, $\omega_S \tau_c \gg 1$ but $\omega_I \tau_c \ll 1$ then

$$\frac{1}{T_{1,M}} = \frac{6}{7} \frac{1}{T_{2,M}}$$

[cf. eqns (9.1) and (9.2)].

Eventually the value of τ_c is such that $\omega_I \tau_c \approx 1$, and we note that $1/T_{1,M}$ then reaches a maximum value, but $1/T_{2,M}$ continues to increase. When $\omega_I \tau_c \gg 1$, then we have that

$$\frac{1}{T_{1,M}} = D_1 \frac{3}{\omega_I^2 \tau_c} \quad \text{and} \quad \frac{1}{T_{2,M}} = D_2(\tau_c)$$

i.e. $1/T_{1,M}$ is proportional to $1/\tau_c$ in contrast with $1/T_{2,M}$ which is proportional to τ_c. The shape of the curves, is, as expected, very similar to those of Figs. 2.7 and 2.12.

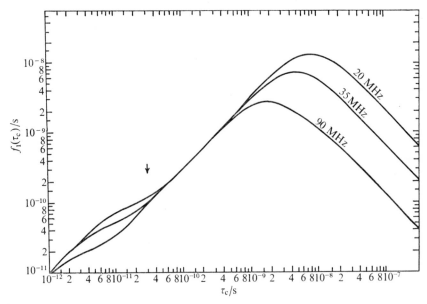

FIG. 9.8. The variation of $f_1(\tau_c)$ with τ_c. The arrow represents the value of τ_c (3×10^{-11} s) for aqueous Mn(II) solutions at 25 °C.

$$f_1(\tau_c) = \frac{3\tau_c}{1+\omega_I^2\tau_c^2}+\frac{7\tau_c}{1+\omega_S^2\tau_c^2}$$

In most cases, when $\omega_S\tau_c \approx 1$, the functional form of the scalar or hyperfine term is such that it contributes little to $1/T_{1,M}$. On the other hand, it does make a significant contribution to $1/T_{2,M}$. Its presence will be to add a constant value to the function in Fig. 9.9 which is essentially proportional to τ_e.

9.6.2. Frequency dependence of $1/T_{1,M}$ and $1/T_{2,M}$

9.6.2.1. τ_c *frequency independent.* The plots of $f_1(\tau_c)$ and $f_2(\tau_c)$ versus τ_c for three different frequencies are shown in Figs. 9.8 and 9.9. We note again the most striking feature that when τ_c is such that $\omega_I\tau_c \gg 1$, the values of $f_1(\tau_c)$ and hence $1/T_{1,M}$ are proportional to $1/\omega_I^2$. The frequency dependence of $f_2(\tau_c)$ is far less marked. The variation of $1/T_{1,M}$ and $1/T_{2,M}$ with frequency, ω_I, are shown schematically in Fig. 9.10a for a fixed value of τ_c. Initially, $1/T_{1,M} = 1/T_{2,M}$ and are independent of the value of ω_I. As ω_I increases so that $\omega_S\tau_c \approx 1$, (for ^{1}H nuclei $\omega_S \approx 660\omega_I$), $1/T_{1,M}$ no longer equals $1/T_{2,M}$. Again we have the situation in which $1/T_{1,M} \to \frac{6}{7} 1/T_{2,M}$ when $\omega_S\tau_c \gg 1$, followed by the frequency independent region until $\omega_I\tau_c \gg 1$, when $1/T_{2,M}$ remains constant but $1/T_{1,M}$ falls as already discussed.

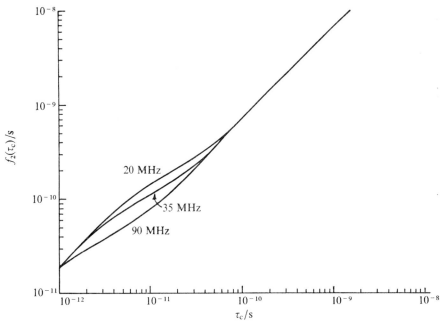

FIG. 9.9. The variation of $f_2(\tau_c)$ with τ_c.

$$f_2(\tau_c) = 4\tau_c + \frac{3\tau_c}{1+\omega_I^2\tau_c^2} + \frac{7\tau_c}{1+\omega_s^2\tau_c^2}$$

9.6.2.2. τ_c frequency dependent. If τ_c is determined by τ_S, then if $\omega_S\tau_v \gtrsim 1$, (see Fig. 9.2) τ_S will be frequency dependent and will increase with increasing ω_I. A plot of the expected frequency dependence of the relaxation rates is shown in Fig. 9.10b. At low values of ω_I the behaviour is identical to that in Fig. 9.10a. This arises because the condition $\omega_S\tau_v \ll 1$ will be fulfilled and τ_S will be frequency independent. If τ_S becomes frequency dependent before

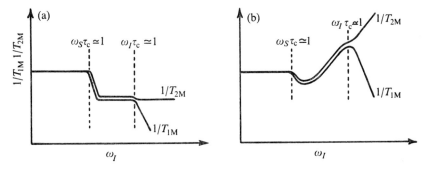

FIG. 9.10. Schematic representation of variation of relaxation rates with frequency. (a) τ_c frequency independent (b) τ_c frequency dependent.

the condition $\omega_S\tau_c(\equiv\omega_S\tau_S) \approx 1$ is reached then the 'fall off' region in Fig. 9.10a will be less pronounced and could disappear. When $\omega_S\tau_S \gg 1$, but still $\ll\omega_I\tau_S$, both $1/T_{1,M}$ and $1/T_{2,M}$ are proportional to $\tau_c(\equiv\tau_S)$ and will increase if τ_S increases. Above $\omega_I\tau_c \approx 1$, $1/T_{2,M}$ continues to increase ($\propto\tau_S$) but $1/T_{1,M}$ falls even more sharply than before.

We can make this approach more quantitative by considering the expressions for τ_S. As τ_S is increased, we reach the situation when $(\omega_S\tau_S)^2 \gg 1$, and thus terms containing $(1+\omega_S^2\tau_S^2)$ may be dropped in the equations for $1/T_{1,M}$ and $1/T_{2,M}$. Thus

$$\frac{1}{T_{1,M}} = D\,\frac{\tau_S}{1+\omega_I^2\tau_S^2} \tag{9.34}$$

and

$$\frac{1}{T_{2,M}} = D\left[\tfrac{2}{3}\tau_c+\frac{\tfrac{1}{2}\tau_S}{1+\omega_I^2\tau_S^2}\right]+C(\tau_S) \tag{9.35}$$

where C and D are constants for a given system.

The expression for $\tau_{1,S}$ may be approximated to

$$\frac{1}{\tau_{1,S}} = \frac{5B\tau_v}{1+2\cdot5\omega_S^2\tau_v^2}.$$

If we consider two limiting cases of $\omega_I^2\tau_S^2 \ll 1$ or $\omega_I\tau_S^2 \gg 1$, in the equations for $1/T_{1,M}$ and $1/T_{2,M}$ we obtain the behaviour shown schematically in Fig. 9.10b. Thus when $\omega_I^2\tau_S^2 \ll 1$, we have, for example:

$$\frac{1}{T_{1,M}} = \frac{\omega_S^2\tau_v}{2BD}+\frac{1}{5BD\tau_v} \tag{9.36}$$

and similarly for $1/T_{2,M}$. So that a plot of $1/T_{1,M}$ or $1/T_{2,M}$ versus ω_S^2 should be a linear function of the square of the frequency. When however $\omega_I^2\tau_S^2 \gg 1$, then $1/T_{2,M}$ is still proportional to ω_S^2 but the expression for $1/T_{1,M}$ is then:

$$\frac{1}{T_{1,M}} = \frac{5BD\tau_v}{\omega_I^2(1+2\cdot5\omega_S^2\tau_v^2)}. \tag{9.37}$$

Since $\omega_I = (\gamma_I/\gamma_S)\omega_S$, we have

$$\frac{1}{T_{1,M}} = \frac{a}{\omega_S^2+b\omega_S^4} \tag{9.38}$$

where a and b are constants. $1/T_{1,M}$ thus falls very rapidly with increasing frequency.

In many Mn(II)–macromolecular complexes, the importance of τ_S in determining τ_c seems well established (see Chapter 10). In such systems, the variations of the relaxation rates with ω_I are found to exhibit a maximum in

$1/T_{1,M}$, and the values of $1/T_{1,M}$ are seen to *decrease* with decreasing frequency. Inspection of Fig. 9.10b shows why such behaviour must be characteristic of a system in which τ_S is important.

9.7. Qualitative deductions from the temperature dependence of the relaxation rates

From Fig. 9.7, we note that the variation of the relaxation rates with $1/T$ may be either negative (in the *slow-exchange* region) or positive in the other regions. In the *fast-exchange* region, the temperature variation of $1/T_{1,P}$ and $1/T_{2,P}$ are determined by those of $1/T_{1,M}$ and $1/T_{2,M}$. Since a change in τ_c is equivalent to a change in temperature, the graphs in the previous sections allow us to tabulate the temperature variations as in Table 9.1.

TABLE 9.1
Variation of relaxation rates with $1/T$ in the fast exchange region

Condition	$\omega_I \tau_c < 1$			$\omega_I \tau_c > 1$		
τ_c	τ_R	τ_M	τ_S†	τ_R	τ_M	τ_S†
$\dfrac{d(1/T_{1,M})}{d(1/T)}$	+	+	±	−	−	∓
$\dfrac{d(1/T_{2,M})}{d(1/T)}$	+	+	±	+	+	±

† The lower sign is applicable if τ_S is frequency dependent.

Thus it may often be possible from experimental observations of $1/T_{1,P}$ and $1/T_{2,P}$ to identify which region fits the experimental results and thus simplify any analysis as we shall see later in the examples involving detailed molecular motion studies. For example, a negative value of $d(1/T_{1,P})/d(1/T)$ could arise from (1) slow-exchange conditions; (2) rapid-exchange conditions with $\omega_I \tau_c < 1$ but with $\tau_c = \tau_S$ (which is frequency dependent); and (3) rapid-exchange conditions with $\omega_I \tau_c > 1$ (and τ_S making no contribution to τ_c or $\tau_c = \tau_S$ which is frequency independent). On the other hand $d(1/T_{2,P})/d(1/T)$ can only be negative if (1) slow-exchange conditions are valid, or (2) rapid-exchange conditions with τ_c determined by τ_S which is frequency dependent.

If the frequency dependence of the relaxation times is also known, then it may be possible to narrow the choice even further. Thus for the cases chosen for $1/T_{1,P}$ above, for example, choice (1) is frequency independent while (2) is strongly frequency dependent. If the scalar term dominates the spin–spin relaxation rate, then the interpretation is still relatively simple and the simple rules in Table 9.1 would still be useful if the correlation times for $1/T_{2,M}$ are τ_S or τ_M. However, if the dipolar and scalar terms are of the same order of magnitude then the treatment may be slightly more complex.

9.8. Relaxation behaviour in some model systems

9.8.1. *Aqueous solutions of* Mn(II) *ions* (*Bloembergen and Morgan*, 1961). Many biochemical systems have a requirement for Mg(II) ions but unfortunately the properties of Mg(II) make it a rather poor probe nucleus. Mn(II), is paramagnetic; it produces marked changes in the relaxation rates of ligands attached to or near it, and also its e.s.r. spectrum can be observed. Since the chemistry of Mn(II) closely resembles that of Mg(II), it is often used as a probe in many biochemical systems. Even where there is no obvious metal requirement, Mn(II) is often used as probe since it has a large magnetic moment (5.9 Bohr magnetons) and thus even at low concentrations, such as those required in biological systems, its use may give measurable effects on the relaxation rates. We shall return to some actual examples later, but for the present, we note that many measurements of the relaxation rates of ^{1}H nuclei in solvent water in Mn(II)–macromolecule complexes have been done. To interpret these measurements, it is a helpful first step to consider the behaviour of pure aqueous Mn(II) solutions. To help illustrate various factors that we

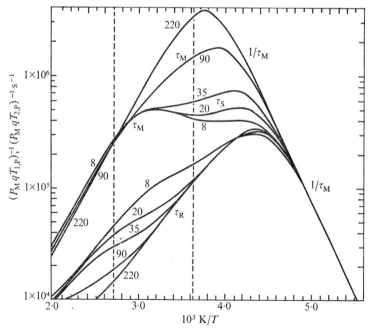

FIG. 9.11. The temperature and frequency variation of the hydrogen relaxation rates in aqueous Mn(II) solutions. The continuous lines are computed from the parameters in Table 9.2 and the relevant correlation times are indicated. The dotted lines indicate the experimentally observable region.

TABLE 9.2

Parameters obtained from analyses of relaxation rates in aqueous Mn(II) solution

B (rad s^{-1})^2	$0{\cdot}1 \times 10^{20}$
τ_v(300 K) (s)	$2{\cdot}1 \times 10^{-12}$
E_v (kcal mole^{-1})	$3{\cdot}9$
τ_M (300 K) (s)	$2{\cdot}3 \times 10^{-8}$
$\Delta H^{\ddagger}$ (kcal mol^{-1})	$8{\cdot}1$
τ_R (300 K) (s)	3×10^{-11}
E_R (kcal mol^{-1})	$4{\cdot}5$
A/h (Hz)	1×10^6†
r (Å)	$2{\cdot}77$
q (co-ordination number)	6

† A more recent determination by Luz and Shulman (1965) gives this value as $6{\cdot}2 \times 10^5$ Hz.

shall encounter in Mn(II)–macromolecule systems, we have computed the theoretical behaviour, over a much wider range of temperature than can be observed experimentally.

The temperature dependence of the paramagnetic ion contributions to the relaxation rates of the water hydrogen nuclei are illustrated in Fig. 9.11. The results are plotted as values of $1/P_M q T_{i,P}$, where $i = 1$ or 2, versus $10^3 K/T$. Quite often, experimental results are plotted in this way since the factor $1/P_M q$ allows the results from different solutions to be 'normalized'. The relaxation behaviour computed is shown for five representative frequencies. The experimentally-observable range is shown within the dotted lines. The interpretation is based on the use of the Solomon–Bloembergen equations for $1/T_{1M}$ and $1/T_{2,M}$, and the equations for chemical exchange [eqns (9.25) and (9.26)]. In this system τ_S is assumed to be given by the Bloembergen–Morgan equation [eqn (9.7)] and outer-sphere relaxation effects are neglected. The parameters which best fit the experimental results are shown in Table 9.2 and it is these parameters which have been used to obtain the continuous curves in Fig. 9.11.

We note first, at low temperatures, the frequency-independent region, characteristic of slow exchange, with $1/T_{1,P}$ and $1/T_{2,P}$ being approximately equal. At higher temperatures we arrive immediately at the conditions of fast exchange and we note that there is no region corresponding to Region II in Fig. 9.7 (i.e. no region for $1/T_{2,P}$ which depends on $\Delta\omega_M$). This is because, in the Mn(II) system, the relaxation in the bound sphere is so efficient that $1/T_{2,M}^2 \gg \Delta\omega_M^2$ whence eqn (9.26) reduces to, (in the absence of any outer sphere relaxation)

$$\frac{1}{T_{2,P}} = \frac{P_M q}{T_{2,M} + \tau_M} \tag{9.39}$$

which is of course similar to the equation for $1/T_{1,P}$.

Thus slow and fast exchange in the Mn(II) system are defined as in Table 9.3. As we have already noted $T_{1,M}$ and $T_{2,M}$ may not be equal and so the transition between fast and slow exchange may apply to different temperature ranges for $1/T_{1,P}$ and $1/T_{2,P}$.

In the fast exchange region ($10^3 K/T <$ ca. 4·4) in Fig. 9.11 the behaviour of $1/T_{2,P}$ is different for the five frequencies shown, and while the values of $1/T_{1,P}$ show very similar behaviour, they are almost an order of magnitude smaller than the values of $1/T_{2,P}$. In this region, the fast-exchange limit:

$$\frac{1}{P_M q T_{2,P}} = \frac{1}{T_{2,M}} \qquad (9.40)$$

and similarly,

$$\frac{1}{P_M q T_{1,P}} = \frac{1}{T_{1,M}}. \qquad (9.41)$$

Thus we are considering the relaxation in the first co-ordination of the Mn(II) ion. The differences in behaviour arise because of the different terms involved

TABLE 9.3

Measurement	Slow exchange	Fast exchange
$1/T_{2,P}$	$T_{2,M} < \tau_M$	$T_{2,M} > \tau_M$
$1/T_{1,P}$	$T_{1,M} < \tau_M$	$T_{1,M} > \tau_M$

in $1/T_{1,M}$ and $1/T_{2,M}$. In particular the value of the hyperfine term dominates $1/T_{2,M}$. The relevant correlation time for the hyperfine term is τ_e, which contains τ_M and τ_S and just which of these dominates in a particular region, is shown on Fig. 9.11. We note that when τ_S dominates, it may appear that $d(1/T_{2,P})/d(1/T)$ may be either positive or negative corresponding to τ_S becoming frequency dependent (i.e. the condition $\omega_S \tau_v \approx 1$ is obtained). The value of τ_S will then increase with increasing frequency until at 90 MHz and higher frequencies its value is such that $\tau_M \ll \tau_S$. Then τ_M dominates τ_e and we have the situation in which $1/T_{2,P}$ is determined by $1/\tau_M$ in the slow-exchange region and τ_M (in $1/T_{2,M}$) in the fast-exchange region. Thus, in this instance, kinetic parameters over a wide temperature range can be obtained.

The relevant correlation time for $1/T_{1,P}$, in the fast-exchange region, is τ_R. The differences in the observed behaviour at different frequencies arise from the dispersion behaviour associated with the term $\omega_S \tau_c (\equiv \omega_S \tau_R$ here) ≈ 1 (see eqn (9.1)).

Of course, at even higher temperatures than those shown, the difference between $1/T_{1,P}$ and $1/T_{2,P}$ would disappear as $1/T_{1,M}$ and $1/T_{2,M}$ both become dominated by τ_M (since this has a larger activation energy than

τ_R and decreases faster) leading to the convergence of the values of $1/T_{1,P}$ and $1/T_{2,P}$ when the condition $\omega_S\tau_c \ll 1$ is met.

We have said above that the main differences between $1/T_{1,M}$ and $1/T_{2,M}$ arise from the scalar contribution to $1/T_{2,M}$. We have to qualify this slightly for this is true only when $\omega_S\tau_e \gg 1$. When this is so the functional form of the scalar term i.e. $[\tau_e/(1+\omega_S^2\tau_e^2)]$ in eqns (9.1) and (9.2), makes little contribution to $1/T_{1,M}$ and $1/T_{2,M}$. (For $A/h \approx 10^6$ Hz it contributes ca. 1% in this example). However $1/T_{2,M}$ also contains a term in $A/\hbar$ (τ_e) which is very significant and hence the difference. When the condition $\omega_S\tau_c \ll 1$ is valid, the contributions of the scalar term to $1/T_{1,M}$ and $1/T_{2,M}$ are equal. In fact, if we recast the equation for $1/T_{2,M}$ [eqn (9.2)] as below this point is more obvious.

$$\frac{1}{T_{2,M}} = D\left(\tfrac{2}{3}\tau_c + \frac{\tfrac{1}{2}\tau_c}{1+\omega_I^2\tau_c^2} + \frac{\tfrac{13}{6}\tau_c}{1+\omega_S^2\tau_c^2}\right) + C\left(\tau_e + \frac{\tau_e}{1+\omega_S^2\tau_e^2}\right) \qquad (9.42)$$

where

$$D = \frac{6}{15}\frac{S(S+1)\gamma_I^2 g^2\beta^2}{r^6} \quad \text{and} \quad C = \tfrac{1}{3}S(S+1)(A/\hbar)^2 \, .$$

In pure aqueous solutions the value of $D = 2\cdot06 \times 10^{15}$ (rad s^{-1})^2 and the ratio of $C/D = 0\cdot022$. Thus the hyperfine (contact) term contributes only because $\tau_e \gg \tau_c$. When $\tau_e \gg \tau_c$ as at high temperatures, then the contribution from the contact term is small. We shall also see later that in many Mn(II)–macro-molecule systems, $\tau_e \approx \tau_c$ and the hyperfine contribution to $1/T_{2,M}$ is very small. As a result of this the separation of τ_R and τ_M is then difficult.

9.8.1.1. *Qualitative deductions from temperature dependence of the relaxation rates in the Mn(II)–aquo complex.* In Section 9.7 we discussed the qualitative deductions that can be made from the observation of the temperature dependence of the relaxation rates. It is often useful to try to obtain some physical insight into the meaning of the experimental results without using elaborate computational methods. For example, let us see how far we can go in interpreting the experimentally-observed region of Fig. 9.11.

An increase in the spin–lattice relaxation rate with increasing temperature can only correspond to the fast-exchange region for $1/T_{1,P}$. The frequency-independent region for $1/T_{2,P}$ (10^3K/$T < 3\cdot0$) must also correspond to the fast-exchange region, but at 8 and 20 MHz, there could be some difficulty in interpreting the region for $1/T_{2,P}$ between 10^3K/$T = 3\cdot4$ and $3\cdot6$. The negative value of $d(1/T_{2,P})/d(1/T)$ could arise from (1) slow exchange, or (2) fast exchange, with τ_S the dominant correlation time. This applies whether $\omega_I\tau_S >$ or <1. Slow exchange may be ruled out by comparison with the results at higher frequencies. (We may note that if the slow exchange region obtains for a given frequency, it must also obtain at the same temperature for any higher frequency. In this case the $1/T_{2,P}$ results at 90 and 220 MHz for instance, are clearly *not* in the slow-exchange region).

Finally, we note the dramatic difference between $1/T_{1,P}$ and $1/T_{2,P}$. If

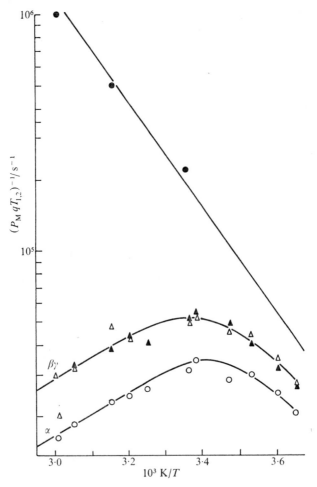

FIG. 9.12. The temperature variation of the paramagnetic contribution to the spin–lattice and spin–spin relaxation rates in aqueous solutions containing 0·1 M ATP and 0·1 mM manganese perchlorate at 86 MHz and pH 8·5. (From Brown *et al.* 1973.)

TABLE 9.4

Parameters for ^{31}P *nuclei in the* Mn–ATP *system*

τ_M (300 K) (s)	$6·5 \times 10^{-6}$
$\Delta H^{\ddagger}$ (kcal mol^{-1})	11·0
$\tau_c = \tau_R$ (300 K) (s)	$6·0 \times 10^{-10}$†
E_R (kcal mol^{-1})	5·0
$r_{\beta\gamma}$ (Å)	3·3
r_{α} (Å)	3·7

† Calculated from hydrogen relaxation rates of the water protons, assuming $\varepsilon_b^* \simeq 10$ and $q^* = 3$. (See page 243.)

only the dipolar term is important in determining both relaxation rates, then such a difference can only come about if the condition $\omega_I \tau_c \gg 1$ obtains. If this is so then $1/T_{1,P}$ should be proportional to $1/\omega_I^2$. This is obviously not so. The only other explanation is that the hyperfine interaction must be contributing to $1/T_{2,P}$.

We note that the more frequencies at which measurements are taken, the less ambiguous the results become. For example, if measurements had only been carried out at 8 MHz, it is difficult to distinguish between the two possibilities mentioned above. So, as a general rule wherever possible the measurements should be performed at more than one frequency, particularly if the results appear to reflect either slow exchange or a negative temperature coefficient for τ_S.

9.8.2. *The* ^{31}P *resonances in* Mn(II)–ATP *solutions* (Brown, Campbell, Henson, Hirst, and Richards, 1973)

The relaxation times of the α, β, and γ ^{31}P nuclei in ATP measured at 86 MHz using Fourier transform techniques, are shown in Fig. 9.12. The behaviour of $1/T_{2,P}$ is that expected for 'slow exchange' over the whole temperature range (i.e. $T_{2,M} < \tau_M$). On the other hand, at high temperatures, fast-exchange conditions are valid for $1/T_{1,P}$, i.e. $T_{1,M} > \tau_M$ where $1/\tau_M$ is the first-order rate constant for the exchange of ATP from the co-ordination sphere of Mn(II). The parameters obtained from an analysis of $1/T_{2,P}$ are simply those relating to τ_M. The remainder of the parameters in Table 9.4 are obtained from an analysis of the spin–lattice relaxation time measurements. The Mn(II) appears to co-ordinate to the β and γ ^{31}P nuclei preferentially. This may indicate some constraint in this system which quite possibly could be absent in macromolecular systems containing Mn(II)–ATP. (e.g. see page 322.)

In this system, the value of $\omega_S \tau_c \gg 1$. Reference to Fig. 9.8 suggests that the spin–lattice relaxation times should be frequency independent as long as the condition $\omega_I \tau_c \ll 1$ is true. Similarly, verification of the slow-exchange region for $1/T_{2,P}$ can be obtained by measurements at different frequencies because $1/\tau_M$ is also frequency independent. However if $1/T_{1,P}$ is frequency independent, it is not possible to obtain an exact value of τ_c and, hence, r. The value of τ_c given in Table 9.4 is determined from the hydrogen relaxation rates of solvent water in the system (see e.g. page 243). Different values of τ_c would lead to different values of r (r being approximately proportional to $\sqrt[6]{\tau_c}$). The ratio of the distances however remains unaltered. This problem of obtaining a value of τ_c is one that we shall frequently encounter when the actual experimental conditions used do not permit a direct determination.

9.8.3. *Aqueous solutions of* Gd(III) *ions* (Dwek *et al.*, 1971)

We include this example because it is a particularly simple illustration of a point already taken for granted. In the fast-exchange region the measurement of relaxation rates over a sufficiently large enough frequency range leads to a unique determination of τ_c. The spin–lattice relaxation rates of aqueous solutions of Gd(III) ions are shown in Fig. 9.13 at 20 and 35 MHz. For this graph, q is assumed to be 10, since its value is not well defined in this system. We note immediately that the values of $1/T_{1,P}$ increase with increasing temperature. Inspection of Table 9.1 and Fig. 9.7 show that this

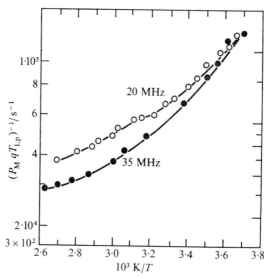

FIG. 9.13. The temperature and frequency variation of the relaxation rates of water hydrogens in aqueous Gd(III) solutions (q is assumed to be 10). (From Dwek *et al.*, 1971.)

can only occur if conditions of rapid exchange apply. The $1/T_{2,P}$ measurements (not shown) were $\leqslant 14\%$ of $1/T_{1,P}$ at all temperatures. This condition can be met by consideration of the dipolar terms only in eqns (9.1) and (9.2) if $\omega_S^2 \tau_c^2 > 1$ but $\omega_I^2 \tau_c^2 < 1$. This implies that the scalar contribution to the relaxation rates can be neglected, i.e. the term $(A/\hbar)\tau_e$ in $1/T_{2,M}$, which is so important in aqueous solutions of Mn(II) ions, is negligible here. The difference with Gd(III) could arise if τ_e is smaller but more probably because $A/\hbar$ itself is small. This latter possibility is not unlikely as the unpaired electrons in Gd(III) are in $4f$-orbitals well shielded from contact interactions with co-ordinated water molecules.

The frequency dependence of the spin–lattice relaxation times disappear between $\omega_S \tau_c > 1$ and $\omega_I \tau_c < 1$ (see Fig. 9.8). For the frequencies 20 and 35 MHz this occurs when $\tau_c \approx 7 \times 10^{-11}$ s. Thus at $10^3\,\mathrm{K}/T = 3.6$ we have determined the value of τ_c. Of course, this treatment does not require that the two curves converge, for the value of τ_c can be obtained from the ratio of the relaxation rates at two or more frequencies.† In this way $\tau_c(300 \text{ K}) \approx 4.5 \times 10^{-11}$ s, and an activation energy for τ_c of ≈ 4 kcal mol^{-1} can be evaluated.

Knowing this value of τ_c it is then possible to calculate the H$_2$O–Gd(III) distance (r) which is ca. 3·1 Å. While the analysis of τ_c is independent of the value of q, this is not so for the experimentally-obtained value of $1/T_{1,M}$ $(\equiv 1/P_M q T_{1,P})$, so that the experimentally-determined value of r will also depend on the value of q. Conversely, if r is known from other measurements q can be obtained.

Note that we do not need to know the explicit nature of τ_c in this example to calculate r, while we shall see later that for various substrates on enzymes we may have to make some assumptions as to the nature of τ_c (see Chapter 10).

In this case the nature of τ_c (which contains τ_M, τ_S, and τ_R) can be obtained since ^{17}O n.m.r. studies suggest a value of 10^{-8} s for τ_M (Reuben and Fiat, 1969) (assuming *whole* water molecules are exchanging) and the observed e.s.r. spectrum suggests a value of ca. 1.5×10^{-10} s for τ_S. Thus the value of τ_c (ca. 10^{-11}) must correspond to rotation of the (H$_2$O)$_n$–Gd(III) complex which is responsible for the relaxation of the water protons.

In this example, we have again neglected outer-sphere relaxation, a debatable point since the lanthanides could well have a definite second co-ordination sphere. Another complication is that the results have been interpreted in terms of a single correlation time τ_R. A more rigorous analysis has been carried out by Reuben (1971a) based on measuring the Gd(III) electron spin–spin relaxation times $\tau_{2,S}$,‡ and then using the parameters obtained from the analysis of the e.s.r. spectrum to calculate τ_{1S}. Using eqns (9.1) and (9.25) the value of τ_R obtained is 7×10^{-11} s at 300 K. The results are shown in Fig. 9.14 and it is noted that the curves at 24·3 and 60·0 MHz cross. Reuben (1971a) attributes this to a frequency dependent $\tau_{2,S}$§ which makes a larger contribution to the correlation time at the lower frequency. This is one of the few examples where such an effect has been observed and analysed and we include it only for completion, and to indicate some of the intricacies one is going to have to be aware of in analysing relaxation data in metal/macromolecular systems.

9.8.4. *Aqueous solutions of* Co(II) *ions and* Ni(II) *ions* (Hauser and Laukien, 1959; Bloembergen and Morgan, 1961; Luz and Shulman, 1965.) The analysis of the relaxation rates of aqueous solutions containing Co(II) has been done on the basis of a fast exchange of water molecules between the bulk and bound sites. A similar analysis applies to Ni(II) but here it appears

---

† Assuming, of course, τ_c is frequency independent.
‡ This is very sensitive to the nature of the anion in the solutions.
§ The experimental points, however can also be well predicted in terms of τ_{1s} only.

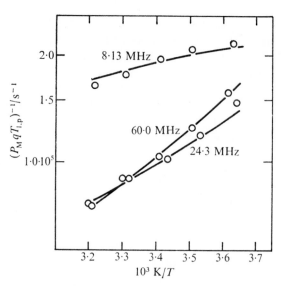

FIG. 9.14. Temperature dependence of the spin–lattice relaxation rate of aqueous Gd(III) solutions; q is assumed to be 10. (From Reuben 1971. © 1971 Amer. Chem. Soc.)

that at low temperatures the exchange is not complete. Neglecting outer-sphere effects and assuming a value for $r \approx 2.8$ Å, a value for $\tau_c \approx 10^{-13}$ s is calculated in each case, using eqns (9.1) and (9.33).† Since τ_R is not expected to be too different from its value in aqueous Mn(II) solutions (10^{-11} s) and an upper limit on τ_M is ca. 10^{-10} s (i.e. diffusion controlled) then for Co(II) and Ni(II) $\tau_c = \tau_S$. The non-observation of an e.s.r. spectrum in either aqueous solution supports this assignment. Several reasons have been proposed to account for the fact that τ_S is so short in these systems [see for, example Ayscough, (1967)].

9.9 Classification of paramagnetic metal ions into relaxation probes and shift probes

The above examples, with Mn(II), Gd(III), Ni(II), and Co(II) indicate that in aqueous solutions conditions of rapid exchange are valid for the solvent proton relaxation rates. This means that the value of $T_{1,M}$ can be obtained and thus the correlation time τ_c. The values of τ_c for the different para-magnetic metal ions appear to form two main classes: (1) Those metals in which $1/\tau_c$ is dominated by $1/\tau_R$, with $\tau_c \approx 10^{-10}-10^{-11}$ s. These include Mn(II), Gd(III), Eu(II), Cu(II), and V(II). (2) Those metals in which $1/\tau_c$ is dominated by $1/\tau_S$, with $\tau_S = 10^{-12}-10^{-13}$ s. These include Co(II), Ni(II), Fe(II), Fe(III), and most of the lanthanide(III) ions.

† Neglecting any outer sphere contributions to the relaxation rates.

Obviously those metal ions in class (1) have a value of $\tau_S > \tau_R$ and so another classification could be on the basis of long (class 1) and short (class 2) electron–spin relaxation times.

The difference in the value of τ_c between the two classes means that for a given system a class (2) metal ion will have a value of $1/T_{1,M}$ that is approximately two orders of magnitude less than that of a class (1) metal ion. Thus class (1) metal ions are obviously the ones to use as *relaxation probes*.

The presence of the paramagnetic ion also results in hyperfine shifts. The successful observation of such shifts will depend on the shift:line-width ratio. The relative inefficiency of class (2) metal ions in causing relaxation, means that the line-widths are correspondingly narrower. Thus class (2) metal ions are often used as *shift probes*.

It is worth stressing that there is nothing 'magical' about this classification. Paramagnetic metal ions give rise to both shifts and changes in relaxation. Just which is most easily observed depends on the experimental conditions. Mn(II), for example, will still give rise to hyperfine shifts but they may not be easily observable. For example, under conditions of rapid exchange, the shift of the bulk water, $\Delta\omega$, in aqueous Mn(II) solutions is given by:

$$\Delta\omega = P_M q \Delta\omega_M.$$

Using eqn (3.3) and a value of $A/h = 1$ MHz, then at an observing frequency of 60 MHz and at 300 K

$$\Delta\nu = 5 \times 10^3 P_M q \text{ (Hz)}$$

In water the co-ordination number of the paramagnetic ion is six, and P_M is given by $[M]/[H_2O]$. $[M]$ is the concentration of paramagnetic ion and $[H_2O] = 55 \cdot 5$ M. Thus in a 1 mM Mn(II) solution

$$\Delta\nu \approx 0 \cdot 5 \text{ Hz}$$

The contribution to the line-width ($\Delta\nu_P$) from the paramagnetic ion is given by

$$\frac{1}{T_{2,P}} = \pi \Delta\nu_P.$$

From Fig. 9.11 we estimate that for $P_M q = 10^{-4}$, $1/T_{2,P} \approx 50 \text{ s}^{-1}$ and thus

$$\Delta\nu_P = 17 \text{ Hz}$$

As a consequence, under these conditions the hyperfine shift is not easily observable. Of course at higher temperatures, the value of $1/T_{2,P}$ is considerably less (see Fig. 9.11) and it may then be possible to observe the shift. Altering the measuring frequency ω_I may also help, since under these conditions (fast exchange) the shift $\Delta\omega$ is directly proportional to ω_I.

References

ABRAGAM, A. (1961). The principles of nuclear magnetism. Clarendon Press. Oxford.

AYSCOUGH, P. B. (1967). Electron spin resonance in chemistry, Methuen, London.

BIRKETT, D. J., DWEK, R. A., RADDA, G. K., RICHARDS, R. E., and SALMON, A. J. (1971). *Eur. J. Biochem.* **20**, 494.

BLEANEY, B. (1972). *J. magn. Reson.* **8**, 91.

BLOEMBERGEN, N. (1957). *J. chem. Phys.* **27**, 572.

—— and MORGAN, L. O. (1961). *J. chem. Phys.* **34**, 842.

BROWN, F. F., CAMPBELL, I. D., HENSON, R. H., HIRST, C. W. J. and RICHARDS, R. E. (1973), to be published.

COHN, M. and LEIGH, J. S. JR., (1962). *Nature (Lond.)* **193**, 1037.

——, ——, and REED, G. H. (1971). *Cold Spring Harbor Symposia on Quantitative Biology*, **36**, 533.

CONNICK, R. E. and FIAT, D. (1966). *J. chem. Phys.* **44**, 4103.

DANCHIN, A. and GUÉRON, M. (1970). *J. chem. Phys.* **53**, 3599.

DWEK, R. A., MORALLEE, K. G., NIEBOER, E., RICHARDS, R. E., WILLIAMS, R. J. P., and XAVIER, A. V. (1971). *Eur. J. Biochem.* **21**, 204.

EISINGER, J., SHULMAN, R. G., and SZYMANSKI, B. M. (1962). *J. chem. Phys.* **36**, 1721.

HAUSER, K. and LAUKIEN, G. (1959). *Z. Physik.* **153**, 394.

HUBBARD, P. S. (1966). *Proc. Roy. Soc.* **A291**, 537.

HUDSON, A. and LEWIS, T. W. E. (1970). *Trans. Faraday Soc.* **66**, 129.

KARGER, J. and PFEIFER, H. (1968), *Ann. des Phys.* **22**, 51.

LEIGH, J. S. JR., (1970). *J. chem. Phys.* **52**, 2608.

LUCKHURST, G. R. (1972). In Electron spin relaxation in liquids, (ed. L. T. Muus and P. W. Atkins) London, Plenum Press.

LUZ, Z. and SHULMAN, R. G. (1965). *J. chem. Phys.* **43**, 3750.

—— and MEIBOOM, S. (1964). *J. chem. Phys.* **40**, 2686.

MILDVAN, A. S. and COHN, M. (1970). *Adv. Enzymol.* **33**, 1.

NAVON, G. (1970). *Chem. phys. Lett.* **7**, 390.

PFEIFER, H., MICHEL, D., SAMES, D. and SPRINZ, H. (1966). *Molec. Phys.* **11**, 591.

REED, G. H. and COHN, M. (1970). *J. biol. Chem.* **245**, 662.

——, and COHN, M. (1972). *J. Biol. Chem.* **247**, 3073.

——, LEIGH, J. S. JR., and PEARSON, J. E. (1971). *J. chem. Phys.* **55**, 3311.

—— and RAY, W. J. JR. (1971). *Biochemistry* **10**, 3190.

REUBEN, J. and COHN, M. (1970). *J. Biol. Chem.* **245**, 662.

——, (1971a). *J. Phys. Chem.* **75**, 3164.

—— (1971b). *Biochemistry* **10**, 2834.

—— and FIAT, D. (1969). *J. chem. Phys.* **51**, 4918.

——, REED, G. H., and COHN, M. (1970). *J. chem. Phys.* **52**, 161.

RUBINSTEIN, M., BARAM, A., and LUZ, Z. (1971). *Molec. Phys.* **20**, 67.

SOLOMON, I. (1955). *Phys. Rev.* **99**, 559.

STERNLICHT, H. (1965), *J. chem. Phys.*, **42**, 2250.

STONE, T. J., BUCKMAN, T., NORDIO, P. L., and MCCONNELL, H. M. (1965). *Proc. Natn. Acad. Sci. (U.S.)* **54**, 1010.

SWIFT, T. J. and CONNICK, R. E. (1962). *J. Chem. Phys.* **37**, 307.

VALLEE, B. L. and WILLIAMS, R. J. P. (1968). *Chem. Britain*, **4**, 391.

WOESSNER, D. (1962). *J. chem. Phys.* **36**, 1.

10

EXAMPLES OF RELAXATION STUDIES OF PARAMAGNETIC METAL–MACROMOLECULE COMPLEXES

10.1. Introduction

INSPECTION of the basic equations [eqns (9.1) (9.2), (9.25), and (9.26)] show that there are essentially two types of information that can be obtained from relaxation measurements, that is, kinetic and structural. Kinetic parameters can be obtained when slow- or intermediate-exchange conditions prevail or when fast-exchange conditions obtain with τ_M as the relevant correlation time in the equations for $1/T_{i,M}$, $(i = 1, 2)$. Under fast-exchange conditions, when this is not so, it may be possible to obtain a lower limit for the value of τ_M.

When conditions of fast exchange apply, the correlation times, the value of the co-ordination number, q, of the ligand, and the distances, r, between the paramagnetic centre and the different ligand nuclei can be obtained by a suitable analysis. In particular, the correlation times can be obtained from a study of the frequency dependence of the relaxation times and then, if q is known, r may be determined. Conversely if r is known, q may be determined, and this has been suggested (Navon, 1970) as a method of determining hydration numbers of paramagnetic ions, particularly Mn(II), when bound to enzymes. This of course involves measurement of the relaxation times for solvent water. Since the concentration of solvent water is 55·5 M, studies of the proton relaxation times have been an attractive starting point when studying macromolecular systems containing paramagnetic ions. Until 1969, however, the importance of obtaining relaxation measurements over as wide a range of temperature as possible and at different frequencies was not realised; this is necessary to aid in an unambiguous interpretation. The work on the relaxation rates of the hydrogen nuclei of water in Mn–RNA solutions by Peacocke, Richards and Sheard (1969) was followed by that of Navon (1970) on Mn–carboxypeptidase A and Mn–pyruvate kinase systems, by Reuben and Cohn (1971) on Mn–pyruvate kinase and by Danchin and Guéron (1971) on Mn–tRNA systems. The list can be extended but we shall draw on these examples primarily to illustrate the main points of the analysis of the relaxation rates of hydrogen nuclei in solvent water. We shall also consider examples concerning the binding of ligands other than water to illustrate how the basic equations can be applied and what importance the conclusions may have on using paramagnetic ions as probes or in 'mapping' on biochemical systems. The careful reader will note small points in all these examples that are perhaps not completely rigorous, nevertheless the basic analyses indicate the method of interpretation of relaxation data in such systems.

10.2. Relaxation rates of the solvent water protons in the Mn(II)–pyruvate kinase and Mn(II)–carboxypeptidase A systems

The metalloenzyme carboxypeptidase A (Mol. wt. 34 000) hydrolyses C-terminal residues, particularly if they contain branched aliphatic or aromatic side chains. The enzyme has one sub-unit and the naturally occurring zinc atom can be replaced by Mn(II). The activity of the manganese enzyme is 35% that of the zinc enzyme.

Pyruvate kinase catalyses the reaction between phosphoenol pyruvate and ADP to form pyruvate and ATP in the presence of obligatory monovalent and divalent ions. In vivo, the divalent ion is Mg(II), but this can be replaced by Mn(II) without any loss in enzymic activity.

In these examples, not all the metal ion is bound to the macromolecules. The Mn(II) was introduced by dialysing the enzyme solutions against a solution containing a known concentration of Mn(II), Navon (1970). The total Mn(II) concentration of the enzyme solution inside the dialysis bag was determined by atomic absorption. From a knowledge of the amount of Mn(II) in the dialysate, the amount bound to the enzyme could be determined. Thus it is not necessary to know how many binding sites for Mn(II) there are on the enzyme. It therefore means that the measured water proton relaxation rates are an average of all the sites. The water proton relaxation rates of the Mn–enzyme solutions are shown in Figs. 10.1 and 10.2. The net relaxation rates $1/T_{1,P}$ and $1/T_{2,P}$ in Figs. 10.1 and 10.2 are the differences between the relaxation rates of the enzyme solution and those of a control solution, (usually taken from the solution outside the dialysis bag, which does not contain an enzyme but otherwise has an identical composition).

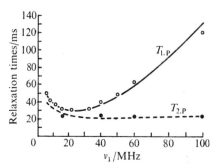

FIG. 10.1. Frequency dependence of the net relaxation times $T_{1,p}$ (○) and $T_{2,p}$ (●), of 6.16×10^{-5} M Mn-pyruvate kinase at 25 °C. The solution contained 0·1 M KCl, 1.5×10^{-4} M excess Mn(II), 0·05 M Tris–HCl buffer pH 7·5. The continuous lines are computed. (From Navon 1970.)

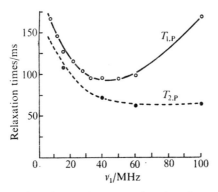

FIG. 10.2. Frequency dependence of the net relaxation times. T_{1p} (○) and T_{2p} (●), of $2·0 \times 10^{-4}$ Mn–carboxypeptidase A at 25 °C. The solution contained 1 M NaCl, $2·0 \times 10^{-4}$ M excess Mn(II), 0·05 M Tris–HCl buffer pH 7·5. The continuous lines are computed. (From Navon 1970.)

The first step in any analysis is to establish whether slow- or fast-exchange conditions apply. This is most conveniently done by an analysis of the temperature dependence of the relaxation rates. In this example, the values of $1/T_{2,P}$ at 100 MHz decreased with increasing temperature. At 100 MHz $T_{2,P}$ is smallest, and since τ_M is independent of frequency, the largest contribution from τ_M is expected at 100 MHz. Since the temperature variation is indicative of fast-exchange behaviour one can conclude that $T_{2,M} > \tau_M$. Since $T_{1,M} \gg T_{2,M}$, fast-exchange conditions obviously apply to the interpretation of the spin–lattice relaxation rates.

We next consider the frequency dependences of the relaxation rates. If there is a unique value of τ_c in $1/T_{1,M}$ or $1/T_{2,M}$, the consideration of the mathematical expressions for the relaxation rates, always suggest that their values should decrease with increasing frequency. The minimum observed in these experiments suggests clearly that τ_c is frequency dependent. Such behaviour as we have already noted in Chapter 9, follows from the frequency dependence of τ_S. Using the basic equations, the fitted theoretical curves are obtained with the parameters listed in Table 10.1 (see also Fig. 10.3). Once τ_c is known, q can be evaluated if r is known. By assuming r was unchanged from its value in the control Mn(II)-aqueous solution at this temperature, the values of q given in Table 10.1 are obtained. The relatively larger error in τ_v for the Mn–pyruvate kinase system arises because of the smaller frequency dependence of $T_{1,P}$ observed in this system.

Several interesting points emerge from the analysis. Firstly the experimental results for $T_{2,P}$ can be reproduced if $\tau_c \approx \tau_e$, and so the contribution of the contact term to $T_{2,M}$ must be relatively small (see p. 205). Then at higher frequencies we note that τ_S increases and its contribution to τ_c is negligibly small. Since $\tau_e \approx \tau_c$, it must follow that both τ_e and τ_c are determined

TABLE 10.1

Best-fit parameter values with their estimated errors for the Mn(II)–*pyruvate kinase and* Mn(II)–*carboxypeptidase complexes (From Navon, 1970)*

	$\dfrac{1}{\tau_r}+\dfrac{1}{\tau_M}$ (s^{-1})	B(rad s^{-1})^2	τ_v(s)	q†
Mn(II)–carboxypeptidase A	$(4\cdot0\pm0\cdot5)\times10^8$	$(3\cdot1\pm0\cdot4)\times10^{19}$	$(7\cdot0\pm0\cdot6)\times10^{-12}$	$1\cdot08\pm0\cdot1$
Mn(II)–pyruvate kinase	$(2\cdot5\pm0\cdot3)\times10^8$	$(0\cdot8\pm0\cdot1)\times10^{19}$	$(14\pm4)\times10^{-12}$	$2\cdot04\pm0\cdot2$

† The estimated errors include errors in the concentration of the bound Mn(II).

mainly by τ_M. The calculation of the values of τ_R, from the Stokes–Einstein expression for the rotation of a rigid sphere in a viscous medium, (see p. 243) which gives values of $1\cdot5\times10^{-8}$ and 1×10^{-7} s for carboxypeptidase (mol. wt. = 36 000) and pyruvate kinase (mol. wt. = 237 000), adds weight to this argument. It has to be assumed though that the manganese ions do not have much greater motional freedom than that of the whole enzyme molecule. This is probably reasonable since the manganese ions are at the active site of these enzymes and are co-ordinated to it through several ligands (see later).

The Mn(II)–pyruvate kinase system has also been studied by Reuben and Cohn (1970). By monitoring the amount of Mn(II) free in solution by use of e.s.r. spectroscopy, they initially studied the stoichiometry and dissociation constant of the Mn(II)–pyruvate kinase complex. The number of binding sites for Mn(II) appeared to be temperature dependent in the range 5–37 °C and approached *four* at the higher temperatures.† They interpreted the results in terms of an equilibrium mixture of two conformational states of the enzyme

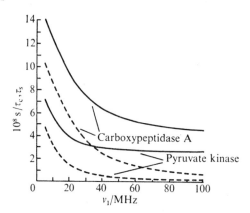

FIG. 10.3. Computed values of $1/\tau_c$ (solid curves) and $1/\tau_s$ (dashed curves) for the results in Figs. 10.1 and 10.2. (From Navon 1970.)

† Pyruvate kinase has four subunits and the analysis assumes that each subunit binds one Mn(II) identically.

one of which (E_A) has a low affinity for Mn(II) and the other (E_B) has *four* 'tight' binding sites. Thus the equilibrium can be written

$$E_A \rightleftharpoons E_B, \qquad K_E = [E_B]/[E_A]$$

If the apparent number of Mn(II) binding sites in the equilibrium mixture is n' then

$$K_E = n'/(4-n') \tag{10.1}$$

The concentration of the high Mn(II) affinity state (i.e. the active state) is given by

$$[E_B] = [E_t]K_E/(1+K_E) \tag{10.2}$$

where $[E_t] = [E_A]+[E_B]$.

By suitable Scatchard plots (see p. 255), values of K_E and K_D can be determined at each temperature and the following relationships were derived:

$$K_{E(KCl)} = \exp(54{\cdot}9-15{\cdot}1\times 10^3/T)$$
$$K_{D(KCl)} = \exp(9{\cdot}58-5{\cdot}73\times 10^3/T)$$
$$K_{E(TMACl)} = \exp(18{\cdot}6-4{\cdot}76\times 10^3/T)$$
$$K_{D(TMACl)} = \exp(14{\cdot}75-7{\cdot}43\times 10^3/T)$$

where TMACl is tetramethyl ammonium chloride. The dissociation constant of the Mn(II)–enzyme complex is thus lower in the presence of the non-activating TMA$^+$ ion.

Using the above relationships to calculate the free and bound Mn(II) concentrations, Reuben and Cohn (1970) measured the temperature dependence of the spin–lattice relaxation rates of the Mn(II)–enzyme complex i.e. $(1/T_{1,P})_b$ (Fig. 10.4). Of course, the spin–lattice relaxation behaviour for pure aqueous Mn(II) solutions has to be known [as in Navon's work (1970)] and allowed for in the analysis.

The analysis of the results illustrates several interesting and helpful points and we shall initially not anticipate Navon's results (1970). The analysis proceeds thus:

(1) The frequency dependence of the relaxation rates (i.e. those at 8.13 MHz being lower than those at 24·3 and 40 MHz) indicates that τ_c must be frequency dependent. As before, this implies that τ_S must be involved in determining τ_c, at least at the lower frequencies.

(2) The frequency dependence of the relaxation rates indicates that they are not dominated by slow-exchange conditions ($\tau_M \gg T_{1,M}$) but τ_M may still make a contribution to $1/T_{1,P}$, i.e. $\tau_M \approx T_{1,M}$. For example, at low temperature there appears to be a slight curvature in the plots of

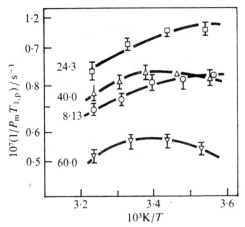

FIG. 10.4. Spin–lattice relaxation rates of the Mn(II)–pyruvate kinase complex at different frequencies. (From Reuben and Cohn 1970.)

relaxation rates versus frequency at 40 MHz and 60 MHz. We can rule out 'slow exchange' as the reason for the negative temperature coefficient at 40 MHz and 60 MHz, since for this to be so $\tau_{\rm M} \gtrsim T_{\rm 1M}$. $\tau_{\rm M}$ is independent of frequency and $T_{\rm 1M}$ is smallest at 24·3 MHz, and yet little curvature in the plots of relaxation rates versus frequency is observed at this frequency. If slow-exchange conditions prevailed one would expect the slow-exchange region to be most manifest at this frequency. Reference to Table 9.1 then suggests that the curvature in the plots at 40 MHz and 60 MHz most probably arises from the condition $\omega_I \tau_{\rm c} > 1$.

Note that the measurement of the spin–spin relaxation rates (e.g. Fig. 10.5) could make this point more obvious since $T_{\rm 1,M} \gg T_{\rm 2,M}$. If slow-exchange

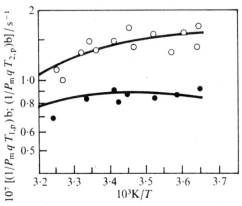

FIG. 10.5. The spin–spin and spin–lattice relaxation rates of the Mn(II)-pyruvate kinase complex at 35 MHz. (From Dwek, Malcolm, Radda, and Richards (1970), unpublished results.)

TABLE 10.2

*Parameters obtained from analysis of relaxation rates
in the* Mn(II)–*pyruvate kinase system*
(*Reuben and Cohn*, 1970)

B (rad^2 s^{-2})	0.146×10^{20}
τ_V (300 K) (s)	6×10^{-12}
E_V (kcal mol^{-1})	1.5 ± 0.1
τ_M (300 K) (s)	$(0.5 \pm 0.1) \times 10^{-8}$
$\Delta H\ddagger$ (kcal mol^{-1})	6.6 ± 0.35

conditions are apparent for $1/T_{1,P}$, they must be equally if not more apparent for $1/T_{2,P}$. This is clearly not the case in Fig. 10.5.

The parameters obtained from the analysis of Reuben and Cohn (1970) are given in Table 10.2. Several interesting comparisons may be made with Navon's (1970) results. The values of τ_M agree reasonably well, but those of B and τ_V differ considerably. With hindsight we may see why this could be so from Navon's results. τ_S becomes important only at low frequencies (Fig. 10.3). Even the large range of frequencies covered by Navon was not sufficient to enable him to obtain very accurate values of B and τ_V. Nevertheless, the actual values of τ_S from both sets of workers are reasonably consistent. The values of τ_c, too, are in good agreement, although the values from the work of Reuben and Cohn are ca. 20% longer on average.

Once τ_c is known then values of q/r^6 can be obtained from eqns (9.1), (9.2), (9.25), and (9.26). In the control aqueous Mn(II) solutions, analysis of Reuben and Cohn gave a value of $r = 2.8$ Å for the Mn—H distance. Using this value in their analysis, $q \approx 2.43$.

The determination of an accurate τ_c value is best accomplished by measurements of both relaxation rates rather than just from the frequency dependence of the spin–lattice relaxation rate. If we use the value of τ_c obtained by Navon (1970), then q is reduced to 2. Thus the results of the two sets of workers are consistent. The determination of hydration numbers, as suggested by Navon, relies on the assumption that r is unchanged from its value in aqueous solutions. Reuben and Cohn (1970), however, did not make the assumption that the value of r is unchanged from that in the pure control solution. They assumed a value of $r = 2.9$ Å and thus obtained a value of $q = 3$.

10.3. Determination of the hydration number in Mn(II)/enzyme systems from measurements of $T_{1,P}$ and $T_{2,P}$

Let us examine more closely the method for determining the hydration number, q. Obviously, if τ_c can be determined from an analysis of the frequency dependence of the relaxation rates, then, if r is known, q can be determined. A better method is obviously one that does not rely on knowing the value of τ_c, for, if τ_c is frequency dependent, obtaining a unique value of τ_c may be difficult. The method below was originally suggested by Navon (1970) and under favourable circumstances requires only *one* measurement of $T_{1,P}$ and $T_{2,P}$ at *one* frequency. The favourable circumstances are that the ratio $T_{1,P}/T_{2,P}$ is greater than ca. 1·5.

10.3.1. *The method*

From the above analysis for the Mn–carboxypeptidase and –pyruvate kinase systems we note that the values of τ_e and τ_c are such that the values of $(\omega_S \tau_e)^2$ or $(\omega_S \tau_c)^2 \gg 1$ and only the terms containing $(\omega_I \tau_c)^2$ or $(\omega_I \tau_e)^2$ in the equations for $1/T_{1,M}$ and $1/T_{2,M}$ have to be retained. The Solomon–Bloembergen equation can then be written as

$$\frac{1}{T_{1,M}} = D\left(\frac{\tau_c}{1+\omega_I^2\tau_c^2}\right) \tag{10.3}$$

and

$$\frac{1}{T_{2,M}} = D\left(4\tau_c+\frac{3\tau_c}{1+\omega_I}\right)+C\tau_e \tag{10.4}$$

where

$$D = \frac{6}{15}\left(\frac{S(S+1)\gamma_I^2 g^2 \beta^2}{r^6}\right) \quad \text{and} \quad C = \tfrac{1}{3}S(S+1)\left(\frac{A}{\hbar}\right)^2$$

If we assume that the values of C and D are the same as for the aquo ion, then $D = 2\cdot06\times10^{15}$ rad^2 s^{-2} and $C/D = 0\cdot022$ (see p. 205). Under conditions of fast exchange

$$\frac{1}{T_{1,P}} = \frac{P_M q}{T_{1,M}} \tag{10.5}$$

$$\frac{1}{T_{2,P}} = \frac{P_M q}{T_{2,M}} \tag{10.6}$$

Thus

$$\frac{T_{1,P}}{T_{2,P}} = \frac{T_{1,M}}{T_{2,M}} = \tfrac{1}{2}+(\tfrac{2}{3}+0\cdot022\tau_e/\tau_c)(1+\omega_I^2\tau_c^2) \tag{10.7}$$

Since $\tau_e \approx \tau_c$, it is reasonable to put the ratio of $\tau_e/\tau_c \approx 1$. Substituting for τ_c in the expression for $1/T_{1,P}$ gives:

$$q = 0\cdot587\times10^{-15}\frac{(T_{1,P}/T_{2,P}-0\cdot5)}{(T_{1,P}/T_{2,P}-1\cdot19)^{\frac{1}{2}}}\frac{\omega_I}{P_M T_{1,P}} \tag{10.8}$$

Since $P_M = N/55\cdot5$, where N is the concentration of paramagnetic species we then have

$$q = 3\cdot26\times10^{-14}\frac{(T_{1,P}/T_{2,P}-0\cdot5)}{(T_{1,P}/T_{2,P}-1\cdot19)^{\frac{1}{2}}}\frac{\omega_I}{N T_{1,P}} \tag{10.9}$$

This expression is still likely to give reasonable answers for q even if the contribution of τ_R to τ_c is comparable with that of τ_S and τ_M, because of the smallness of the term containing τ_e/τ_c. It is clearly most sensitive when the ratio of $T_{1,P}/T_{2,P}$ differs significantly from 1·19. This is likely to be so at high magnetic fields.

The value of D in this experiment is calculated assuming a value of $r \approx 2·74$ Å. This differs somewhat from the free-ion value for the Mn—H distance of 2·8 Å in aqueous solutions of Mn(II).† The differences may arise because there are difficulties in making up accurately known concentrations of Mn(II), and any errors in the concentrations will reflect in the value of r obtained. Also systematic machine errors could result in different workers obtaining different values of the relaxation rates which would also lead to differing values of r. To minimize these discrepancies, $1/T_{1,M}$ should be measured from a control solution of the Mn(II) ions and assuming

$$\tau_c \approx 3 \times 10^{-11} \text{ s}$$

at 25 °C, a value of D calculated for the aqueous solution using the Solomon–Bloembergen equation for $1/T_{1,M}$. This value of D is then used in the calculation of the hydration number.

When the value of τ_c is frequency independent, there is a further method that is often used for the determination of q. As we have noted above, under conditions of fast exchange, it is sufficient to write

$$\frac{1}{P_M q T_{1,P}} = \frac{1}{T_{1,M}} = D\left(\frac{\tau_c}{1+\omega_I^2\tau_c^2}\right) \tag{10.10}$$

(i.e. terms in $\omega_S^2\tau_c^2$ are negligible).

We may write this equation as:

$$T_{1,P} = \frac{1}{P_M q D\tau_c}(1+\omega_I^2\tau_c^2). \tag{10.11}$$

If the product $\omega_I^2\tau_c^2 \gtrsim 1$ then a graph of $T_{1,P}$ versus ω_I^2 will be linear with an intercept of $1/P_M q D\tau_c$ and a slope of $\tau_c/P_M q D$. The product (slope × intercept) gives $(1/P_M q D)^2$ while the ratio gives $1/\tau_c^2$. If the value of D is known, then q can be calculated. Additionally, the value of τ_c can be evaluated. If τ_c is frequency dependent, there will of course be significant deviations from linearity.

As an example of this method we may take the values of $T_{1,P}$ for the Mn(II)–pyruvate kinase system. Above 40 MHz, τ_c is independent of frequency. (Fig. 10.3). A plot of $T_{1,P}$ versus ω_I^2 is shown in Fig. 10.6 from which values of $\tau_c \approx 3·0 \times 10^{-9}$ s and $q \approx 2$ are obtained. (The value of D is assumed to be the same as above.) The slight discrepancy in the value of τ_c may result because only four experimental points were used in this calculation and because the method is only really accurate when $\omega_I^2\tau_c^2 \gg 1$.

† Using a value of $r = 2·8$ Å, $D = 1·8 \times 10^{15}$ rad² s⁻².

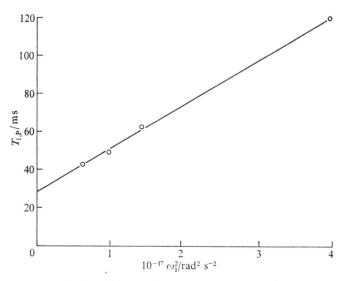

FIG. 10.6. Determination of q and τ_c for the Mn(II)–pyruvate kinase complex, from a plot of $T_{1,P}$ versus ω_I^2. Data from Fig. 10.1.

The danger of using this method is that it is not always possible to know *a priori* whether τ_S is important. We include it because experimental results of spin–lattice relaxation times are often presented as plots of $T_{1,P}$ versus ω_I^2.

10.4. Relaxation of the solvent water protons in some Mn(II)–creatine kinase-nucleotide complexes

Creatine kinase (mol. wt 80 000) has two apparently identical sub-units and catalyses the reaction.

$$\text{ATP}^{4-} + \text{Creatine} \underset{\longleftarrow}{\overset{M^{2+}}{\rightleftharpoons}} \text{ADP}^{3-} + (\text{Phosphocreatine})^{2-} + \text{H}^+$$

where M^{2+} may be Mg^{2+}, Mn^{2+}, Co^{2+}, or Ca^{2+}.

These examples are included to illustrate the general feature that τ_S is important in determining τ_c at low frequencies, and to indicate that outer-sphere relaxation effects may be important in certain systems. The conclusions are based on an analysis of the water proton relaxation rates.

In contrast to the Mn(II)–pyruvate kinase and Mn(II)–carboxypeptidase systems, the spin–lattice relaxation rates of the water protons in the binary Mn(II)–creatine kinase complex only show very slight increases compared with those of control solutions containing no enzyme. It is probable, but not definitely established, that this slight increase may reflect non-specific binding of the metal to the enzyme. However addition of ADP or ATP to Mn(II)–creatine kinase solutions gives relatively large increases in the water proton relaxation rates providing evidence for formation of the ternary

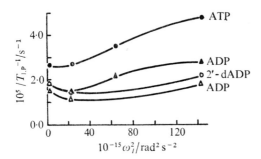

FIG. 10.7. A comparison of the frequency dependence of the molar relaxation time of E–Mn–nucleotide complexes of creatine kinase. The open symbols represent an experiment comparing ADP ($\triangle$) 0·107 mM and 2′-dADP ($\square$) 0·106 mM. Both solutions contained 0·1 mM $MnCl_2$, 8·25 mg per ml of enzyme in 0·05 M N-ethylmorpholine–HCl buffer, pH 8·0; $T = 16$ °C. The closed symbols represent an experiment in the same buffer with enzyme concentration 6 mg per ml comparing ADP ($\blacktriangle$) 0·108 mM with ATP ($\bullet$) 0·096 mM, $T = 22$ °C. (From Reed, Diefenbach, and Cohn 1972.)

complex. (see p. 278). The ternary complex probably has a substrate bridge between enzyme and metal (Cohn, 1970 and references therein). The frequency dependence of the water proton relaxation times of some representative enzyme–Mn(II)–nucleotide complexes are shown in Figs. 10.7 and 10.8. (Reed *et al.*, 1972).

The plots of $T_{1,P}$ versus ω_I^2 show a slight minimum indicative of the behaviour expected if τ_S makes a significant contribution to the relaxation rates at low frequencies. In each of these experiments, there is a significant amount of free metal–nucleotide complex in addition to the enzyme–metal–nucleotide complex and so only qualitative deductions can be made. The temperature dependence of the ternary creatine kinase–Mn(II)–ADP complex is also qualitatively similar to that observed in the Mn(II)–pyruvate kinase system.

An appreciable change in the spin–lattice relaxation rate and a large

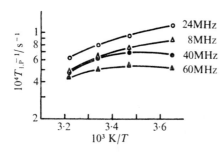

FIG. 10.8. Temperature dependence (ranging from 1–37 °C) of molar relaxation time of the ternary creatine kinase–Mn–ADP complex at four frequencies. The solutions contained 0·1 mM $MnCl_2$, 0·12 mM ADP, 9 mg per ml of enzyme in 0·05 M N-ethylmorpholine–HCl buffer, pH 8·0. (From Reed *et al.* 1972.)

decrease in reactivity of the 'essential' SH groups of the enzyme occurs on addition of creatine to solutions containing creatine kinase, Mn(II), and ADP. However Milner–White and Watts (1971) have shown that the reactivity of the SH groups in the abortive quaternary complex is a function of the particular anion present in solution. The effect of three anions on the water proton spin–lattice relaxation rates are shown in Table 10.3. (Reed and Cohn, 1972). The anions acetate, chloride, and nitrate are representative of

TABLE 10.3

Anion effects on PRE of the quaternary complex

The solutions contained $MnCl_2$, 0·1 mM, ADP, 0·2 mM; creatine, as indicated; and enzyme, 10 mg per ml; HEPES–KOH, 50 mM, pH 8·0. $T = 1$ °C. Measurements were carried out at 24·3 MHz.

Added anion	Concentration of anions (mM)	Creatine (mM)	$1/T_{1,P}$† (s^{-1})
None		30	4·81
(I) Acetate	10	50	4·57
	20	50	4 51
(II) Chloride	10	50	3·19
	20	50	2·79
(III) Nitrate	1	30	2·02
	2	30	2·00

† $1/T_{1,P}$ is the paramagnetic contribution to the longitudinal relaxation rate.

three classes of anions. The three classes are based on (1) their affinity for the abortive quaternary complex and (2) the magnitude of the decrease in the water proton relaxation rate. For example, nitrate has the highest affinity and causes the largest decrease in the relaxation rates, while anions like acetate have little effect on the relaxation rates and relatively low affinities for the quaternary complex.

The frequency dependence of the water solvent relaxation time on the quaternary complex in the presence of nitrate is shown in Fig. 10.9. We note immediately that the linear plot of $T_{1,P}$ versus ω_I^2 strongly suggests that τ_c is not frequency dependent. From the intercept and the slope (see 221), values of $\tau_c \approx 10^{-8}$ s can be calculated, (the exact value depends on the temperature). Assuming the usual value of D in eqn (10.11) the hydration number is calculated to be less than one! Similar effects are observed in the presence of chloride and acetate ions (Reed and Cohn, 1972) and the 'apparent' hydration numbers q' and values of τ_c at 0 °C are listed in Table 10.4.

The low hydration number suggests that *no* water molecules in the first co-ordination sphere contribute significantly to the observed relaxation rate.

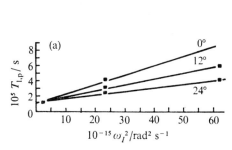
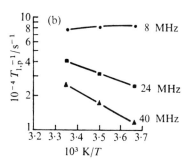

FIG. 10.9. (a) Frequency dependence of molar relaxation time of the quaternary creatine kinase–Mn–ADP–creatine complex at three temperatures ranging from 0–24 °C. (b) Same data, $1/T_{1,P}$ plotted as a function of temperature. The solutions contained 0·1 mM $MnCl_2$, 0·2 mM ADP, 30 mM creatine, 0·2 mM KNO_3, and 7·7 mg per ml of enzyme in 0·05 M HEPES–KOH buffer, pH 8·0 (From Reed *et al.* 1972.)

(It does not necessarily mean that there are none, for they may not be undergoing chemical exchange at a rate fast enough to affect the relaxation rates). If the observed behaviour results from outer sphere relaxation it is then necessary to choose a model to interpret the results. However the effects may originate from hydrogens of the protein itself undergoing exchange with the solvent water or from protons in water molecules which are held rigidly to the protein.† In both these cases, the use of the Solomon–Bloembergen treatment is justified rather than the diffusion model discussed on p. 190.

TABLE 10.4

Variation of τ_c and q' for the quaternary abortive complex in presence of anions

The solutions contained $MnCl_2$, 0·1 mM; ADP, 0·2 mM; anion, concentrations as indicated (potassium salts); creatine, 30 mM in the nitrate experiment, 80 mM in the chloride and acetate experiments; creatine kinase, 7·7 mg per ml; HEPES–KOH buffer, 50 mM, pH 8·0. $T = 0$ °C.

| | $T_{1,P}$ at frequencies of | | | | |
| | 8·13 MHz | 24·3 MHz | 40·0 MHz | τ_c | |
Anion	($10^3 \times$ s)	($10^3 \times$ s)	($10^3 \times$ s)	($10^9 \times$ s)	q'†
Nitrate, 2 mM	121	415	876	12	0·31
Chloride, 50 mM	138	320	532	7·9	0·36
Acetate, 10 mM	136	203	332	5·0	0·55

† q' is the ratio of the observed value of the paramagnetic contribution to the relaxation rate, $1/T_{1,P}$, to the value expected for 1 water molecule in the first co-ordination sphere of Mn(II).

† The estimated value of τ_R for these systems is ca. $2·6 \times 10^{-8}$ s.

Some support for this interpretation comes from the temperature dependence of the solvent water spin–lattice relaxation times with the quaternary complex in the presence of nitrate. The negative temperature coefficients at 40 and 60 MHz do not result from slow-exchange conditions. The arguments follow a familiar pattern. τ_M is frequency independent, and if it dominated completely the relaxation rate, then no frequency dependence would be observed. If there is a significant contribution from τ_M, i.e. if $\tau_M \approx T_{1,M}$ its effect would be most marked at the highest value of $1/T_{1,P}$, for here $T_{1,M}$ would be lowest. At 8·13 MHz the temperature coefficient is, if anything, positive. The observed behaviour probably arises from the condition $\omega_I \tau_c \approx 1$ at 8·13 MHz and correspondingly greater than unity at higher frequencies. Thus at 8 MHz, from $\omega_I \tau_c \approx 1$ we obtain $\tau_c \approx 2 \times 10^{-8}$ s.

What then can we say about the nature of τ_c? Probably in this example there is a significant contribution from τ_R and possibly a distribution of values for τ_M from different hydrogens in the different sites. The linearity of the plot suggests that τ_S is unimportant and thus its value must be longer than 2×10^{-8} s. The possibility of including contributions from diffusion of water molecules relative to the Mn(II)–complex has not been considered, but it is possible that one of the intermediate outer-sphere cases mentioned in Chapter 9 applies. When one considers all the variables, more detailed analyses than those presented here do not seem worthwhile.

As a corollary, the resonances of some of the anions themselves, particularly ^{14}N and ^{35}Cl, can be studied and it is then possible, in conjunction with studies on the various substrates, to build up a picture of the transition-state analogues which may provide an insight into the mechanism of the action of the enzyme. We shall return to this theme later.

10.5. Some conclusions from the analyses of the water relaxation rates in Mn(II)–enzyme complexes

Perhaps the most important result of the previous examples is that τ_c contains a significant contribution from τ_S. This is demonstrated most easily by measurements of the frequency dependence of the spin–lattice relaxation times under conditions of fast exchange, which reveal a minimum in plots of $T_{1,P}$ versus ω_I, characteristic of systems where τ_S is important. In the study of the solvent water, we noted in Mn(II)–pyruvate kinase and Mn(II)–carboxypeptidase that τ_M is also important in determining τ_c, particularly at high frequencies, and that τ_R may be neglected. Clearly the neglect of τ_R is not always justified and in smaller macromolecules, e.g. lysozyme (mol. wt. ca. 14 000) for which τ_R is estimated as ca. 5×10^{-9} s, it could make a significant contribution to τ_c.

When we considered the Solomon–Bloembergen equations in Chapter 9, we also considered some complications that could occur. In the light of the above examples, we may examine some of these.

The difficulty of whether to use $\tau_{1,S}$ or $\tau_{2,S}$ does not arise for the values of τ_c and τ_e are such that the terms containing $\omega_S^2\tau_c^2$ or $\omega_S^2\tau_e^2$ in eqns (9.1) and (9.2) may be neglected. Thus the value of $\tau_{2,S}$ does not enter into the analysis. It is interesting to compare the values of B and τ_v derived for the Mn(II)–enzyme systems in the previous section with those for aqueous Mn(II) solutions. The values of q may indicate that the Mn(II) in the enzymes is quite well embedded in the protein and the similarity of the B and τ_v values with those in aqueous solution, might indicate a relaxation mechanism for the electron spin that does not necessarily involve impact of solvent molecules which distort the zero field splitting.

The effect of anisotropic rotation is unlikely to be unimportant in the Mn(II)–carboxypeptidase and Mn(II)–pyruvate kinase systems. For anisotropic averaging, any rotational motion would have to be faster than τ_M and τ_S, and this seems intuitively to be unlikely. We note however that the analysis of a system with a hydration number of $q = 5$ could give an apparent value of $q = 1\cdot6$ if anisotropic averaging occurs (i.e. $1-R = 8/25$; see page 186 ff).

For Mn(II) the g-value is always likely to be isotropic. This is because Mn(II) is an S-state ion which has no orbital angular momentum and the anisotropies in g arise for first-row transition metals mainly from spin–orbit coupling. It is possible that Mn(II) could bind to a group such as SH and that mixing of orbital angular momentum from the sulphur atom could result in spin–orbit coupling and an anisotropic g-value. A crude calculation however shows that this situation is extremely unlikely.

The interpretation of *water proton relaxation* rates in the quaternary Mn(II)–creatine–ADP–creatine kinase complex, suggests that outer-sphere relaxation is not necessarily negligible in these systems. However, in the Mn(II)–creatine systems only small enhancement of water relaxation rates compared with aqueous Mn(II) solutions are observed. In macromolecular systems, the observed relaxation rate generally will depend on the value of q and τ_c. It may be a relatively safe generalization to state that τ_c does not vary very greatly for the different systems and that, when only small relaxation rate enhancement in macromolecular systems are observed, outer-sphere effects could be important. These may result from exchangeable protons on the protein or water attached to amino acid groups near to the paramagnetic metal site.

10.6. The binding of various ligands to Co(II) human carbonic anhydrase (Taylor, Feeney, and Burgen 1971)

The human enzyme has an essential difference from the bovine in that it contains an SH group. We have already discussed in Chapter 6 some of the properties of and an example of relaxation measurements on the zinc bovine carbonic anhydrase. The zinc can be replaced reversibly by several divalent ions, but only Co(II) gives a retention of similar catalytic and inhibitor-binding properties.

The example is included to illustrate the type of exchange behaviour that we have already mentioned in Chapter 9. In this example only a limited number of measurements of line widths were undertaken and we shall see that important information on the exchange behaviour for the binding of various anionic ligands can be obtained. The values of $1/T_{2,P}$ were obtained from measurements, of the line-width at half-height, $\Delta\nu_P$. The usual expression links these variables:

$$\pi\Delta\nu_P = 1/T_{2,P} \qquad (10.12)$$

(where $\Delta\nu_P$ is in Hz). Under the experimental conditions of Taylor *et al.* (1971) all the Co(II) was bound to the enzyme, and further it could be arranged that in every case at least 97 % of the ligand was bound to the enzyme.

The temperature dependences of the hydrogen and fluorine relaxation rates with various carboxylate inhibitors are shown in Figs. 10.10–10.12. The ^{19}F relaxation rates and that for the hydrogen in the formate inhibitor have a region at lower temperatures indicative of slow-exchange behaviour. Measurements at a different frequency, (although few), seem to indicate that at higher temperatures the relaxation rates are frequency dependent. This behaviour corresponds to Region II of Fig. 9.7. It will be recalled that in

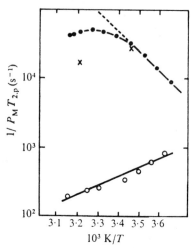

FIG. 10.10. Influence of temperature on the transverse relaxation rate for formate and acetate complexes of cobalt(II)–human carbonic anhydrase C, pH* 7·6. (Samples were made up in D_2O; pH* values are meter readings.) (●) 400 mM formate in the presence of 0·1 mM enzyme at 100 MHz; and (×) at 60 MHz; (○) 240 mM acetate in the presence of 1·5 mM enzyme at 100 MHz. (From Taylor *et al.*, 1971. © 1971 Amer. Chem. Soc.)

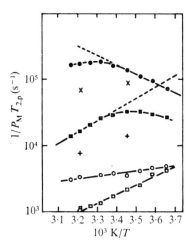

FIG. 10.11. Influence of temperature on the transverse relaxation rate of ^1H and ^{19}F for mono- and di-fluoroacetate complexes of cobalt(II)–carbonic anhydrase C, pH* 7·6. (●) ^{19}F in $CH_2FCO_2^-$ at 94·2 MHz; (■) in $CHF_2CO_2^-$ at 94·2 MHz; (○) ^1H in $CH_2FCO_2^-$ at 100 MHz; (□) ^1H in $CHF_2CO_2^-$ at 100 MHz; (×) ^{19}F in $CH_2FCO_2^-$ at 56·4 MHz; (+) ^{19}F in $CHF_2CO_2^-$ at 56·4 MHz. (From Taylor *et al.*, 1971. © 1971 Amer. Chem. Soc.)

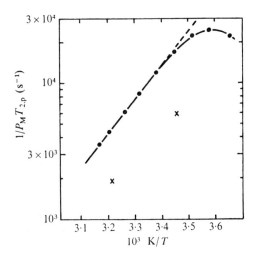

FIG. 10.12. Influence of temperature on the transverse relaxation rate of ^{19}F for 400 mM trifluoroacetate with 0·40 mM cobalt(II) isozyme C, pH* 7·6 at (●) 94·2 MHz and (×) 56·4 MHz. (From Taylor *et al.*, 1971. © 1971 Amer. Chem. Soc.)

the limit when $1/\tau_M^2 \gg \Delta\omega_M^2 \gg 1/T_{2,M}$

$$\frac{1}{P_M T_{2,P}} = \tau_M \Delta\omega_M^2.\dagger \qquad (10.13)$$

if outer-sphere contributions are neglected. If these conditions obtain at two frequencies, then the ratio of the relaxation times should be given by the ratio of the square of the frequencies (in this case 2·87). The results in Table 10.5 show that at high temperatures this is approximately the case. The discrepancies may arise from either significant outer-sphere contributions or because the above *limiting* conditions are not quite fulfilled.

TABLE 10.5

Frequency dependence of the transverse relaxation rate for various carboxylate ligands with cobalt(II)-carbonic anhydrase C

Ligand	Nucleus	$10^3 K/T$	A	B	A/B
Formate†	^1H	3·46	$2·95 \times 10^4$	$2·46 \times 10^4$	1.20
		3·22	$4·7 \times 10^4$	$1·76 \times 10^4$	2·67
Monofluoroacetate‡	^{19}F	3·46	$1·37 \times 10^5$	$8·5 \times 10^4$	1·61
		3·22	$1·75 \times 10^5$	$7·2 \times 10^4$	2·45
Difluoroacetate‡	^{19}F	3·46	$3·3 \times 10^4$	$1·39 \times 10^4$	2·37
		3·22	$1·82 \times 10^4$	$7·9 \times 10^3$	2·30
Trifluoroacetate‡	^{19}F	3·46	$1·7 \times 10^4$	$5·9 \times 10^3$	2·89
		3·22	$4·9 \times 10^3$	$1·75 \times 10^3$	2·80

† A, $1/P_M T_{2,P}$ at 100 MHz; B, $1/P_M T_{2,P}$ at 60 MHz.
‡ A, $1/P_M T_{2,P}$ at 94·2 MHz; B, $1/P_M T_{2,P}$ at 56·4 MHz.

Confirmation of the above exchange behaviour can, in principle, be obtained from measuring the 'bulk' chemical shifts of the various ligand resonances. For example, the behaviour of the chemical shift, $\Delta\omega_A$, of the 'bulk' formate ^1H nucleus is shown in Fig. 10.13, and gives the expected behaviour. *Provided* conditions are arranged so that rapid exchange prevails then the bound chemical shift, $\Delta\omega_M$, can be obtained from the relationship:

$$\Delta\omega = P_M \Delta\omega_M$$

In general, Co(II) has an anisotropic g-value and thus it is likely that $\Delta\omega_M$ contains contributions from both contact and pseudo-contact interactions and so some care is obviously necessary in interpreting its value. Although, in this example, assuming the carboxyl group co-ordinates to the Co(II), there are likely to be several chemical bonds between the hydrogens and

† Strictly speaking the equation should contain the symbol q—the number of inhibitor molecules (or ligands) bound. Often this symbol is omitted in equations. In such cases it is assumed that $q = 1$.

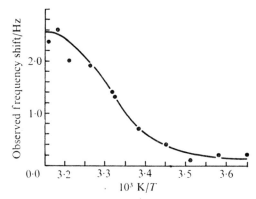

FIG. 10.13. Frequency shift difference, $\Delta\omega$, of the formate hydrogen for solution containing 400 mM formate and 0·10 mM cobalt(II)–carbonic anhydrase C, pH* 7·6. Chemical shifts are measured relative to an internal reference, dioxan, and the recorded values are differences observed in the absence and presence of p-carboxybenzene sulphonamide (p-CBS). (From Taylor *et al.*, 1971. © 1971 Amer. Chem. Soc.)

fluorine and the Co(II) and thus pseudo-contact interactions would dominate. Alternatively, if τ_M is known, then $\Delta\omega_M$ can be obtained from the equation for the transverse relaxation rate, $1/T_{2,P} = \tau_M\Delta\omega_M^2$. Of course, the best way is to use all the equations, thus obviating the need for using only the limiting forms of the equations under conditions where they may not be strictly valid. Once an exchange phenomenon has been verified, examination of the concentration dependence of the apparent exchange rate can establish whether ligand substitution is affected by bimolecular or higher-order processes. For example, in this case the following processes are possible:

$$E\text{—}I \underset{}{\overset{k_{-1}}{\rightleftharpoons}} E^+ + I^-$$

and

$$E\text{—}I + I^{-*} \rightleftharpoons EI^* + I^-$$

where E represents the enzyme and I the ligand. In the second process the attacking ligand influences the dissociation rate of the leaving ligand and the apparent exchange rate should be concentration dependent. For formate, the exchange rate is concentration independent and the second process does not contribute to the exchange rate as determined by n.m.r. The dissociation of the leaving group, as a unimolecular process, is assumed to control turnover of the ligand to the enzyme and these rates are listed in Table 10.6. In all cases $\Delta H\ddagger$ is ca. 9 kcal mol^{-1}. The values of τ_M (at 298 K) for the various inhibitors are also listed in Table 10.6. For the fluorinated inhibitors, these were calculated by suitable extrapolation of the relaxation data as indicated by the dotted lines in Figs. 10.10–10.12.† The values of K_1 in Table

† For the trifluoroacetate, the temperature dependence of the relaxation data in the fast-exchange region represents essentially the activation energy for τ_M, since the temperature variation of $\Delta\omega_M$ is very small (varying approximately as $1/T$).

TABLE 10.6

Estimated association and dissociation rate constants for various carboxylate anion complexes of cobalt(II)–carbonic anhydrase C, 25 °C

Ligand	$K_I(M^{-1})$†	$k_1(M^{-1}\,s^{-1})$‡	$k_{-1}\,(s^{-1})$§
Formate	$6\cdot5\times10^3$	$3\cdot9\times10^8$	$6\cdot0\times10^4$
Monofluoroacetate	$1\cdot1\times10^3$	$2\cdot2\times10^8$	$2\cdot0\times10^5$
Difluoroacetate	$1\cdot9\times10^3$	$1\cdot9\times10^8$	$1\cdot0\times10^5$
Trifluoroacetate	$1\cdot3\times10^3$	$2\cdot0\times10^8$	$1\cdot5\times10^5$

† Intrinsic affinity calculated from the apparent affinity at pH 7·6 (Taylor and Burgen, 1971).
‡ Calculated from k_{-1} and K_I.
§ Determined from life times measured by n.m.r. and extrapolation to 25 °C.

10.6 were obtained from the relationship:

$$K_I = \frac{[EI]}{[E][I]} = \frac{k_{+1}}{k_{-1}} \tag{10.14}$$

since $1/\tau_M$ may be identified with k_{-1}.

Finally, we mention the control experiments. In principle, these could be carried out with the diamagnetic enzyme, in which Zn(II) replaces Co(II). Taylor *et al.*, (1971), however, used a more sophisticated approach. By addition of sufficient of the inhibitor *p*-carboxybenzenesulphonamide, the carboxylate inhibitors were displaced and thus the reported relaxation rates must reflect those for specific sites on the enzyme. Any of the inhibitors not displaced by the *p*-carboxybenzenesulphonamide will still contribute in the 'control experiment'.

The most startling conclusion about these rates is that the Co(II) enzyme exhibits ligand association rates k_1 which are between two and three orders of magnitude greater than octahedral aquo-Co(II). Whatever the cause, this gives a selective advantage for catalysis to the enzyme–metal co-ordination site. A comparison with the sulphonamide inhibitors is interesting in that the values of the associative rate constant k_1 are approximately the same for the two sets of inhibitors but the dissociation rates of sulphonamides are ca. $10^{-1}\,s^{-1}$, much slower than for the anionic inhibitors. The longer residence of the sulphonamides might be a result of the hydrophobic interaction involving the phenyl ring which may contribute to the stabilization of the complex (see p. 136).

10.7. The binding of sulphacetamide to bovine carbonic anhydrase (Lanir and Navon, 1972)

We noted previously that the zinc ion in the metallo-enzyme carbonic anhydrase can be replaced reversibly by various divalent metal ions including

Mn(II). Moreover the activity of the Mn(II)–enzyme is only ca. 4·5% that of the zinc enzyme. The dissociation constant, K_D, for the Mn(II)–enzyme complex is ca. 400 μM at 25 °C and pH 7. These values of K_D may be contrasted with those for the Zn(II)– and Co(II)–enzyme which are several orders of magnitude lower. In fact, the usual procedure for removing the zinc from the enzyme is by dialysis against o-phenanthroline at acid pH for about ten days!

The n.m.r. spectrum of sulphacetamide consists of three groups of resonances, assigned to the acetyl hydrogens, and the hydrogens *ortho* and *meta* to the sulphonamide group (see p. 134). The addition of this inhibitor to the Mn(II)–enzyme results in large broadening of its resonances. By measuring the change in line-width as a function of inhibitor concentration, the dissociation constant of the inhibitor–metallo-enzyme complex K_I can be obtained. The relevant equation:

$$\frac{1}{\Delta \nu_P} = \frac{\pi(T_{2,M}+\tau_M)}{[E]_0}(K_I+[I])$$

has been derived previously (p. 138). $\Delta \nu_P$ in these experiments is the difference between the solution containing the enzyme inhibitor and the Mn(II) ions and a similar solution of the inhibitor containing an identical Mn(II) concentration but no enzyme.

In most n.m.r. experiments of this type, the ratios of inhibitor to enzyme concentrations are usually large and it is advisable to check the possibility of any non-specific binding being responsible for the observed line broadening or other effects. Lanir and Navon (1972) showed that, in their experiments, the sulphacetamide was indeed bound to the active site by displacement using the specific inhibitors toluene-p-sulphonamide. This was monitored by measuring the narrowing of the line-width of the sulphacetamide methyl resonances as a function of the concentration of the second inhibitor.

Before we discuss the temperature variation of the relaxation rates we have to be able to calculate the fraction of the total inhibitor concentration bound to the enzyme at any temperature. This is given by:

$$P_M \approx \frac{[EI]}{[I_0]} = \frac{[E]_0}{K_I+[I]_0}. \tag{10.16}$$

Thus the temperature dependence of K_I must be known and this can be obtained from the above method carried out at several temperatures.

The temperature dependences of $1/P_M T_{2,P}$ for the three resonances are shown in Fig. 10.14. As usual the first stage in the analysis is to determine whether or not conditions of fast exchange obtain. The temperature dependences of the relaxation rates of all three resonances and particularly the methyl protons of the acetyl group have a region in which the relaxation rate

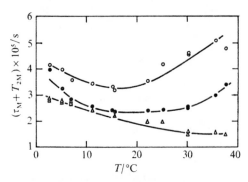

FIG. 10.14. Temperature dependence of $(T_{2\mathrm{M}}+\tau_{\mathrm{M}})$ of sulphacetamide bound to Mn(II)–bovine carbonic anhydrase. (○) Phenyl hydrogens *meta* to the sulphonamide group. (●) Phenyl hydrogens *ortho* to the sulphonamide group. (△) Methyl protons of the acetyl group. (From Lanir and Navon 1972. © 1972 Amer. Chem. Soc.)

decreases on decreasing the temperature. Reference to Table 9.1, indicates that this will occur if 'slow-exchange' behaviour is becoming important, or fast-exchange conditions obtain with τ_S having an apparent 'negative activation energy'. Clearly τ_{M} is not completely dominant, i.e. $\tau_{\mathrm{M}} \gg T_{2,\mathrm{M},i}$, (where $T_{2,\mathrm{M},i}$ are the respective relaxation times of the three groups of resonances) for τ_{M} is independent of frequency and is obviously the same for all protons and thus the values of $1/T_{2,\mathrm{P}}$ should all be identical. If initially, though, it is assumed that slow-exchange conditions apply only to the methyl group, then the value of τ_{M} can be determined and thus the contribution of $T_{2,\mathrm{M}}$ to the relaxation rates of the hydrogens of the phenyl group can be evaluated. However, when this is done there still appears to be a minimum in the plots of $T_{2,\mathrm{P}}$ for the phenyl group hydrogens. Thus it is likely that both of the above possibilities for the temperature dependence of the relaxation behaviour must be considered. In principle, it is possible to evaluate τ_{M} and the separate $T_{2,\mathrm{M}}$ values from the measurements so far described. However, in order to calculate the values of the Mn(II)–ligand distances from $T_{2,\mathrm{M}}$, it is necessary to know the value of the correlation time. At one frequency this may be done by measuring the ratio of $T_{1,\mathrm{P}}/T_{2,\mathrm{P}}$ and the analysis used by Lanir and Navon (1972) proceeds as follows:

(1) $T_{1,\mathrm{P}}$ was measured (using the progressive saturation technique) for each group of hydrogens.
(2) The ratio

$$\frac{T_{1,\mathrm{P}}}{T_{2,\mathrm{P}}} = \frac{T_{1,\mathrm{M}}+\tau_{\mathrm{M}}}{T_{2,\mathrm{M}}+\tau_{\mathrm{M}}} \qquad (10.17)$$

and the quantity

$$\frac{1}{P_{\mathrm{M}}T_{2,\mathrm{P}}} = \frac{1}{T_{2,\mathrm{M}}+\tau_{\mathrm{M}}}$$

were measured for each of the three hydrogen groups. *Three* equations of the form of eqn (10.18) are obtained

$$\frac{T_{1,M}}{T_{2,M}}[(T_{2,M}+\tau_M)-\tau_M]+\tau_M = \frac{T_{1,P}}{T_{2,P}}(T_{2,M}+\tau_M). \tag{10.18}$$

In the three equations there are only two unknowns τ_M and $T_{1,M}/T_{2,M}$ and thus both may be obtained. The ratio $T_{1,M}/T_{2,M}$ is the same for each group of hydrogens as shown in (3).

(3) The previous examples indicate that, in macromolecular systems, the values of τ_c are such that terms in $\omega_S^2\tau_c^2$ in the Solomon–Bloembergen equations may be neglected. By *assuming* that the hyperfine terms in $1/T_{2,M}$ are small then, from the dipolar parts of the Solomon–Bloembergen equations we have:

$$\frac{1}{T_{1,M}} = \frac{2}{15}\frac{\mu^2\gamma_I^2}{r^6}\left[\frac{3\tau_c}{1+\omega_I^2\tau_c^2}\right] \tag{10.19}$$

$$\frac{1}{T_{2,M}} = \frac{1}{15}\frac{\mu^2\gamma_I^2}{r^6}\left[4\tau_c+\frac{3\tau_c}{1+\omega_I^2\tau_c^2}\right] \tag{10.20}$$

from which we obtain

$$\frac{T_{1,M}}{T_{2,M}} = \tfrac{7}{6}+\tfrac{2}{3}(\omega_I^2\tau_c^2). \tag{10.21}$$

If τ_c is the same for each group of hydrogens and if the hyperfine interaction is indeed negligible then $T_{1,M}/T_{2,M}$ should be a constant. From the value of τ_c (Fig. 10.15) the various metal–ligand distances can be calculated, using eqn (10.19) or (10.20).

The experimental calculated parameters for the Mn(II)–sulphacetamide complex are given in Table 10.7. By taking the average values of the distances and assuming that the sulphonamide nitrogen is directly bound to the metal ion, the model in Fig. 10.16 represents *a unique conformation*. However, as Lanir and Navon (1972) point out, other structures in which

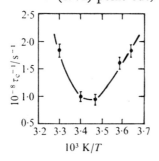

FIG. 10.15. Temperature dependence of τ_c^{-1}. (From Lanir and Navon 1972. © 1972 Amer. Chem. Soc.)

TABLE 10.7

Experimental and calculated parameters for sulphacetamide bound to Mn(II)–carbonic anhydrase B (From Lanir and Navon, 1972. © 1972 Amer. Chem. Soc.)

(°C)	CH₃	$T_{1,P}/T_{2,P}$ H_{ortho}†	H_{meta}†	$\tau_M \times 10^5$ (s)	$T_{1,M}/T_{2,M}$	$\tau_c \times 10^9$ (s)	CH₃	Distance in Å from Mn(II) H_{ortho}†	H_{meta}
30	1·95	5·1	6·1	1·35	9	5·45±0·5	4·5	5·8	6·8
21	3·4	9·9	14·5	1·68	28	10·1±1·0	4·6	5·8	6·8
15	3·9	8·1	13·3	1·94	32	10·9±1·1	4·6	5·3	6·4
6	1·9	3·1	4·7	2·35	11	6·2±0·6	4·6	5·5	6·3
2	1·9	3·9	3·5	2·53	9	5·45±0·5	4·8	5·75	6·3

† H_{ortho} and H_{meta} refer to the phenyl hydrogen position in relation to the sulphonamide group in the inhibitor.

the metal ion is not directly bound to the sulphonamide nitrogen are also consistent with the observed results and can not be ruled out.

In this example the possibility of rapid rotation of the methyl group is not considered although it probably occurs. With hindsight we can see that because of the distances involved the angle of the cone formed by the rotation of the methyl group about the paramagnetic ion–methyl group axis is very small (see Fig. 9.5). The reduction in dipolar coupling is thus negligible.

The value of $1/\tau_M$ can, as usual, be equated with the dissociation rate constant k_{off}. By use of the relationship $K_s = k_{off}/k_{on}$, a value of $k_{on} = 9 \times 10^6$ M^{-1} at 25 °C is derived. This value is very similar to the binding rate constants for various unsubstituted sulphonamides to the zinc enzyme. The independence of the value of k_{on} for the various sulphonamides and for the different metals can be reconciled with a mechanism for complex formation that involves as the initial step, only the hydrophobic interaction of the benzene ring with the enzyme. As we saw in the example on p. 132, such an interaction has, in fact, been detected by n.m.r. spectroscopy.

Finally, in the analyses we turn to the temperature dependence of τ_c (Fig. 10.15). The value of τ_M is (ca. 10^{-5}) much longer than that of τ_c (ca.10^{-8})

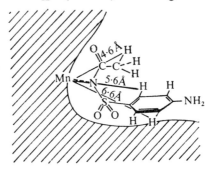

FIG. 10.16. Possible structure of the carbonic anhydrase–sulphacetamide complex. (From Lanir and Navon 1972. © 1972 Amer. Chem. Soc.)

and thus

$$\frac{1}{\tau_c} = \frac{1}{\tau_S} + \frac{1}{\tau_R}.$$

(10.22)

The shape of the curve for $1/\tau_c$ versus $10^3 K/T$ (Fig. 10.15) could arise from $1/\tau_c$ being controlled by $1/\tau_R$ at high temperatures and $1/\tau_S$ at low temperatures. The minimum would arise because τ_S has an opposite temperature dependence to τ_R, as discussed initially. If we assume that, at the minimum, $1/\tau_c$ contains equal contributions from $1/\tau_R$ and $1/\tau_S$ then the value of either of these is ca. 2×10^8 s. This value of τ_R can be compared with the value of ca. 1.5×10^8 s estimated for the tumbling time of the whole molecule; this adds support to this interpretation. The value of τ_S also allows us to obtain some information on τ_v. For a 'negative' temperature coefficient for τ_S $\omega_S \tau_v < 1$. Thus $\tau_v < 1/\omega_S$ which at 60 MHz means that $\tau_v < 2.5 \times 10^{11}$ s (which fits well with the values for other systems discussed previously). However, it is worth stressing that in this particular example it is not necessary to know the explicit nature of τ_c to calculate the distances. The τ_c was determined directly by the $T_{1,P}/T_{2,P}$ measurements of the substrate resonances. We shall meet some examples later where these measurements are not always carried out and the value of τ_c is obtained in an indirect manner. In such cases it *is* necessary to know or to make some assumptions about the explicit nature of τ_c.

The above example illustrates a very important point in that each group of hydrogens can exhibit a different region of chemical exchange. Thus the methyl hydrogens of the acetyl group exhibit mainly 'slow exchange' behaviour. The phenyl hydrogens *ortho* to the sulphonamide group show intermediate behaviour while the *meta*-phenyl hydrogens are on the fast–intermediate/borderline of chemical exchange. The behaviour clearly depends on the relative magnitudes of τ_M and $T_{2,M}$ for each group. The nearer the group to the Mn(II), the smaller the value of $T_{2,M}$ for that group. If fast-exchange conditions apply to this group they must of course apply to all other groups. Conversely, if slow-exchange conditions apply to the group furthest from the Mn(II), they must apply to all other groups. In some experiments it may only be convenient to check this condition with one or two groups and it may be helpful to bear these simple criteria in mind. It is often useful, too, to remember that even if slow-exchange conditions do apply *upper limits* for r can be obtained (since $\tau_M > T_{2,M}$) if a value of τ_c is assumed (or determined for another group in the same molecule).

10.8. The binding of several inhibitors to Mn(II)–carboxypeptidase A (Navon, Shulman, Wyluda, and Yamane, 1968)

This was one of the original examples of distance calculations and is included mainly for historical reasons and also because it presents a useful exercise for the interested student. The high resolution spectra of various inhibitors changed in the presence of Mn(II)–carboxypeptidase A.

For the strongest inhibitor, indoleacetic acid, the value of the linewidths and thus $P_M T_{2,P}$ was found to be the same for all hydrogens. One explanation is that slow exchange conditions apply to all the hydrogens. Navon *et al.* (1968) confirmed that this was the case by variable-temperature studies, in which the value of $1/T_{2,P}$ was shown to decrease with increasing temperature. An upper limit on the Mn—H distances of the inhibitor can be obtained from the inequality $T_{2,M} < (T_{2,M} + \tau_M)$; however a value of τ_c is required Reference to Fig. 10.3 suggests that the value of $\tau_{1,S}$ at 60 MHz is ca. 7×10^{-9} s.

TABLE 10.8

Experimental values† of $P_M(T_{2,P}) = (T_{2,M} + \tau_M)$ for Mn(II)–carboxypeptidase inhibitor complexes and calculated values of r and τ_M.

| Inhibitor | $(T_{2,M} + \tau_M)$ (s) | | Upper limits of r (Å) | | τ_M (s) |
	—CH$_2$—	Other hydrogens	—CH$_2$—	Other hydrogens	
Indoleacetate	$2 \cdot 1 \times 10^{-4}$	$2 \cdot 0 \times 10^{-4}$	8·7	8·7	$2 \cdot 0 \times 10^{-5}$
C(CH$_3$)$_3$CH$_2$CO$_2^-$	$7 \cdot 4 \times 10^{-5}$	$1 \cdot 5 \times 10^{-4}$	7·4	8·2	$7 \cdot 4 \times 10^{-4}$
Br·CH$_2$·CO$_2^-$	$2 \cdot 3 \times 10^{-5}$		6·1		$2 \cdot 3 \times 10^{-5}$
CH$_3$·O·CH$_2$·CO$_2^-$	$4 \cdot 5 \times 10^{-6}$	$7 \cdot 2 \times 10^{-6}$	4·5	5·3	$< 4 \cdot 5 \times 10^{-6}$

† Assumed to be at 60 MHz, although the original paper does not make this clear.

and thus, combining this with $\tau_R = 1 \cdot 5 \times 10^{-8}$, gives the value of τ_c as ca. $0 \cdot 33 \times 10^{-8}$ s. This value has been used in the expression for $T_{2,M}$ to calculate the upper limits given in Table 10.8. Strictly speaking, the value of $\tau_{1,S}$ could well be different for each inhibitor, but we have ignored this complication.

In the case of t-butylacetic acid, the value of $(T_{2,M} + \tau_M)$ for the t-butyl group is longer than that of the methylene hydrogens. Perhaps this is because co-ordination to the Mn(II) occurs with the carboxylate group attached directly to the metal, so that slow-exchange conditions do not completely apply to the C(CH$_3$)$_3$ hydrogens, because a larger value of r gives a larger value of $T_{2,M}$ i.e. $T_{2,M(C(CH_3)_3)} \sim \tau_M$. The temperature variation of the relaxation times for the CH$_2$ hydrogens indicated that slow-exchange conditions apply Using a value of τ_M, $= 7 \cdot 4 \times 10^{-5}$ s, then $T_{2,M(CH_3)_3)} = 7 \cdot 6 \times 10^{-5}$ s, yielding a value of $r = 7 \cdot 4$ Å for the t-butyl hydrogens. (The upper limit for r in the table is obtained by assuming $T_{2,M(C(CH_3)_3)} = 1 \cdot 5 \times 10^{-4}$ s.) This value of r should be compared with a value of ca. 7·6 Å obtained from a model where the inhibitor has its carboxylate group attached directly to the Mn(II) ion.

For methoxyacetate, the weakest of the inhibitors here, fast-exchange conditions do apply and an upper limit for the distance of the Mn(II) to the methylene hydrogens is calculated as 4·5 Å, compared with a value of 4·8 Å from molecular models. The value of 5·3 Å for the methyl hydrogens is only consistent with the molecular model of the carboxylic acid in the first co-ordination sphere for a bent configuration. Thus it would seem from the experimental results that these inhibitors bind to Mn(II)–carboxypeptidase with the free carboxylate group in the first co-ordination sphere of the Mn(II) ion. However, the X-ray model (Reeke, Hartsuck, Ludwig, Quiocho, Steitz, and Lipscomb, 1967) shows that for the substrate, the carbonyl of the penultimate residue, rather than the terminal carboxyl group, is bound to the metal.

10.9. The thallium–pyruvate kinase (PK)–Mn(II) and thallium–(PK)–phosphoenolpyruvate–Mn(II) complexes (Reuben and Kayne, 1971)

This is an important example for it provided the first direct evidence for a change in the distance between the two metal ion activators, Tl(I) and Mn(II), of pyruvate kinase (PK) on addition of the substrate, phosphoenolpyruvate. It also extended the use of Tl(I) as a probe for monovalent cations such as K(I). Previous measurements of hydrogen and phosphorous resonances by Manners, Morallee, and Williams (1970) on diamagnetic Tl(I)–nucleotide complexes had indicated the potential use of Tl(I) as a probe for K(I), but the results by Reuben and Kayne (1971) were the first Tl(I) n.m.r. measurements in a biological system.

The addition of Mn(II) to a solution of $TlNO_3$ containing pyruvate kinase results in a broadening of the ^{205}Tl resonance. This broadening is dramatically increased by addition of phosphoenolpyruvate. By use of the progressive saturation technique, the value of the spin–lattice relaxation rate was obtained. In all cases the paramagnetic-ion contribution to the spin–spin relaxation rate, $1/T_{2,P}$, was much greater than that for the spin–lattice relaxation rate, $1/T_{1,P}$. In this work variable-temperature measurements were not made but from the inequivalence of $1/T_{1,P}$ and $1/T_{2,P}$ it was concluded that, at least for $1/T_{1,P}$, fast-exchange conditions are valid. The value of $1/T_{2,P}$ then allows upper limits for the residence time of ^{205}Tl on the enzyme to be obtained by writing $\tau_M = T_{2,M}$ in the usual equation:

$$\frac{1}{T_{2,P}} = \frac{P_M}{T_{2,M} + \tau_M} \tag{10.23}$$

For the thallium–enzyme–Mn(II) complex $\tau_M < 1·6 \times 10^{-5}$ s. (Actually from the dissociation constant of the enzyme–thallium complex and assuming a diffusion controlled association step, τ_M can be estimated as 10^{-7} s).

There next arises the problem of choosing the correlation time to calculate the thallium–manganese distances. As usual this may be τ_R, τ_M, or $\tau_{1,S}$. τ_R can be estimated as 10^{-7} s (see p. 242) and from above τ_M is probably of similar magnitude. As with the previous example, the value of $\tau_{1,S}$ can be

17

estimated from analysis of the water solvent relaxation rates in the Mn(II)–pyruvate kinase system. At the frequency of these experiments (ca. 24 MHz) it is ca. 0.94×10^{-8} s. The values calculated for the thallium–manganese distance is 8·2 Å.

For the Mn(II)–pyruvate kinase–phosphoenolpyruvate complex, the value of $\tau_{1,S}$ is unknown. Reuben and Kayne (1971) state that an analysis of the e.s.r. spectrum of this complex and the Mn(II)–pyruvate kinase complex indicate that the values of τ_S are of the same order of magnitude. Using a value of $\tau_S = 0.94 \times 10^{-8}$ s the Mn(II)–thallium distance in the phosphoenol–pyruvate complex is 4·9 Å.

In many examples such as this there may be insufficient instrumentation available to carry out a complete analysis, such as we have seen in some of the preceding examples. In such cases ambiguities in interpretation are bound to arise. In this example Reuben and Kayne (1971) point out that the expected T_{1P}/T_{2P} ratio is 2·52 assuming fast-exchange conditions and only considering the dipolar contributions to $1/T_{1,P}$ and $1/T_{2,P}$. The *observed* ratio is 10·1 for the ternary complex and 13·8 for the quaternary complex. Two explanations can be put forward to explain the discrepancy: (1) fast-exchange conditions are not valid for $1/T_{2,P}$ or (2) fast-exchange conditions do apply but there may be a significant hyperfine contribution to $1/T_{2,M}$. Reuben and Kayne (1971) favoured the latter explanation since, in the absence of the enzyme, similar differences in $1/T_{1,P}$ and $1/T_{2,P}$ were obtained, but the conclusion is obviously tentative.

It is interesting to compare the expected paramagnetic effects on the relaxation rates of the alkali metals with those for ^{205}Tl. For a solution

TABLE 10.9

Some properties of nuclei of alkali metals (and thallium-205) and expected paramagnetic relaxation effects due to interaction with manganese(II) in the presence of pyruvate kinase

Nucleus	Spin, I	Natural abundance (%)	Relative sensitivity at constant field relative to $H = 1.000$	$(\gamma_I/\gamma_{205\text{Tl}})^2$	$1/T_{1,P}$† (s^{-1})	$1/T_{2,P}$‡ (s^{-1})
^{7}Li	$\frac{3}{2}$	92·57	0·294	0·4535	3·6	9·2
^{23}Na	$\frac{3}{2}$	100	0·0927	0·2100	2·28	5·74
^{39}K	$\frac{3}{2}$	93·08	5.08×10^{-4}	0·0065	0·1	0·25
^{87}Rb	$\frac{3}{2}$	27·2	0·177	0·3215	3·0	7·6
^{133}Cs	$\frac{7}{2}$	100	0·047	0·0516	0·7	1·8
^{205}Tl	$\frac{1}{2}$	70·48	0·192	1	5·1	12·9

† The relaxation rates are calculated at constant field (equivalent to a resonance frequency of 24·3 MHz for Tl), with $\tau_s = 0.94 \times 10^{-8}$ s, for a solution containing 0·1 M alkali ion, 42·6 μM pyruvate kinase, and 0·94 mM Mn(II).

‡ Assuming dipolar contributions only.

containing 0·1 M thallium nitrate, 42·6 μM pyruvate kinase and 0·92 Mn(II) the *observed* values of $1/T_{1,P}$ and $1/T_{2,P}$ are 5·1 s^{-1} and 51·4 s^{-1} respectively. If we assume that dipolar contributions only were significant in determining $1/T_{2,P}$, then, for a value of $\tau_c = 0·94 \times 10^{-8}$ s, $T_{1,P}/T_{2,P} = 2·52$ whence $1/T_{2,P} = 12·9$ s^{-1}. For the alkali metals, we shall make the assumption that only dipolar effects contribute. The values of $1/T_{1,P}$ obtained are listed in Table 10.9. We note that to a first approximation they may be obtained by multiplying the values for ^{205}Tl by $(\gamma_I/\gamma_{205_{Tl}})^2$, but a more exact calculation takes into account the functional dependence of the correlation times in the expression for $1/T_{1,M}$, i.e. terms like $\tau_c/(1+\omega_I^2\tau_c^2)$. The actual situation is however not likely to be so simple, because all the other nuclei except ^{205}Tl possess quadrupole moments. It is possible that any binding to an enzyme may alter the electric-field gradients sufficiently so that the resonance become so broad as to make the detection of any additional paramagnetic effects very difficult.

10.10. Methods for obtaining τ_c

We have seen that one of the major problems in the calculation of distances is obtaining an accurate value of τ_c, and we end this section on molecular motion by summarizing various methods for its determination.

10.10.1. *From measurements of* $T_{1,M}$ *and* $T_{2,M}$

Under conditions of *fast* chemical exchange:

$$\frac{1}{P_M q T_{i,P}} = \frac{1}{T_{i,M}}$$

(10.24)

$$i = 1, 2.$$

If $1/T_{i,P}$ is measured for each nucleus, it is possible to evaluate τ_c from the ratio $T_{1,M}/T_{2,M}$ making the *assumption* that the hyperfine terms contribute little to the relaxation rates. We have from eqns (9.1) and (9.2)

$$\frac{T_{1,M}}{T_{2,M}} = \frac{4+3/(1+\omega_I^2\tau_c^2)+13/(1+\omega_S^2\tau_c^2)}{6/(1+\omega_I^2\tau_c^2)+14/(1+\omega_S^2\tau_c^2)}.$$

(10.25)

A plot of this ratio at two frequencies is shown in Fig. 10.17 and we note that this method is really only sensitive for values of $\tau_c \geqslant 1/\omega_I$.

This is probably the best method and should be used whenever possible. The assumption of fast chemical exchange should be checked where possible for we shall see later that in some examples, e.g. the ^{31}P resonances of AMP in the Mn(II)–AMP complex, that fast exchange is valid only for the $1/T_{1,P}$ measurements. (See also Section 9.8.2.)

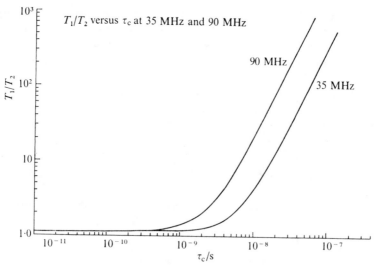

FIG. 10.17. Plot of T_1/T_2 versus τ_c at 35 and 90 MHz. Note that this ratio is only really sensitive to the value of τ_c when $\tau_c \geqslant \omega_I$.

10.10.2. *From the frequency dependence of the spin–lattice relaxation rates*

We have seen that it is possible to obtain a value of τ_o from the frequency dependence of the spin–lattice relaxation rates but the computation is difficult if τ_o is frequency dependent. If τ_o is frequency independent then from eqn (10.11) we have a straightforward analysis since

$$T_{1.P} = \frac{1}{P_M q \, D\tau_o} (1 + \omega_I^2 \tau_o^2) \tag{10.26}$$

and as we have already noted the ratio of slope/intercept is $1/\tau_o^2$.

This method obviously only works well when $1 \gg \omega_I^2 \tau_o^2$. It has the drawback that to obtain accurate τ_o values several frequencies are required and that *a priori* it may not be possible to know that τ_o is really frequency independent.

10.10.3. *By use of Stokes' law*

In macromolecular systems, whatever the mechanism determining τ_o, it is unlikely to be longer than the rotational diffusion time of the whole molecule. Thus an *upper* limit on τ_o (and thus r) can be obtained.

The rotational diffusion time for the whole molecule can be calculated from Stokes' Law, viz.

$$\tau_R = \frac{4\pi\eta a^3}{3kT} \tag{10.27}$$

where η is the viscosity of the solvent, a the effective radius of the complex, k the Boltzmann constant and T the absolute temperature. We can rewrite eqn (10.27) as

$$\tau_R = M\bar{V}\eta/RT \tag{10.28}$$

where M is the molecular weight of the complex, $\bar{V}$ is the partial specific volume (≈ 0.72 cm^3 g^{-1} for most enzymes). There is a slight drawback in using eqn (10.28) in that the value of M makes no allowance for any hydration of the enzyme, i.e. in solution it is the effective value of a that is required. Equation (10.27) is often referred to as the Stokes–Einstein equation.

10.10.4. *From e.s.r. spectra*

In some cases it is possible to observe the e.s.r. spectrum of the paramagnetic complex and values of $\tau_{2,S}$ can sometimes be estimated from the spectrum. Since $\tau_{1,S} > \tau_{2,S}$ we may obtain a lower limit for $\tau_{1,S}$. If $\tau_{1,S}$ is the relevant correlation time then we can calculate a *lower* limit for r.

10.10.5. *From measurements of the relaxation rates of the solvent water in the system*

This method has as its basis the assumption that the correlation time for the water hydrogens, τ_0, in the first hydration sphere of the paramagnetic ion is the τ_c applicable to the other nuclei under observation. The correlation time may be determined by measuring $(T_{1,P}/T_{2,P})$ as in method 1 or by use of the relaxation enhancement of the water hydrogens caused by the addition of the macromolecule, ligand, etc. The basis of the method is discussed in Chapter 11 on Proton Relaxation Enhancement and only a brief outline will be given here. We shall illustrate the calculation initially for the Mn(II)–AMP complex and then consider what extra considerations have to be taken into account for macromolecular systems.

The method involves measuring the relaxation rates of the water hydrogens of the metal complex and comparing these with the relaxation rates of an aqueous solution of the metal. If we *assume* that fast exchange of the water hydrogens must be valid in both cases, then a charactertistic enhancement parameter ε_b for the metal complex is defined as (see p. 249).

$$\varepsilon_b = \frac{q^* \, T_{1,M}}{q \, T_{1,M}^*} \tag{10.29}$$

where the asterisk† denotes the relaxation rate of water hydrogens in the complex. q^* and q are the appropriate hydration numbers of the metal ion. The following further assumptions are then made.

(1) The M—OH$_2$ distance is unchanged in the metal complex from its value in pure aqueous solution.

---

† The use of an asterisk in this section and throughout Chapter 11 denotes parameters relating to the presence of a macromolecule. Strictly speaking this convention should have been used in the earlier sections of this chapter—but it seemed unnecessary.

(2) The hyperfine contribution to $1/T_{1.M}^*$ and $1/T_{1.M}^*$ can be neglected. We then obtain from eqns (9.1) and (10.29)

$$\varepsilon_b = \frac{q^* f(\tau_c^*)}{q\ f(\tau_c)} \tag{10.30}$$

where

$$f(\tau_c) = \frac{3\tau_c}{(1+\omega_I^2\tau_c^2)} + \frac{7\tau_c}{(1+\omega_S^2\tau_c^2)}.$$

A similar equation applies for $f(\tau_c^*)$. To solve eqn (10.30) for $f(\tau_c^*)$ we need to know q^*, q, and $f(\tau_c)$. $f(\tau_c)$ may be obtained from an analysis of the relaxation rates of the pure aqueous solution of the paramagnetic ion for which q is usually known, e.g. for Mn(II), $q = 6$ and τ_c (300 K) $= 3 \times 10^{-11}$ s. The value for q^* is clearly less than q, for metal binding to the ligand to occur and either the value has to be guessed or found by other types of measurements (e.g. from the $T_{1,P}/T_{2,P}$ ratio see p. 241). This obviously is one of the drawbacks of the method.

From the analyses of relaxation rates of solvent water protons in section 10.2 we note that $f(\tau_c^*) > f(\tau_c)$ which provides the physical basis for the enhancement. We return to this and the methods of measuring enhancement parameters in more complex systems in Chapter 11. For the present, we note that the value of ε_b will be frequency (and temperature) dependent and Fig. 10.18 shows plots of ε_b versus τ_c^* at two frequencies, for the Mn(II) system using $q = 6$, $\tau_c = 3 \times 10^{-11}$ s and $q^* = 4$. Because there is a maximum in the curve, obtaining a unique value of τ_c^* may be difficult and we may

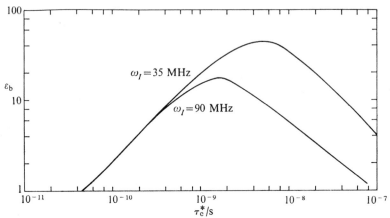

FIG. 10.18. Variation of ε_b with correlation time at 35 and 90 MHz for Mn(II) systems, at 298 °C assuming $q = 6$, $q^* = 4$ and $\tau_c = 3 \times 10^{-11}$ s. Note that the curves have a maximum when $\tau_c^* = 1/\omega_I$.

require measurements of ε_b at different frequencies or measurements of the spin–spin relaxation rates.

The experimental value of ε_b for the Mn(II)–AMP complex is 2. If we assume that the Mn(II) co-ordinates to the AMP via two oxygen molecules then $q^* = 4$ and a value of $\tau_c^* \approx 9 \times 10^{-11}$ s is obtained. In this example it is reasonable to assume that the value of τ_c^* is, in fact, the rotational correlation time of the complex. Thus if Mn–AMP distances are required, the above value of τ_c^* may be used in the analysis of the AMP relaxation rates Notice that method 1 is really insensitive in this range of τ_c^* values for most conventional frequencies.

There are considerable dangers in applying the same method to the analysis of ligand relaxation rates in ternary metal–macromolecule–ligand complexes. We have seen that in the examples of analyses that at high frequencies the correlation time for the relaxation behaviour of the relaxation rates of water hydrogens contains a considerable contribution from τ_M^*. Ligand exchange rates are generally slower than hydrogen exchange rates and thus the value of τ_c^* calculated in this way may be totally inapplicable to the ligand nuclei. The relevant τ_c^* for the ligand nuclei will contain τ_S^* and/or τ_R^*. However for the water hydrogens, τ_S^* is mainly important at low frequencies. Thus the method could be used to determine a τ_c^* at low frequencies which would be more applicable to the ligands. (It must be remembered though that τ_S^* may be frequency dependent.)

This method has been widely used, despite the obvious difficulties, and it is for this reason that we have given such a detailed discussion. In general, if the value of τ_c^* determined in this way contains a significant contribution from τ_M^* for the water hydrogens, the values of τ_c^* for the ligand nuclei will be too small, and consequently values of metal ligand distances in macromolecules obtained in this way will represent lower limits.

Appendix

Chemical-exchange spin decoupling

Many of the experiments described in this Chapter involve measuring the broadenings of resonances caused by the addition of paramagnetic ions. In cases where the resonances are coupled to a second nucleus such that spin–spin multiplets are present, the addition of paramagnetic ions may actually lead to an apparent narrowing of the multiplet.

Consider two nuclei both of spin $I = \frac{1}{2}$. The spectrum of each nuclei will exhibit a doublet splitting, with a spin–spin coupling constant J_{AX}. If nucleus A is closer to the paramagnetic ion than is X, we then have that $1/T_{1,M}^A > 1/T_{1,M}^X$ and of course $1/T_{1,P}^A > 1/T_{1,P}^X$. The value of $1/T_{1,p}$ depends on the concentration of added paramagnetic ion. If the addition of a paramagnetic ion results in the condition $1/T_{1,P}^A < J_{AX}$ then effective spin decoupling of B from A will occur. (In other words, the nuclei B will only sense the average value of the field associated with A, see p. 363.)

The condition $1/T_{1,P}^A < J_{AX}$ may be achieved in stages. Initially, on adding the paramagnetic ion there is a broadening of the X resonances, followed by coalescence and then narrowing, as the concentration of paramagnetic ion increases. The effects are thus very similar to chemical exchange and the term 'chemical-exchange spin decoupling' has been given to this phenomenon (Frankel, 1970) to distinguish it from conventional spin-decoupling experiments in which a second radio-frequency is employed.

The analysis involves substituting $1/2T_{1,M}^A$† for $1/\tau_M$ in the usual equations for chemical exchange (Carrington and McLachlan, 1967). Of course the nature of the situation varies with the complexity of the multiplet. However, we include this as a cautionary note that 'unexpected' line-widths of some resonances can occur if spin–spin multiplets are present.

The problem may be easily overcome by conventional spin decoupling experiments thereby removing the multiplet before addition of the paramagnetic ion. Alternatively, in the absence of conventional decoupling, measurements may also be carried out at either of the two extremes, i.e. when the splittings are completely resolved or when 'chemical-exchange' decoupling is complete.

Chemical-exchange spin decoupling is in general unimportant in measurements of spin–lattice relaxation rates.

References

Carrington, A. and McLachlan, A. D. (1967). Introduction to Magnetic Resonance—Harper and Row. (New York & London). Chapter 12.

Cohn, M. (1970). Quart. Rev. Biophys. 3, 61.

Danchin, A. and Guéron, M. (1970). J. chem. Phys. 53, 3599.

Dwek, R. A., Malcolm, A. D. B., Radda, G. K., and Richards, R. E. (1970). Unpublished results.

Frankel, I. S. (1970). J. chem. Phys. 50, 943.

Lanir, A. and Navon, G. (1972). Biochemistry, 11, 3536.

Manners, J. P., Morallee, K. G., and Williams, R. J. P. (1970). Chem. Comm. 965.

Milner-White, E. J. and Watts, D. C. (1971). Biochem. J. 122, 727.

Navon, G. (1970), Chem. Phys. Letts. 7, 390.

——, Shulman, R. G., Wyluda, B. J., and Yamane, T. (1968). Proc. natn. Acad. Sci. 60, 86.

Peacocke, A. R., Richards, R. E., and Sheard, B. (1969). Molec. Phys. 16, 177.

Reed, G. H. and Cohn, M. (1972). J. biol. Chem. 247, 3073.

——, Diefenbach, H., and Cohn, M. (1972). J. biol. Chem. 247, 3066.

Reeke, G. N., Hartsuck, J. A., Ludwig, M. L., Quiocho, F. A., Steitz, T. A., and Lipscomb, W. N. (1967). Proc. natn. Acad. Sci. 58, 6220.

Reuben, J. and Kayne, F. J. (1971). J. biol. Chem. 246, 6227.

—— and Cohn, M. (1970). J. biol. Chem. 245, 662.

Taylor, P. W. and Burgen, A. S. V. (1971). Biochemistry, 10, 3859.

——, Feeney, J., and Burgen, A. S. V. (1971). Biochemistry, 10, 3866.

---

† The probability of a nucleus ($I = \frac{1}{2}$) changing its spin state is $1/2T_{1,M}^A$. The paper by Frankel (1970) ignores the factor 2.

11

PROTON RELAXATION ENHANCEMENT

11.1. Introduction

THE examples given in Chapter 10 show that the relaxation rates of the solvent water protons in the first hydration sphere of a paramagnetic ion may be increased when the metal ion is bound to a macromolecule. The analysis shows that the main reason for the enhancement of the relaxation rates results from changes in the correlation time for the dipolar interaction τ_c^*. In this Chapter we consider more fully some of the uses of measuring the solvent water proton relaxation rates. This is a particularly attractive starting point for studying macromolecules since the concentration of water molecules is ca. 55·5 M and thus there are few problems with sensitivity of the spectrometer.

The presence of the macromolecule (E) to which the metal (M) can bind introduces the equilibrium:

$$M + E \rightleftharpoons ME.$$

Thus we now have to consider water protons in three environments: (1) those in the bulk; (2) those bound to the 'free' paramagnetic ion (designated the f-site) which has a concentration M_f; and (3) those bound to the paramagnetic ion on the macromolecule (designated the b-site) which has a concentration M_b. If we assume that M_b and M_f are small (as is the case experimentally), then the concentration of water molecules at each of these two sites is small compared with the concentrations of water hydrogens in the bulk solvent. We may then neglect any chemical exchange of water hydrogens between the two paramagnetic ion sites, and consider only chemical exchange between the bulk solvent at each of these two sites. In such a case, the paramagnetic contribution to the observed relaxation rates will be the sum of the effects in the two sites, i.e.

$$\frac{1}{T_{i,P}^*} = \left(\frac{1}{T_{i,P}^*}\right)_f + \left(\frac{1}{T_{i,P}^*}\right)_b \tag{11.1}$$

where $i = 1$ or 2 and f and b refer to the two environments. The asterisk denotes the presence of the macromolecule. $1/T_{i,P}^*$ is related to the observed relaxation rate $1/T_{i,\text{obs}}^*$ by the usual equation:

$$\frac{1}{T_{i,P}^*} = \frac{1}{T_{i,\text{obs}}^*} - \frac{1}{T_{i,(0)}^*} \tag{11.2}$$

where $1/T_{i,(0)}^*$ is the relaxation rate of the water protons in a similar solution containing the enzyme but no paramagnetic ion, i.e. the diamagnetic control.

In order to proceed further, it is necessary to have explicit forms for $1/T^*_{i,P}$. We shall restrict the discussion to the case where there is a definite hydration sphere for which the Solomon–Bloembergen equations describe correctly the relaxation behaviour. We then need to know the fraction of water molecules in each environment. These are given by

$$\frac{M_f q}{N_P} ; \frac{M_b q^*}{N_P}$$

where q and q^* are the number of water molecules co-ordinated in the first hydration sphere of the metal ion when free and when bound to the macromolecule. N_P is the total concentration of water molecules in the system (55·5 M).

The explicit form for $1/T^*_{1,P}$ is from eqns (9.2) and (9.25)

$$\frac{1}{T^*_{1,P}} = \left(\frac{1}{T^*_{1,\text{obs}}} - \frac{1}{T^*_{1,(0)}} \right) = \frac{M_b q^*}{N_P} \left(\frac{1}{T^*_{1,M} + \tau^*_M} \right)_b + \frac{M_f q}{N_P} \left(\frac{1}{T^*_{1,M} + \tau^*_M} \right)_f \quad (11.3)$$

where b and f refer to the two environments.

The corresponding equation, when no macromolecule is present is, of course,

$$\frac{1}{T_{1,P}} = \frac{1}{T_{1,\text{obs}}} - \frac{1}{T_{1,(0)}} = \frac{M_f q}{N_P} \left(\frac{1}{T_{1,M} + \tau_M} \right) \quad (11.4)$$

where $M_t = M_b + M_f$. (Similarly, equations can be written for $1/T_{2,P}$ and $1/T^*_{2,P}$.) It is noted that outer-sphere relaxation has been neglected. If only the value of $1/T^*_{1,P}$ is required, any outer-sphere contribution will not affect any analysis. However when equations such as those above containing explicit forms for $1/T^*_{1,P}$ are written outer-sphere terms should, strictly speaking, be included.

11.2. Relaxation enhancement

Since relaxation rates tend to be large it is more convenient to refer these rates to a corresponding solution of the paramagnetic ion in which the macromolecule is absent. Thus it is usual to define the relaxation enhancement factor, ε^*, for a solution as

$$\varepsilon^*_i = \frac{1/T^*_{i,P}}{1/T_{i,P}} = \frac{1/T^*_{i,\text{obs}} - 1/T^*_{i,(0)}}{1/T_{i,\text{obs}} - 1/T_{i,(0)}} \quad (11.5)$$

where $i = 1$ or 2 and the asterisk indicates the presence of a macromolecule. In the following discussion, for simplicity, we shall only consider the enhancement of the spin–lattice relaxation times. (Similar considerations will also apply to the enhancement of the spin–spin relaxation time.)

From the use of eqns (11.3) and (11.4) in eqn (11.5) we obtain:

$$\varepsilon^* = \frac{M_b}{M_t}\varepsilon_b + \frac{M_f}{M_t}\varepsilon_f \tag{11.6}$$

where M_t represents the total ion concentration. ε_f is given by

$$\varepsilon_f = \frac{(T_{1,M}+\tau_M)}{(T_{1,M}^*+\tau_M^*)} \tag{11.7}$$

and is assumed to be unity. The predominant assumption here is that the microviscosity of the regions surrounding the free paramagnetic ions is identical in the solutions in which macromolecules are present or absent. ε_b is given by

$$\varepsilon_b = \frac{q^*}{q} \cdot \frac{(T_{1,M}+\tau_M)}{(T_{1,M}^*+\tau_M^*)} \tag{11.8}$$

ε_b is called the *characteristic enhancement of the metal–macromolecule complex* and corresponds physically to the situation of measuring an enhancement of a solution in which all the metal is bound to the macromolecule.

Equation (11.8) may be rewritten as:

$$\varepsilon^* = x_b\varepsilon_b + x_f\varepsilon_f \tag{11.9}$$

where x_b and x_f are the mole fractions of bound and free manganese.

Thus we see that the enhancement of the water protons depends on the amount of metal ion bound to the macromolecule and thus, like any other spectroscopic technique where a change in a measured parameter, A, is observed in the presence of different amounts of B, the dissociation constant for AB can be obtained.‡ We note that if eqn (11.9) is valid, no further assumptions are needed in order to obtain binding parameters and, of course, it is not necessary to know the explicit cause of the enhancement. However, there is an advantage of this technique over most others, for the characteristic enhancement of AB, ε_b, contains information on the molecular motion of the water hydrogens around the metal binding site. Changes at the metal site from addition of substrates or from variations in pH or temperature, for example, may cause changes in the relaxation rates of the water protons, most conveniently expressed in the enhancement parameter.

11.2.1. The nature of ε_b

Clearly, for binding $q^* < q$ and if this were the only parameter to change, then there would be a decrease in the relaxation rate of the water protons in the presence of the macromolecule. (Indeed if it can be shown that no other

‡ Assuming that the change in A reflects the binding of B directly rather than (say) some conformational change in the macromolecule.

factors change then this could be a useful method for estimating co-ordination numbers.) As we have seen in Chapter 10 with macromolecules this is not, in general, the case. The remainder of the expression for ε_b involves $(T^*_{1,M})$, (τ^*_M), and the corresponding terms for the free aquo-ions. We proceed by considering two limiting cases to apply in eqn (11.8)

(1) $(T^*_{1,M}) \gg (\tau^*_M)$ (fast exchange)

(2) $(\tau^*_M) \gg (T^*_{1,M})$ (slow exchange).

Fast exchange. The expression for ε_b is given by eqn (11.8)

$$\varepsilon_b = \frac{q^*}{q} \frac{(T_{1,M}+\tau_M)}{(T^*_{1,M})} \tag{11.10}$$

If we write eqn (9.1) as:

$$\frac{1}{T_{1,M}} = \frac{Kf_1(\tau_c)}{r^6}, \tag{11.11}$$

where $f_1(\tau_c) = 3\tau_c/(1+\omega_I^2\tau_c^2)+7\tau_c/(1+\omega_S^2\tau_c^2)$ and K is a constant for each particular nucleus then we may combine eqns (11.10) and (11.11) to obtain:

$$f_1(\tau^*_c) = \varepsilon_b \cdot \frac{q}{q^*} \cdot \frac{r^6}{K} \left(\frac{1}{T_{1,M}+\tau_M} \right) \tag{11.12}$$

If the M—OH$_2$ distance is known (or *assumed* to be unchanged from that in the pure aqueous solution) then $f_1(\tau^*_c)$ can be calculated, if q^* is estimated, since $q/(T_{1,M}+\tau_M)$ can be measured directly. In the majority of cases for solutions of metal ions at laboratory temperatures (q.v. Chapter 10) the relaxation rate of the water protons is such that $T_{1,M} \gg \tau_M$, i.e. fast-exchange conditions apply to the pure aquo-ion. Thus we obtain:

$$f_1(\tau^*_c) = \varepsilon_b \frac{q}{q^*} f_1(\tau_c) \tag{11.13}$$

which is the same as eqn (10.30).

We note from eqn (11.13) that, if $f(\tau^*_c) \gg f(\tau_c)$ then $\varepsilon_b > 1$. The reason that $f(\tau^*_c)$ may be greater than $f(\tau_c)$ is well illustrated by reference to the analysis of the Mn(II)–pyruvate kinase system discussed in some detail in Chapter 10. The analysis of the relaxation data for the complex allowed the calculation of the correlation times shown in Table 11.1, in which those for the free aquo ion are included for comparison. In the aquo ion the value of τ_c is determined by that of τ_R (the shortest correlation time). In the complex τ^*_c is determined by τ^*_S and τ^*_M—and τ_R no longer makes a contribution. Of course the mechanism responsible for relaxation does not *have* to change in

TABLE 11.1

Comparison of the correlation times for the relaxation behaviour of aqueous
$Mn(II)$ *solutions and solutions of* $Mn(II)$–*pyruvate kinase at* 25 °C *and* 20 MHz

	Aqueous Mn(II)‡ (s)	Mn(II)–pyruvate kinase† (s)	
τ_R	3×10^{-11}	$\sim 10^{-7}$	τ_R^*
τ_M	$2 \cdot 5 \times 10^{-8}$	4×10^{-9}	τ_M^*
τ_S§	$\sim 1 \times 10^{-8}$	$\sim 0 \cdot 7 \times 10^{-8}$	τ_S^* §
τ_c	3×10^{-11}	$2 \cdot 56 \times 10^{-9}$	τ_c^*
τ_e	$7 \cdot 2 \times 10^{-9}$	$2 \cdot 56 \times 10^{-9}$	τ_e^*
q	6	2	q^*

† From Navon (1970)
‡ From Bloembergen and Morgan (1961)
§ Frequency dependent

order that an enhancement may be observed. With enzymes of lower molecular weight than pyruvate kinase, it is quite likely that τ_R^* could still be important in determining τ_c^*. Of course in metal ligand complexes e.g. Mn(II)–nucleotide complexes the relevant correlation time is almost certainly that of rotation.

Slow exchange. To establish this question unambiguously necessitates measurements of the frequency and temperature dependence of both the relaxation times. One possible example to which it might apply, is Cr(III) which is inert kinetically. For the aquo-ion the half-life for exchange of *whole* water molecules is ca. 40 h at room temperature. Although the proton exchange rate is much faster (ca. 10^5 s^{-1} at 25 °C) slow-exchange conditions still apply (Luz and Shulman, 1965).

11.2.2. *Experimental questions for the measurement of* ε_b

We stress again that there is nothing magical in using an enhancement parameter *per se*, rather than the relaxation rates of the various solutions. The advantage we have mentioned: that of the enhancements being small numbers as opposed to the relaxation rates which are large numbers. However, experimentally, the use of enhancement parameters enables any systematic errors in the spectrometer and any errors in the metal ion concentration to be minimized. To obtain binding parameters the explicit nature of ε_b is unimportant, but it is obviously desirable to try to work under conditions where the largest enhancements can be observed. Thus from an experimental view three questions must be answered: (1) does ε_b vary with frequency? (2) what is the maximum value of ε_b at a given frequency? and (3) is it preferable to measure the spin-lattice or spin–spin relaxation enhancement?

The answers are found by inspection of eqn (11.12) and the corresponding one for $f_2(\tau_c^*)$. For convenience we will consider only the limiting cases of fast exchange applying to the water molecules in the two environments of the metal ion.

Essentially the answers have already been discussed, for ε_b reflects the ratio of the spin–lattice relaxation times. If $f_1(\tau_c^*)$ is independent of frequency then any frequency dependence of ε_b will arise from that of $f_1(\tau_c)$. In many systems, we have seen that τ_S^* makes a significant contribution to $f_1(\tau_c^*)$ and thus will be frequency dependent. As an example, consider the frequency dependence of the paramagnetic contribution to the spin–lattice relaxation times of solvent water of the ternary complexes of the Mn(II)–formyl–tetra-hydrofolate synthetase (Reed, Diefenbach, and Cohn, 1972). For comparison, we also show the frequency dependence of the pure aqueous Mn(II) solutions. A plot of the observed enhancement, ε^* versus frequency has a maximum at 24·13 MHz. In Fig. 11.1 the enhancements of the ATP and ADP complexes at 8·13 MHz are very similar while they differ considerably at 24·3 MHz. Thus some care is necessary in interpreting the values of ε_b or ε_t etc. at only one frequency. In the phosphorylase system, discussed later, we shall consider a scheme for the AMP activation of the enzyme based on the value of the enhancement parameter of solvent water hydrogens. Clearly, there can be dangers in such an approach unless the relaxation behaviour is studied in some detail. In this last example it is possible from other techniques to confirm the scheme (see pp. 273 and 294).

In general, however, the maximum value of the spin–lattice relaxation times occurs when $\tau_c^* = 1/\omega_I$ and thus $f_1(\tau_c^*)$ will have a maximum at this value (see Fig. 10.18). As an example of the magnitude of $(\varepsilon_b)_{max}$, consider the Mn(II) probe, for which $\tau_c = 3 \times 10^{-11}$ s (at 298 K) and $q = 6$ for the aquo-ion. If we assume $q^* = 5$ in a complex then $(\varepsilon_b)_{max} = 85$ at 20 MHz and 26 at 90 MHz. If τ_c is increased, e.g. by lowering the temperature, then the maximum value of ε_b is reduced correspondingly. Thus the higher the temperature the larger the possible maximum value of ε_b. By contrast the spin–spin relaxation rates do not have a maximum value, but will continue to increase as τ_c^* increases. This means that the relaxation rate enhancement one should measure depends on the nature of the metal ion and also the value of τ_c^*. Experimentally, the spin–lattice relaxation times are probably easier to measure. As an aside we note that if values of ε_b etc. are going to be used to calculate correlation times (see pp. 243–245) it is probably advisable to check if $\tau_c^* >$ or $< 1/\omega_I$. This can be done simply by measuring the ratio of the relevant water spin–spin and spin–lattice relaxation rates (see Chapter 10).

If we assume that $\tau_c^* < 1/\omega_I$ then for Mn(II) the spin–spin relaxation enhancement will be less than those of the spin–lattice relaxation enhancement. This is because in aqueous Mn(II) systems $1/T_{2,M} \gg 1/T_{1,M}$ since $\tau_e \gg \tau_c$ (see Chapter 9). The scalar term thus dominates $1/T_{2,M}$ and while τ_e changes relatively little in the macromolecular complex, τ_c may change considerably (see Table 11.1). The presence of the significant scalar term in the aquo ion

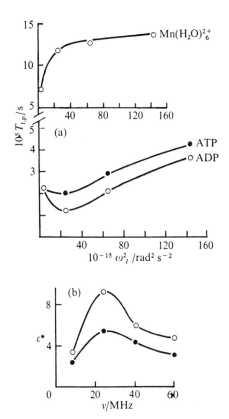

FIG. 11.1(a). Frequency dependence of the paramagnetic contribution to the molar relaxation times $T_{1,P}$ of the ternary E–Mn–nucleotide complexes of formyltetrahydrofolate synthetase. (b) ε^* as a function of frequency v; the upper curve ($\bigcirc$) refers to ADP and the lower curve ($\bullet$) to ATP. The solutions contained 0·1 mM $MnCl_2$, 0·5 M KCl, 0·2 M mercaptoethanol, 0·4 mM nucleotide, and enzyme (19·5 mg per ml) in 0·05 M HEPES–KOH buffer, pH8·0; $T = 21\ ^{\circ}C$. (From Reed et al., 1972.)

has the effect of 'masking' the dipolar enhancement. On the other hand, for a metal ion like Gd(III) there is little scalar interaction and in pure aqueous solutions $1/T_{1,M} \simeq 1/T_{2,M}$ and thus measurements of either relaxation rate should give almost the same enhancement (as long as $\tau_c^* < 1/\omega_I$).

11.3. Binding studies

The study of the proton relaxation of the water molecules around the paramagnetic ion in the presence of different amounts of macromolecules or ligands allows the determination of the stability constant for the paramagnetic ion–macromolecule or paramagnetic ion–ligand complex. These complexes are termed binary complexes, and have an enhancement parameter, ε_b. The addition of a substrate to form the metal–macromolecule–substrate complex gives a ternary complex, (characterized by an enhancement parameter, ε_t), while the further addition of a substrate will form a quaternary complex (enhancement parameter, ε_q). In this section we shall deal first in some detail with the binary complexes and the determination of their binding parameters, and indicate some of the problems that are encountered in corresponding studies with ternary and higher complexes.

11.3.1. *Binary complexes*

11.3.1.1. *The equilibrium.* The interaction of a paramagnetic ion with a macromolecule, or enzyme is described by the equilibrium

$$M + nE \rightleftharpoons ME$$

where M and E represent metal and macromolecule or enzyme, respectively. If there are n equivalent and identical binding sites for M per protomer, then the effective concentration of E can be written as nE where E represents the concentration of e nzyme molecules. The dissociation constant, K_D is

$$K_D = \frac{[M_t][nE_f]}{[ME]}. \tag{11.14}$$

Since

$$[M_t] = [M_f] + [M_b]$$

where the subscripts t, f, and b refer to total, free, and bound metal or macromolecule (enzyme), respectively we have:

$$K_D = \frac{nE_f([M_t] - [M_b])}{[M_b]}, \tag{11.15}$$

where $[nE_f]$ is the concentration of free enzyme binding sites.

11.3.1.2. *The enhancement parameter ε_b.* The hydrogen relaxation enhancement of the bound form ε_b, is related to the observed enhancement ε^* by eqn (11.9)

$$\varepsilon^* = x_b\varepsilon_b + (1 - x_b)$$

where x_b is the mole fraction of bound metal. By use of eqn (11.15) x_b is given by

$$x_b = \frac{[M_b]}{[M_t]} = \frac{[nE_f]}{K_D + [nE_f]}. \tag{11.16}$$

11.3.1.3. *Calculation of n and K_D by e.s.r. spectroscopy.* In some favourable cases, e.g. Mn(II), the concentration of free ion, can be measured by e.s.r. spectroscopy. Generally, the bound metal does not contribute significantly to the signal when free metal is present. Noting that

$$[nE_f] = [nE_t] - [M_b] \tag{11.17}$$

use of eqn (11.15) then allows K_D to be determined directly. If the number of binding sites is unknown this can be determined by measuring the free metal concentration and then plotting the data according to the methods of Hughes and Klotz or Scatchard, using the following equations:

$$\frac{1}{\bar{\mu}} = \frac{K_D}{[nM_t]} + \frac{1}{n}$$ (11.18)

or

$$\frac{\bar{\mu}}{[M_t]} = \frac{1}{K_D}(n-\bar{\mu})$$ (11.19)

where $\bar{\mu}$ is the average number of metal ions bound per molecule of enzyme, $[M_b]/[E_t]$, and n is the number of metal binding sites (assumed to be equivalent and independent). Plots of $1/\bar{\mu}$ versus $1/[M_t]$, or $\bar{\mu}/[M_t]$ versus $\bar{\mu}$ give K_D and n. The experiment is usually carried out by titration of a fixed concentration of enzyme with metal. For this reason this titration is sometimes referred to as an *M-titration*.

11.3.1.4. *Calculation of ε_b by e.s.r. and proton relaxation enhancement.* The proton relaxation enhancement of the bound form, ε_b, can sometimes be obtained by measuring the concentration of free metal by e.s.r., and the observed enhancement, ε^*, by n.m.r. From section 11.5.1.2 we note that

$$\varepsilon_b = \frac{\varepsilon^*-1}{x_b} + 1$$

where $x_b = ([M_t]-[M_f])/[M_t]$.

11.3.1.5. *Calculation of* n, K_D, *and* ε_b *from PRE data.* The variables n, K_D, and ε_b can be obtained directly from the PRE measurements by two methods which involve the titration of a fixed concentration of M with enzyme (*E-titration*). This gives nK_D and ε_b. The titration of a fixed concentration of enzyme with metal (M titration) gives K_D and n.

E-titration(I).   From eqns (11.16) and (11.9)

$$x_b = \frac{[nE_f]}{K_D+[E_f]}$$

and

$$x_b = \frac{\varepsilon^*-1}{\varepsilon_b-1}$$

so that

$$\frac{1}{\varepsilon^*-1} = \frac{K_D+[nE_f]}{[E_f](\varepsilon_b-1)}$$ (11.20)

and

$$\frac{1}{\varepsilon^*-1} = \frac{1}{[nE_f]}\cdot\frac{K_D}{(\varepsilon_b-1)}+\frac{1}{(\varepsilon_b-1)}.$$ (11.21)

18

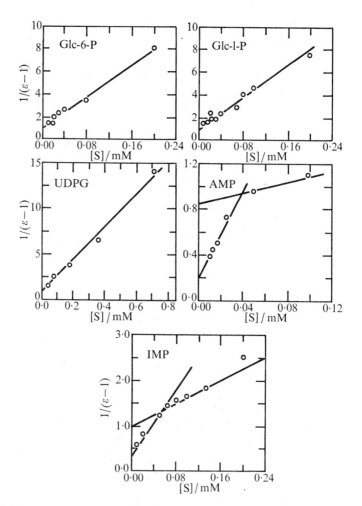

FIG. 11.2. Form of plot of E titrations of Mn(II) with various ligands, s, at pH $= 8\cdot5$ and 22 °C. ([Mn(II)] $= 10^{-4}$ M).

At high concentrations, $[E_f] \approx [E_t]$ and a plot of $1/(\varepsilon^*-1)$ versus $1/[E_t]$ will yield a straight line of slope $K_D/n(\varepsilon_b-1)$ and intercept $1/(\varepsilon_b-1)$. If $[E_f]$ differs substantially from $[E_t]$, the values derived from a linear extrapolation of such a plot will not be accurate.

E-titration(II). An alternative form of eqn (11.21) which has often been used is

$$\frac{1}{\varepsilon^*} = \frac{1}{\varepsilon_b} \cdot \frac{K_D}{[nE_t]+K_D/\varepsilon_b} + \frac{[nE_t]}{[nE_t]+K_D/\varepsilon_b} \tag{11.22}$$

The following approximations are now made

$$[nE_t]+K_D/\varepsilon_b = [nE_t]$$

and that

$$[nE_t]/([nE_t]+K_D/\varepsilon_b) = 1$$

so that eqn (11.22) becomes

$$\frac{1}{\varepsilon^*} = \frac{1}{[nE_t]}\cdot\frac{K_D}{\varepsilon_b}+\frac{1}{\varepsilon_b}. \qquad (11.23)$$

Examples of E-titrations. In some experiments it is possible to arrange conditions so that, to a very good approximation $[E_t] = [E_b]$. For example, in the binding of Gd(III) and Mn(II) to lysozyme (Dwek, Ferguson, Williams, and Xavier, 1973), $K_D \approx 10$ mM and a typical titration involves 100 μM of metal with enzyme varying from ca. 1 to 6 mM. In such cases E-titration(I) will give fairly accurate values of ε_b and K_D. It can also be used for the binding of ligands such as AMP, Glc-1-P, IMP, Glc-6-P, UDPG to metal ions, e.g. Mn(II). In a fairly typical case (Birkett, Dwek, Radda, Richards, and Salmon, 1971) the concentration of substrate or ligand, $[S_t]$, is varied over the range 2–100 mM at a $[Mn(II)]$ of 100 μM, so that the above approximation holds and plots of $1/[S_t]$ against $1/(\varepsilon^*-1)$ are linear. Some typical examples are shown in Fig. 11.2.

(The plots for AMP and IMP are biphasic, the linear portions at high nucleotide concentrations probably corresponding to higher complexes or some form of stacking.)

By way of contrast let us consider the case of Mn(II) binding to phosphorylase b (Birkett *et al.*, 1971). The value of ε_b and K_D varies with pH. Using the parameters in Table 11.2 for two pH values we can calculate which form of E-titration is most valid at the different pH values. From Table 11.2 we see that at pH = 8·5, form (II) can be used but at pH = 7·0 only form (I)

TABLE 11.2

Validity of approximations made for double-reciprocal plots

pH	$[E_t]$† (μM)	$[E_t]$ (μM)	$[E_t]+K_D/\varepsilon_b$ (μM)	$[E_t]/([E_t]+K_D/\varepsilon_b)$
8·5	150	111	126	0·88
8·5	100	71	86	0·83
8·5	50	34	49	0·70
7·0	150	134	187	0·72
7·0	100	88	141	0·62
7·0	50	43	96	0·45

At pH = 8·5, the metal ion is 100 μM, $\varepsilon_b = 11$ and $K_D = 170$ μM.
At pH = 7·0, the metal ion is 50 μM, $\varepsilon_b = 5\cdot4$ and $K_D = 285$ μM.
† It is assumed that $n = 1$ per protomer.

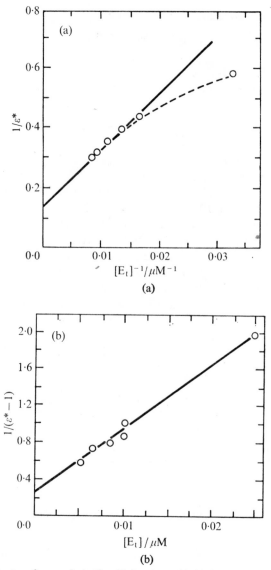

FIG. 11.3. The two forms of plotting E-titrations. (a) 10^{-4} M Mn(II) with phosphorylase b at pH 8·5, the dotted line is computed using the values in Table 11.1; (b) 5×10^{-5} M Mn(II) with phosphorylase b at pH = 7. (Adapted from Birkett *et al.*, 1971.)

will give reasonable estimates (Fig. 11.3(a) and (b)). So some care is obviously required in processing the data. Just which method should be used depends very much on the values of K_D and ε_b. Both procedures should be used to obtain preliminary estimates of ε_b and K_D followed by an iterative procedure to check the values obtained.

11.3.1.6. *Iterative procedure for calculating* K_D. For simplicity let us assume $n = 1$. Equation (11.15) may then be written as:

$$K_D = \frac{[E_f]([M_t]-[E_t]-[E_f])}{([E_t]-[E_f])}. \tag{11.24}$$

From this, $[E_f]$ and thus $[M_b]$ can be calculated assuming the value for K_D obtained from the limiting slope of either plot (see pp. 256 and 257). Using the value for ε_b obtained from the intercept, ε^* can be calculated from

$$\varepsilon_b = \frac{[M_b]}{[M_t]}\varepsilon_b + \frac{[M_f]}{[M_t]}\varepsilon_f. \tag{11.25}$$

By use of this procedure, the values of K_D and ε_b can be adjusted until consistency is obtained over the full range of concentrations for the E-titration.

M-titration. This involves a titration of a fixed concentration of enzyme with metal. From suitable analysis K_D and the number of binding sites can be obtained. In this method the amount of bound and free metal as well as the total enzyme concentration must be measured. The form of the equations has already been given in Section 11.3.1.3. Sometimes the free metal concentration can be measured by e.s.r. However, the accuracy of the e.s.r. technique is such that it is often better to calculate $[M_b]$ from the proton relaxation enhancement data. From eqn (11.9)

$$[M_b] = \left(\frac{\varepsilon^*-1}{\varepsilon_b-1}\right)[M_t] \tag{11.26}$$

so that, if we know ε_b from an E-titration, the amount of free and bound metal can be calculated from the observed enhancement. The data can then be plotted according to eqns (11.18) and (11.19).

An example of an M-titration using this technique is illustrated in Fig. 11.4, for the binding of Mn(II) to phosphorylase a. From an E-titration, ε_b is 18·5, and using this value, together with the observed enhancements as discussed above, the Scatchard plot gives the value of $K_D \approx 150\,\mu M$ and $n \approx 1$ (Dwek, Radda, Richards, and Salmon, 1972).

Cautionary note (Deranleau, 1969): For most binary complexes, the difficulties in evaluating any two parameters from one titration alone (e.g. K_D and ε_b from an E-titration and K_D and n from an M-titration) are the same as those involved in the determination of an equilibrium constant and extinction coefficient for a binary complex by any spectrophotometric titration procedure. For example, if the only method to be employed is an E-titration then ca. 75% of the saturation curve should be covered for good accuracy in ε_b and K_D. This is not always possible and in this case the use of e.s.r. independently where possible (or some other technique) to determine K_D is needed. The E-titration is then fitted using this value of K_D.

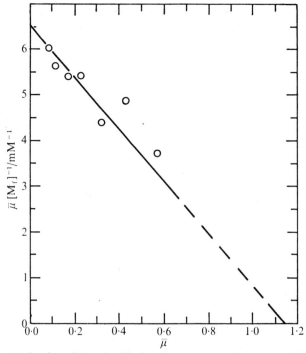

FIG. 11.4. M-titration of Mn(II) with phosphorylase a at pH 8·5 and 22 °C. (Enzyme concentration = 160 μM.) (Adapted from Dwek *et al.* 1972.)

11.3.2. *Ternary complexes*

For ternary complexes the evaluation of the parameters is slightly more complex than in the binary case because of the several coupled equilibria that exist. In a solution containing enzyme, ligand metal the following equilibria and enhancement parameters must be considered

$$K_D = \frac{[E_f][M_f]}{[ME]} \cdots\cdots \frac{Enhancement}{Parameter} \varepsilon_b \qquad (11.27)$$

$$K_1 = \frac{[M_f][S_f]}{[MS]} \cdots\cdots\cdots\cdots\cdots \varepsilon_s \qquad (11.28)$$

$$K_2 = \frac{[MS][E_f]}{[MES]} \cdots\cdots\cdots\cdots\cdots \varepsilon_t \qquad (11.29)$$

$$K_3 = \frac{[ME][S_f]}{[MES]} \cdots\cdots\cdots\cdots\cdots \varepsilon_t \qquad (11.30)$$

$$K_A = \frac{[ES][M_f]}{[MES]} \cdots\cdots\cdots\cdots\cdots \varepsilon_t \qquad (11.31)$$

$$K_S = \frac{[E_f][S_f]}{[ES]} \qquad (11.32)$$

where ε_b, ε_S, and ε_t are the enhancements of the complexes ME, MS, and MES respectively, and S represents ligand (substrate or modifier).

We note that $K_1 K_2 = K_3 K_D = K_A K_S$. Two general cases may be distinguished. In the first, ligand and metal bind at separate sites on the enzyme so that K_2, eqn (11.29), may be disregarded. The binding of ligand and metal may not, however, be independent as there may be heterotropic effects mediated through the ternary structure of the protein. If this is so, K_A will represent the binding of metal to a different form of the enzyme which exists only when the ligand is bound. The second general case is where the ligand–metal complex (MS) binds as such to the enzyme and this may, or may not, involve a metal bridge. In this case, the equilibrium represented by K_2 must be considered. Conservation equations for the above set of equilibria are

$$[S_t] = [S] + [ES] + [MS] + [EMS] \qquad (11.33)$$

$$[E_t] = [E] + [ES] + [EM] + [EMS] \qquad (11.34)$$

$$[M_t] = [M] + [EM] + [MS] + [EMS]. \qquad (11.35)$$

The observed relaxation enhancement for a solution containing the three components of the equilibrium mixture is given by

$$\varepsilon^* = \sum_i [M_i] \varepsilon_i / [M_t]$$

$$= \frac{[M_f]}{[M_t]} \varepsilon_f + \frac{[ME]}{[M_t]} \varepsilon_b + \frac{[MS]}{[M_t]} \varepsilon_S + \frac{[MES]}{[M_t]} \varepsilon_t. \qquad (11.36)$$

If the values of K_1, K_D, and K_S are known together with the corresponding enhancement factors, it is possible, in principle, for a trial value of K_2 or K_3, to calculate the equilibrium composition of a solution-containing enzyme, metal ion, and substrate. However, such a calculation involves simultaneous solutions of eqns (11.27)–(11.32). The non-linear form of these equations dictates a numerical solution and one method that can be used is the computer adaptation of the standard successive approximation method for complex equilibria (Reed, Cohn, and O'Sullivan, 1970). In cases in which K_S is known, titration data can be subjected to a non-linear least-squares fit to eqn (11.36). The scatter in ε_t calculated from ε^* at each data point is minimized by the use of K_2 as a variable parameter. When K_S is unknown, the data can be analysed according to the scheme outlined in Fig. 11.5. (A computer program is given in Appendix 11.1.)

These procedures have been used to evaluate the values of ε_t and the binding constants in the Mn(II)–ADP–creatine kinase and the Mn(II)–ATP–adenylate kinase systems. The original methods for determining K_2 and ε_t were based on graphical solutions of eqn (11.36) and relaxation measurements were originally restricted to the region where $[\text{ADP}_t] \leqslant [\text{Mn}_t]$† in an attempt to avoid significant contributions from the E–S equilibrium. The theoretical curves in Fig. 11.6 indicate the degeneracy of the possible solutions of the

† Since ADP competes with Mn–ADP for the same site.

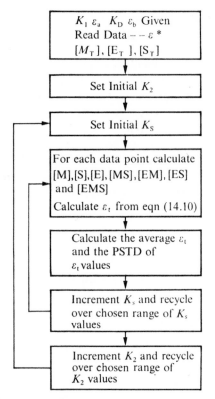

FIG. 11.5. Outline of computer program for nonlinear least squares fitting of PRE data; K_2 and K_S are varied. PSTD, per cent standard deviation. (From Reed *et al.* 1970.)

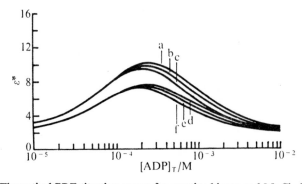

FIG. 11.6. Theoretical PRE titration curves for creatine kinase and $MnCl_2$ (0·1 mM) with ADP. Curves (a), (b), and (c) for creatine kinase 0·19 mM; curves (d), (e), and (f) for creatine kinase 0·12 mM (active sites). $K_1 = 0·03$ mM, $\varepsilon_s = 1·7$, $K_D = 0·50$ mM, $\varepsilon_b = 6·5$ for all curves. Curves (a) and (d): $K_2 = 0·067$ mM, $K_S = 0·10$ mM, $\varepsilon_t = 19.5$. Curves (b) and (e): $K_2 = 0·116$ mM, $K_S = 0·174$ mM, $\varepsilon_t = 24·0$. Curves (c) and (f): $K_2 = 0·17$ mM, $K_S = 0·25$ mM, $\varepsilon_t = 28·5$. (From Reed *et al.* 1970.)

parameters to the experimental results for creatine kinase for $[ADP_t] \leqslant [Mn_t]$. For example, for some parts of the titration data in Fig. 11.6 it is not possible to distinguish between an $\varepsilon_t = 28\cdot5$ and $\varepsilon_t = 9\cdot5$.

In order to cover more of the saturation curve, higher ADP and enzyme concentrations can be used. However, in the creatine–kinase system there is an added complication, that of competition between free ADP and Mn–ADP for the same enzyme site. By using trial values of the parameters to fit the data, it is possible to predict concentration regions where the enhancement

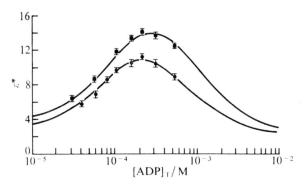

FIG. 11.7. PRE titration data for creatine kinase and $MnCl_2$ (0·10 mM) with ADP. Upper curve: $[E_t] = 0\cdot389$ mM active sites; lower curve, $[E_t] = 0\cdot205$ mM active sites. Temperature 24 °C. Solid curves drawn with $K_D = 0\cdot5$ mM, $\varepsilon_b = 6\cdot5$, $K_1 = 0\cdot03$ mM, $\varepsilon_s = 1\cdot7$, $K_2 = 0\cdot064$ mM, $\varepsilon_t = 20\cdot5$, $K_S = 0\cdot112$ mM. Values of K_2, K_S, and ε_t taken from minimum % standard deviation (2·9) in regression analysis. (From Reed *et al.*, 1970.)

data is optimal for distinguishing between various families of K_2, K_S, and ε_t (see for example, Fig. 11.7).

In this example the equilibrium constants K_1, K_2, and K_S are the same order of magnitude and the unique solution is difficult. Similar problems would be probably encountered if K_1 is very much larger than the other two. However, it should be noted that the difficulties encountered with the determination of binding constants by this technique are similar to those of any experimental methods that can be used to investigate the equilibria. Nevertheless, despite large uncertainties in the K values, reasonable limits can be set on the value of ε_t. The value of ε_t may be useful in determining the correlation time of the ternary complex (to be used in distance calculations—see p. 274).

Despite the rigorous method, there is the possibility of using graphical solutions provided that any approximations made are noted and, where possible, tested. Some examples which are amenable to graphical solutions are: (1) when the substrate saturates the macromolecule; or (2) when the substrate has all the metal ion bound to it.

From eqns (11.36) and (11.27)–(11.32), it is possible to derive eqn (11.37)

$$\varepsilon^* = \frac{\dfrac{[E_t]}{K_D}\varepsilon_b + \dfrac{[ES]}{K_A}\varepsilon_t + \dfrac{[S_t]}{K_1}\varepsilon_S}{1 + \dfrac{[E_t]}{K_D} + \dfrac{[ES]}{K_A} + \dfrac{[S_t]}{K_1}} \tag{11.37}$$

An example of the first case is in phosphorylase b where the Mn(II) ion is added as a probe (see page 268), and the effector, AMP, is saturating. In such cases, the concentration of E_f becomes very small and the terms containing it in eqn (11.37) can be disregarded. Further it is then likely that $[S_t]/K_1 < 1$ and, if $[E_t] > 1$, eqn (11.37) may be written as

$$\frac{1}{\varepsilon^*} = \frac{1}{[ES]}\frac{K_A}{\varepsilon_t} + \frac{1}{\varepsilon_t}. \tag{11.38}$$

A plot of $1/\varepsilon^*$ versus $1/[E_t]$ (assuming that $[E_t] = [ES]$) will be a straight line of intercept $1/\varepsilon_t$ and slope K_A/ε_t. If however the ligand concentration needed to saturate the enzyme is $\approx K_1$, then it may not be valid to ignore the last terms in eqn (11.37) and an iterative procedure can then be used as below.

11.3.2.1. *Iterative procedure for ternary complexes.* Assuming the enzyme is saturated with substrate, eqn (11.38) is used to obtain trial values of K_A and ε_t. Putting $[S_t] = [S_t]$ we have:

$$K_1 = \frac{[S_t]([M_t]-[MS]-[MES])}{[MS]} \tag{11.39}$$

and

$$K_A = \frac{([ES]-[MES])([M_t]-[MES]-[MS])}{[MES]} \tag{11.40}$$

Assuming that $[ES] = [E_t]$, eqns (11.39) and (11.40) can be solved for $[MS]$ and $[MES]$ and thus $[M_t]$ and by use of the value of ε_t from eqn (11.38), ε^* can be predicted from eqn (11.37). This procedure is really equivalent to the first procedure for solving eqn (11.37) mentioned above.

The *second example* involving a graphical solution is the binding of Mn–ATP to phosphofructokinase (Jones, Dwek, and Walker, 1972). Here, under the reported conditions, the binding of Mn(II) to the enzyme is very weak and ε_b is very small (<1) so that the term in $[ME]$ in eqn (11.36) can be ignored. It is then simple to derive eqn (11.41)

$$\frac{1}{(\varepsilon^*-\varepsilon_s)} = \frac{1}{(\varepsilon_t-\varepsilon_s)}\frac{K_2}{[E_t]} + 1 \tag{11.41}$$

which is similar to eqn (10.21) and can be treated in the same way.

Both of the above examples allow simplification of the mathematical techniques but unfortunately there is no *one* way of using graphical solutions. If rigorous computing is not to be used, then the graphical procedure chosen will depend on the system studied.

11.3.3. *Quaternary and higher complexes*

Here we have to consider the simultaneous presence of two ligands. In such a solution there will be six metal complexes MS, MS′, ME, MES′, MES, MES′S with corresponding enhancements ε_s, ε_s', ε_b, ε_t', ε_t, ε_q.

In general, we may write

$$\varepsilon^* = \sum_i \frac{[M_i]}{[M_t]} \varepsilon_i. \tag{11.42}$$

The enhancement of the solution is given by

$$\varepsilon^* = \frac{[M_t]}{[M_t]} \varepsilon_t + \frac{[MS]}{[M_t]} \varepsilon_s + \frac{[MS']}{[M_t]} \varepsilon_s' + \frac{[ME]}{[M_t]} \varepsilon_b +$$
$$+ \frac{[MES]}{[M_t]} \varepsilon_t + \frac{[MES']}{[M_t]} \varepsilon_t' + \frac{[MESS']}{[M_t]} \varepsilon_q \tag{11.43}$$

Of course the binary and ternary complex constants must be known, but the solution of the equation for ε_q is now very difficult even using computation techniques and conditions must be arranged so that some simplification can be made. For example, if S and S′ are both present in sufficient concentration to saturate the enzyme, the fourth, fifth, and sixth terms in the above equation may be neglected.

For higher complexes than quaternary, the problem becomes even more difficult and obtaining the enhancement parameters and binding constants becomes a very complex task. At the simplest level, however, even differences in the values of the enhancement parameters, with different substrates, can yield quite useful information on the structure and function of an enzyme. An example of this is for the AMP activation of phosphorylase b. Here the relaxation rates of the bound water hydrogens, are measured and the enhancements are used to monitor *changes* in the environment of the bound metal ion, as a function of ligand and temperature. The different enhancement parameters shown in Fig. 11.12 are assumed to arise from different conformational states of the enzyme. However caution is advisable before assuming that the enhancement parameters are state functions and where possible other evidence is useful to help justify any schemes proposed on the basis of these enhancement parameters. In this particular instance the scheme can be confirmed by e.s.r. measurements on a spin-label probe (see page 294).

We have discussed proton relaxation enhancement in some detail and it is obviously a convenient method of studying various types of binding. The problems associated with multiple equilibria are common to all forms of spectroscopy from which binding parameters may be determined. The question then arises as to which systems will be most productive to study by this method in the future. Clearly many of the enzymes already mentioned, particularly the kinases, which have a definite metal requirement are an obvious choice. However there are indications that the study of regulatory or allosteric enzymes will prove very productive. In these systems there may be conformational transitions that may not necessarily involve extensive rearrangements of the macromolecule and may therefore not be detected by studies of 'bulk' physical properties such as optical rotatory dispersion, circular dichroism, etc. More localized transitions often can be detected relatively easily by the use of probes which by virtue of their spectroscopic properties are sensitive to relatively minor changes in their environment. Of course, X-ray crystallography can, in principle, also detect these changes but the collection of data may be time consuming and uneconomical. N.m.r. spectroscopy may afford a relatively quick method of doing this and also has the advantage that the solution is studied. For this reason, we will deal in some detail with the phosphorylase system as an example of the points we have already encountered and as an indication of what kinds of experiments can be done. All the range of experiments mentioned for diamagnetic systems will be applicable but with paramagnetic systems there is the advantage of relating the observed effects to a definite centre and removing much of the ambiguity found with diamagnetic systems.

11.4. Phosphorylase b and a

The enzyme, phosphorylase, catalyses the reaction:

$$(\text{glycogen})_n + \text{HPO}_4^{2-} \rightleftharpoons (\text{glycogen})_{n-1} + \text{glucose-1-phosphate}.$$

Although the reaction is represented as an equilibrium, *in vivo* phosphorylase functions solely in the direction of glycogen breakdown. The glucose-1-phosphate is subsequently degraded by the glycolytic enzymes to pyruvic acid, with the production of energy-rich ATP phosphate bonds. Because the demands for energy, and hence glycogen breakdown, vary enormously, it is hardly surprising that phosphorylase turns out to be a control or regulatory enzyme.

There are two forms of rabbit muscle phosphorylase, the inactive 'b' form, which is the major form in the resting state of the muscle (and which may be activated by liganding with AMP), or the active 'a' form. The essential difference between the b and a forms is that the latter is phosphorylated on one specific serine residue per sub-unit. The conversion of phosphorylase b to phosphorylase a is accomplished via another enzyme, phosphorylase kinase, in the presence of Mg^{2+} and ATP. There are thus two methods of activating phosphorylase, either from an increase in AMP levels (a reflection of the decrease in ATP levels) or by chemical modification. In this latter process a small increase in the level of a hormone, adrenaline, leads to a sequence of reactions which results in the activation of phosphorylase kinase and then conversion of phosphorylase b to the a-form (Fig. 11.8).

Most of the regulatory enzymes contain several subunits and another major form of biological regulation is by the binding of a ligand (termed the effector or regulator) to one subunit in such a way that it not only affects the activity of that subunit but also the activity of the others. (The best known example is haemoglobin, where the binding of the first oxygen molecule

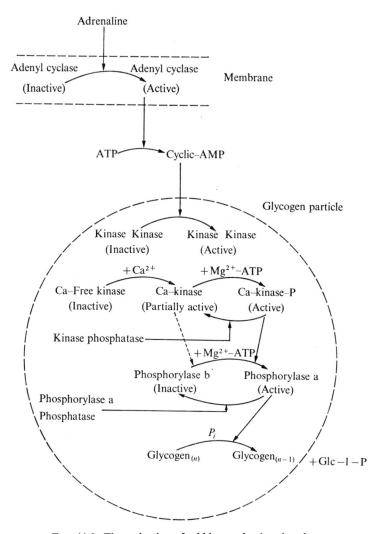

Fɪɢ. 11.8. The activation of rabbit muscle phosphorylase.

increases the affinity of the other three subunits towards oxygen.) The binding of effectors is not described, in general, by the hyperbolic binding equation but usually follows a sigmoidal binding curve.

Although phosphorylase b has no divalent metal-ion requirement, Mn(II) ions can apparently be specifically introduced into the protein without any measurable effect on the activity of the enzyme (Birkett *et al.*, 1971). The basic subunit on phosphorylase b has a mol. wt. of 100 000 and a summary of the various states of aggregation is shown in Fig. 11.9. Tetramer formation of phosphorylase b occurs in the presence of AMP and it is found that divalent ions especially Mn(II) enhance this tetramer formation, a point to which we shall later make further reference. The use of a fluorescent probe suggests that the value of K_D for Mn(II) from the enzyme sites involved in tetramer formation is ca. 6 mM.

The PRE studies (E- and M-titrations) show that there are two equivalent and independent manganese binding sites per oligomer (200 000), i.e. one metal ion per subunit. The variation of K_D and ε_b (at 35 MHz) with pH for this binding is shown in Fig. 11.10. The first point to make is that all the values of $K_D \ll 6$ mM. The nature of these experiments was such that low concentrations of Mn(II) ions were used and it is important to note that the metal binding sites detected by this technique are quite different from those involved in the aggregation as detected by fluorescence. Presumably it would be possible to detect these too by PRE but undoubtedly at the high concentrations of Mn(II) required for such experiments there may be many non-specific sites which might complicate the analysis.

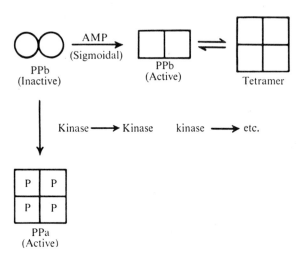

FIG. 11.9. States of aggregation of phosphorylase. PPb is phosphorylase b and PPa is phosphorylase a.

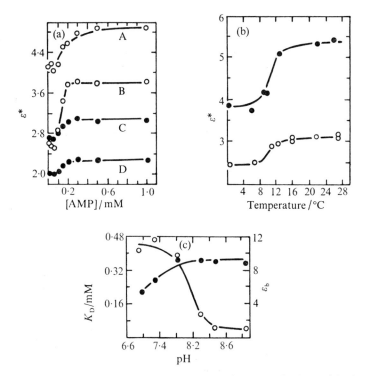

FIG. 11.10(a). Effect of increasing AMP concentration on ε^*, in the Mn(II)–phosphorylase b system at pH = 8·5. (A) 10^{-4} M Mn(II) at 19 °C. (B) 10^{-4} M Mn(II) at 1 °C. (C) 10^{-3} M Mn(II) at 19 °C. (D) 10^{-3} M Mn(II) at 1 °C.
(b) Temperature variation of ε^* in the Mn(II)–phosphorylase b system. The enzyme concentration is 102 μM and pH = 8.5. (c) Variation of ε_b and K_D with pH for the Mn(II)–phosphorylase b system. $T = 22$ °C. (○), K_D; (●) ε_b. (From Birkett et al. 1971).

The next point we note from Fig. 11.10 is that the variation of K_D with pH for Mn(II) implies a group with a pK_a value of ca. 8·1 while the pH titration for ε_b implies a group with a pK_a ca. 7·4. There are several reasons why this difference may occur. For example, the pH variation of K_D could reflect competition between H$^+$ ions and Mn(II) for the same ligand site on the enzyme or ionization of a group on the ligand remote from Mn(II). Either of these factors could influence ε_b although the former is much more likely to do so. Also a change in ε_b could result from a conformational change in the enzyme which might not alter K_D. The assignment of any pK_a value to a particular amino acid residue is obviously hazardous in the absence of the knowledge of the crystal structure. So many factors could contribute to an 'atypical' pK_a. The value of these experiments in such cases is really to suggest the optimum pH at which the relaxation and thus binding studies can be carried out, concomitant with enzyme activity and maximum enhancement effects. Most of the experiments which are described below were performed at pH ca. 8·5, where the value of K_D is such that for reasonably accessible concentrations of this enzyme (ca. 100 μM/subunit) a significant fraction of Mn(II) is bound (see p. 257 for an example).

The temperature variation of the spin–lattice relaxation times of water hydrogens in a solution containing Mn(II) and phosphorylase b shows a discontinuity at ca. 13 °C. The temperature variation of the enhancements (at two different pH values) show this too (Fig. 11.10b), although such plots can be misleading, for the discontinuities could arise from the behaviour of the relaxation rates of the control solution. A similar temperature transition has been detected by the change in the pattern of chemical reactivities of the SH groups (Birkett et al., 1971) and by kinetic and binding studies (Kasten-schmidt, Kastenschmidt, and Helmreich, 1968a and b). M-titrations below 13 °C again show that one Mn(II) ion per subunit is bound and that the value of K_D is hardly altered. However the value of ε_b drops from 11·5 (above 13 °C) to 6·3. When taken together with all the other data, it seems likely that the high- and low-temperature forms of phosphorylase b represent two different conformations of the enzyme. (We shall have recourse to note later that this temperature transition is still observed in the presence of either AMP, Glc-1-P, or both.)

The addition of the effector AMP to a solution containing phosphorylase b and Mn(II) gives an increase in the observed enhancement (Fig. 11.10a). The binding is clearly sigmoidal and a computer analysis of these curves (Salmon, 1972) similar to that in the Appendix gives values of $\varepsilon_t = 15$; $K_A \approx 150$ μM; and a Hill coefficient of $n = 1·4$. (The maximum value of n in any system is equal to the total number of subunits.) Confirmation of the values of K_A and ε_t is obtained from an analysis of suitable E titrations. At 1 °C, ε_t for the Mn–AMP–phosphorylase b complex is 10 (cf. ε_b (1 °C) = 6·3) and again K_A and K_D are approximately similar. The change in ε_t from ca. 15 to 10 on decreasing the temperature occurs at ca. 13 °C.

Obviously such binding studies can be extended. For example, the sigmoidal character of the AMP binding curve is still apparent in the presence of the substrates, glucose-1-phosphate and glycogen. Glycogen which prevents the ligand-induced aggregation (Kastenschmidt et al., 1968a) has little effect on the AMP enhancement and it again seems likely that the value of ε_t represents a different conformational state of the enzyme.

The binding of the substrate, glucose-1-phosphate, to phosphorylase b can be analysed by suitable plots (Birkett et al., 1971) and values of ε_t of 5·9 and $K_D \approx 70$ μM (at 22 °C) are obtained. On this basis and on studies of the reactivities of the SH groups in the presence of glucose-1-phosphate (Birkett et al., 1971), a third conformation of the enzyme is postulated in the presence of Glc-1-P. A fourth conformation can be postulated by examining the quaternary complex AMP–glucose-1-phosphate–Mn(II)–enzyme. Analysis of the binding results is of course difficult in any quaternary system, but by using saturating concentrations (for the enzyme) of glucose-1-phosphate and AMP it was concluded that ε_q is in the range 7–10 at 22 °C and ca. 8 at low temperatures. Some of the difficulties in obtaining a unique fit are apparent from Fig. 11.11 and this figure also serves as a warning against placing too much

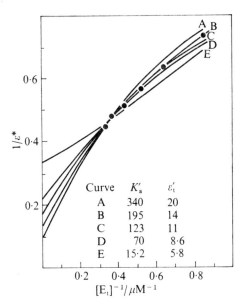

Curve	K'_a	ε'_t
A	340	20
B	195	14
C	123	11
D	70	8·6
E	15·2	5·8

FIG. 11.11. Titration of Mn^{2+} with phosphorylase b in the presence of 1 mM AMP and 10 mM Glc-1-P. Circles experimental points. Curves (A) to (E) are calculated using K_1 (Glc-1-P) = 34 mM, ε_S (Glc-1-P) = 2·0, K_1 (AMP) = 3·5 mM, ε_S (AMP) = 2·0 and K'_a and ε'_t as indicated. (From Birkett *et al.* 1971.) K'_a is the dissociation constant of Mn^{2+} from the Mn-AMP-Glc-1-P-enzyme complex.

reliance on graphical solutions in ternary and higher systems without the type of analyses discussed on pp. 260–263.

Taking into account all the different conformational states of the enzyme and the behaviour at low temperatures it is possible to postulate eight different states for phosphorylase b. These are summarized in Fig. 11.12.

The value of ε_b in the *phosphorylase a*–Mn(II) complex (ca. 18) (Dwek *et al.*, 1972) is closer to that of the b-form in the presence of AMP (i.e. the ternary complex) than that of the simple binary unliganded Mn(II)–phosphorylase b. In addition, AMP does not alter the enhancement of phosphorylase a. This provides strong support for the conclusion that the change in ε, for phosphorylase b, on activation by AMP, reflects a conformational change. The value of ε_b for the two enzymes is consistent with the expectation that phosphorylation of phosphorylase b 'locks' the enzyme in a conformation

19

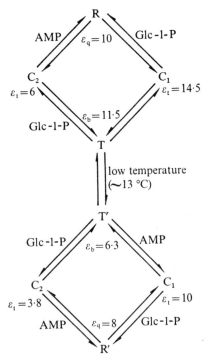

FIG. 11.12. Proposed scheme for the allosteric transitions in phosphorylase b. T represents the inactive conformation of the enzyme in the absence of ligands and R that of the active enzyme in the presence of both AMP and Glc-1-P. C_1 and C_2 represent the conformations in the presence of AMP and Glc-1-P respectively. The forms T', R', C_1', and C_2' represent the corresponding conformations at low temperatures. The respective enhancement parameters from PRE studies are indicated. (From Birkett et al. 1971.)

similar to that observed in the presence of saturating concentrations of AMP (see Fig. 11.9). The temperature transition at ca. 13 °C is also observed with phosphorylase a, again suggesting a conformational equilibrium similar to that in phosphorylase b. This too is to be expected if it is remembered that this transition occurs in the absence of AMP (see page 270).

The addition of glucose-1-phosphate to phosphorylase a gives a decrease in relaxation rates of the ternary complex compared with the binary complex. Its value, $\varepsilon_t = 10.9$, is similar to that for the quaternary complex of phosphorylase b suggesting that the assignment of the fourth conformation to the phosphorylase b quaternary complex is reasonable.

Finally, we mention for completion that as with any spectroscopic method much information can be obtained from suitably-designed competition experiments (as long as the two substances competing have different enhancement effects on the enzyme). For example, there might be two substrates competing for the same site, with different ε_t values, or competition between two metals, e.g., Mg(II) (diamagnetic $\varepsilon_b = 0$) and Mn(II). Competition between these two metals has in fact been shown for phosphorylase b (Brooks, Dwek, and Radda, 1972).

11.5. Some conclusions regarding the empirical use of enhancements

In the above phosphorylase system, the conformational scheme postulated for the activation of phosphorylase b by AMP has been confirmed, to some extent, by the measurements of the phosphorylase a complexes. It also relies on other physical techniques (Birkett et al., 1971). To assume that the values of ε_b, ε_t, and ε_q etc. represent different conformations is to give a similar weighting to the importance of the enhancement parameters as, for example, the extinction coefficient has in optical spectroscopy. In fact a comparison between the two is interesting. Both are independent of sample size and concentration, and both are dependent upon frequency. However extinction coefficients are not usually very temperature dependent, in contrast to the enhancement parameters which can show a marked temperature dependence. Changes in the co-ordination number of the metal ion, or even in the value of τ_c^* lead to the changes in enhancement and great caution is always necessary before interpreting the measurements as being 'state functions' of a particular system. This is because n.m.r. is sensitive to very small energy changes (and thus changes in protein structure) which may have little to do with the overall structure–function relationships of the protein. The success in the phosphorylase system could be fortuitous, for the determination of accurate enhancement parameters in systems where multiple equilibria are present is difficult. Nevertheless, it is a surprising feature that the values of the enhancement parameters reflect the different conformations of the enzyme at all, when it is considered that the enzyme has no divalent metal ion requirement.

With these reservations it is interesting to re-examine the classification of Mildvan and Cohn (1970) of the $1:1:1$ co-ordination schemes of ternary complexes of enzyme, manganese, and substrate. Four co-ordination schemes are postulated and the classification is based on a comparison of the relative enhancements the of ternary complexes (ε_t) and those of the binary Mn–E(ε_b) complexes (Table 11.3).

Type I enzymes, (substrate bridge) show little enhancement in binary complexes of manganese and enzyme but addition of a nucleotide substrate produces a relatively large enhancement. This suggests that the substrate is necessary to bring the manganese (and its hydrated water) into the enzyme environment via a substrate bridge complex, and this seems a very reasonable interpretation. Type II enzymes have significant enhancements in the Mn–E binary complexes, which are reduced on addition of substrates. On the other hand, type III enzymes have $\varepsilon_b = \varepsilon_t$ and this could arise, for example, if the

TABLE 11.3

Correlation of theoretical co-ordination scheme and empirical enhancement behaviour

Co-ordination scheme	(A) Substrate bridge E—S—M	(B) Metal bridge E—M—S or E—M⟨S	(C) Enzyme bridge S—E—M
Enhancement behaviour	Type I ($\varepsilon_b < \varepsilon_t$)	Type II ($\varepsilon_b > \varepsilon_t$)	Type III ($\varepsilon_b = \varepsilon_t$)

substrate binds to the enzyme at a site remote from the metal and does not affect the metal hydration sphere, or through fortuitous compensation of various factors. Again though we note that such a scheme is open to many ambiguous interpretations. For example, a change in enhancement parameter between a binary and ternary complex could result from a change in relaxation mechanism. In order to make the scheme more rigorous, relaxation measurements as a function of temperature and frequency should be carried out. Also ambiguities can arise if there is compulsory, rather than random, binding of substrate and metal to the enzyme. Despite these drawbacks, it will be interesting in the future when sufficient enzymes have been studied, to see what generalizations, if any, will emerge.

Conformational schemes, however, tell us little about the extent of any changes in enzyme structure or the interaction between the different ligand sites. For this the relative positions of the sites have to be known. As we have seen in Chapter 10, the use of paramagnetic probes also affords a means of obtaining this information.

11.6. The use of PRE data in the calculation of paramagnetic ion–ligand distances

The changes in relaxation rates of a particular nucleus arising from the presence of a paramagnetic ion is related to the distance between them. The calculation of the actual distance requires the following information: (1) a knowledge of all the relevant binding constants describing the equilibria in solution; (2) that fast-exchange conditions apply to the particular resonance under study; (3) the value of the correlation time characterizing the relaxation rates of the particular nucleus.

While variable temperature studies of the nuclear relaxation rates can show whether fast-exchange conditions are valid, PRE studies can often help in the two other points. Firstly, PRE studies often afford a relatively quick and easy method for obtaining binding constants. Secondly, in some cases, there is the possibility of calculating the correlation times from the characteristic enhancement values (see p. 243). Below we consider two examples of the observation of ligand nuclei in Mn(II) complexes and the analysis of their relaxation rates in which the PRE studies make an important contribution.

11.6.1. *The* Mn(II)–AMP *complex at* pH $= 8\cdot5$ (*Bennick, Campbell, Dwek, Radda, Price, and Salmon*, 1971; *Henson*, 1972)

The individual resonances of some selected protons of AMP are shown in Fig. 11.13 and we observed that the addition of Mn(II) gives a broadening of these resonances.

The calculation of the dissociation constant of the complex may be carried out by monitoring the water proton relaxation rates of solutions containing Mn(II) as the AMP concentration is varied, i.e. an E-titration.

The characteristic enhancement for the Mn(II)–AMP complex is ca. 2 and, as we have discussed on p. 243, this enables a value of τ_c to be calculated. In this example $\tau_c = 9 \times 10^{-11}$ s and this corresponds to τ_R in this system.

The temperature variation of both relaxation rates indicates that fast-exchange conditions apply to the protons and thus using the above value of τ_c the distances listed in Table 11.4 are obtained. Incidentally, the distances calculated from either relaxation rate are the same, indicating that for the protons the scalar (or hyperfine) interaction is unimportant. This is perhaps expected, if the Mn(II) is not directly co-ordinated to the hydrogens but co-ordinates to the phosphate, for the scalar interaction is transmitted through chemical bonds and is attenuated rapidly.

For the ^{31}P resonance there is quite a difference in the relaxation rates. Their temperature dependences are shown in Fig. 11.14, and we note immediately that for $1/T_{2,\mathrm{P}}$ slow exchange conditions are valid because in the case of ^{31}P there is a very large hyperfine contribution to $1/T_{2,\mathrm{M}}$. Although in this instance no correction was made for the temperature variations of the dissociation constant an approximate value of $\Delta H^{\ddagger}$ in the Mn–AMP complex is ca. 14 kcal mole^{-1} and the value of τ_{M} (298 K) is ca. $1\cdot1 \times 10^{-6}$ s. For the spin–lattice relaxation rates, fast-exchange conditions do obtain and using a value of $\tau_c \approx 9 \times 10^{-11}$ s, this gives an Mn(II)–^{31}P distance of 3·0 Å.

The slow-exchange conditions which apply to the ^{31}P *resonance* spin–spin relaxation rates could still obtain on binding to a macromolecule. For fast exchange, the condition $T_{2,\mathrm{M}} \gg \tau_{\mathrm{M}}$ must apply. If τ_{M} increases on binding, then this, coupled with any increase in τ_c, will tend to exaggerate the initial inequality $\tau_{\mathrm{M}} \gg T_{2,\mathrm{M}}$. Relatively large increases in the distances would then be required for fast-exchange conditions to be valid.

The calculation of the absolute Mn(II)–ligand nuclei distances depends on the exact value of the correlation time, which is difficult to determine in this system except by PRE studies. Nevertheless, the ratio of the distances does not depend on the value of τ_c (with the proviso that the same τ_c is equally applicable to every resonance). In principle, it is possible to obtain the

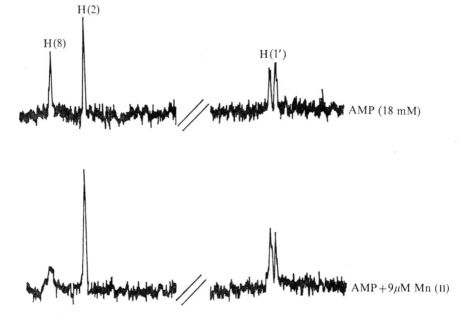

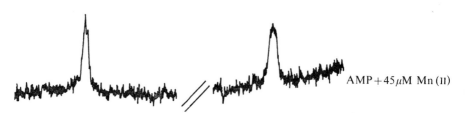

FIG. 11.13. The line broadening of some AMP protons caused by the addition of Mn(II). (The hydrogen nearest to the metals is broadened the most.) Conditions pH = 8·5; 22 °C; and 90 MHz.

TABLE 11.4

Distance of Mn(II) *from various nuclei on* AMP

Nucleus	Distance (Å)
H(8)	3·4
H(2)	4·9
H(1′)	5·1
^{31}P	3·0

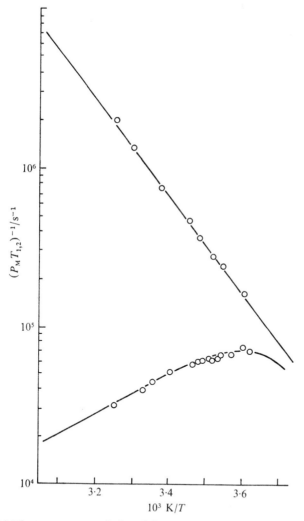

FIG. 11.14. The temperature variation of the paramagnetic contributions to the ^{31}p nuclear relaxation rates in aqueous solutions of AMP (0·1 M) containing $MnCl_2$ (0·1 mM) at pH 8·5 and 86 MHz. (From Henson, 1972.)

relaxation rates of all the nuclei and obtain the conformation of the Mn(II)–AMP complex. In solution there may be many conformations of the complex, and the distances determined will, therefore, represent averages. Because of the $1/r^6$ dependence in the relaxation rates, the conformations with the smallest values of r will obviously contribute most to the average. Additionally, a certain amount of caution is obviously required if there is not a unique site for the Mn(II). For example, if only a small percentage of Mn(II) coordinates to the nitrogen atom near H(8) or to the OH groups of the ribose ring this could have a dominant effect on some of the nuclear relaxation rates.

One of the main problems in the calculation of distances which we have stressed is the determination of the correlation times τ_c. However it is worth noting that relaxation measurements on *four* nuclei in a rigid molecular frame are sufficient to locate the position of the metal ion relative to this frame. For example in AMP, ^{13}C relaxation measurements on C(2), C(4), C(6), and C(8) would be sufficient. Alternatively combinations involving H(2) and H(8) could be used. Similarly procedures can be used for the ribose ring. Thus the conformation of the complex could be evaluated without any knowledge of τ_c. The idea simply is that the analysis of the relaxation rates of a nucleus places that nucleus, relative to the paramagnetic ion, on the surface of a sphere distance r from the centre. In a rigid system in which the inter-relationships of the nuclei are known, the intersection of four such spherical surfaces must be at the probe. Obviously it is always better to take as many reference points as possible and oversolve the problem, but a computer search may be involved in order to find the best solution for the structure. In general, the systems we shall deal with are undersolved either because of lack of resolution, for example in assigning *all* the hydrogens on the ribose ring in AMP or because of sensitivity problems, in measuring (say) the ^{13}C relaxation rates.

11.6.2. *Some complexes of creatine kinase*

This example illustrates some of the difficulties in distance determination, often encountered with substrates such as ATP, ADP, etc. which bind very strongly to Mn(II). We shall discuss the results that have been obtained for the Mn(II)-distances in the active complex of creatine kinase, viz. Mn(II)–ADP–creatine–creatine kinase (Leigh, 1971; Cohn, Leigh, and Reed, 1971).

Initially we start from the proton relaxation enhancement results for the Mn(II)–ADP–creatine kinase complex. The value of the enhancement of the ternary complex (ε_t) is much greater than that of the binary complex (ε_b) suggesting that the substrate is necessary to bring the manganese into the enzyme environment via a substrate bridge complex. We need this information because attempts to determine the conformation of Mn–ATP in the enzyme complex have been largely unsuccessful (Leigh, 1971; Cohn *et al.*, 1971), and it is instructive to consider why this is so. The success of such experiments depends on two factors: (1) the relative concentrations of the Mn–ATP and Mn–ATP–Enzyme complexes; and (2) the difference in relaxation rates between the two complexes. The relative concentrations can be obtained from a knowledge of the equilibrium constants. By use of the same notation as previously (p. 260) we may write:

$$\frac{K_1}{K_A'} = \frac{K_8}{K_2} = \frac{[ATP_t]}{[Mn\ ATP]} \times \frac{[Mn\ ATPE]}{[ATP\ E]}. \tag{11.44}$$

Under the usual experimental conditions the concentration of ATP is in large excess over the enzyme thus $[ATP_t] = [ATP_{total}]$. The concentration of the various paramagnetic species is also in general quite small. If we assume a Mn–H(8) distance of 3·4 Å (see Table 11.1) then for this proton it can be calculated that $1/T_{2,M} = 80\ 000\ s^{-1}$. (The correlation time required for this calculation is estimated as $1\cdot1 \times 10^{-10}$ s from PRE data with $\varepsilon_b = 2$ and assuming $q^* = 3$.) it can also be calculated that addition of 7·5 μM Mn(II) to a 20 mM solution of ATP will give a broadening of 10 Hz of the H(8) resonance. Thus, as we have generally seen in all these examples, small concentrations of Mn(II) can result in very large effects. This usually means that [Mn–ATP] and [Mn–ATPE] $\ll$ [ATP-E] and thus it is reasonable to assume that [ATP-E] $\approx$ [E$_t$]. Implicit in the above equation is that ATP and Mn–ATP compete for the same site, which is certainly the case for creatine kinase.

The observed relaxation rate for each resonance is given by

$$\frac{1}{(T_i^*)_{obs}} = \frac{(P_M)_t q'}{(T_{i,M}^*)_t} + \frac{(P_M)_b q'}{(T_{i,M}^*)_b} + \frac{(P_M)_{b'} q'}{(T_{i,(0)}^*)_{b'}} + \frac{1}{T_{i,(0)}^*} \tag{11.45}$$

$$i = 1, 2$$

where $(P_M)_t$ and $(P_M)_b$ refer to the fractions of ATP bound to the Mn(II) in the free solution and on the enzyme and $(P_M)_{b'}$ is the fraction of ATP bound to the enzyme which does not have Mn(II) bound to it. q' is assumed to be unity in each case, i.e. only one ATP molecule is bound per subunit. $1/T_{i,(0)}^*$ is the relaxation rate of the nucleus on the free ATP in the solution and $1/(T_{i,(0)}^*)_{b'}$ is the corresponding relaxation rate in the presence of the enzyme. A complete determination of the four components then requires four distinct measurements. Even if $(P_M)_b \approx (P_M)_t$, the question of whether we will observe an effect will depend on the ratios of $(T_{i,M}^*)_t$ and $(T_{i,M}^*)_b$. On the enzyme we have seen that the value of τ_c^* is greater than that in the free solution. If r is unchanged then $1/(T_{2,M}^*)_b \gg 1/(T_{2,M}^*)_t$ and provided the concentration ratios are favourable an effect on the line-width will be observed. If the term in $1/(T_{2,M}^*)_b$ is not much greater than the sum of the other terms—the nature of its calculation (involving the subtraction of three terms from $(1/T_i^*)_{\text{obs}}$ each with some experimental error) can lead to considerable uncertainties.

All the above discussion only emphasizes the difficulties that can sometimes occur and stress to the reader the need for being critical in reading the literature regarding the determination of distances in such cases.

The failure by Leigh *et al.* (1971) to observe an effect probably results from a combination of all the possibilities mentioned above and of course the fact

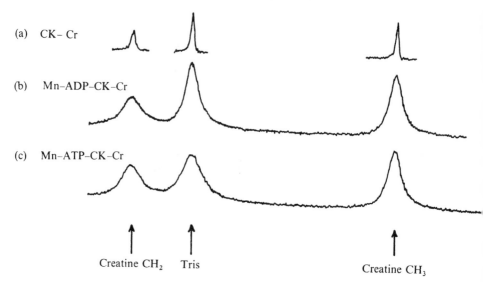

(a)   CK–Cr

(b)   Mn–ADP–CK–Cr

(c)   Mn–ATP–CK–Cr

Creatine CH$_2$ Tris   Creatine CH$_3$

FIG. 11.15. Creatine spectra (50 mM and at 60 MHz). (a) in the presence of creatine kinase only (16 mg ml); (b), abortive quaternary complex formed by addition of Mn–ADP (0·6 mM); (c), equilibrium mixture formed by addition of an eightfold excess of phosphocreatine to the quaternary complex at pH 6·8. (From Leigh, 1970.)

that the Mn–ATP distances may actually be longer in the enzyme complex than in the binary complex. All that can be concluded is that the metal ion does not move appreciably closer to any of the observed nuclei in the ternary complex.

When the substrate does not bind Mn(II) so tightly then the experiment is much easier since experimental conditions can be so arranged that the first term in eqn (11.45) is negligible or very small in comparison with the second. Such is the case for the Mn–creatine complex which has a dissociation constant of ca. 670 mM (*cf.* K_D for Mn–ATP of ca. 10 μM). The effects of the metal on the hydrogens of creatine in the abortive Mn–ADP–creatine kinase–creatine complex are easily observed (Fig. 11.15). Leigh (1970) reports that in a sample containing creatine (65 mM), creatine kinase (1 mM), and Mn(II) (0·2 mM), the relaxation rates of the creatine hydrogens are not very different from Mn(II)–creatine solutions in the absence of enzyme. The addition of ADP (1 mM) results in a fourfold increase in the observed relaxation rates for the creatine hydrogens. This implies that the nucleotide is necessary to bind the Mn(II) to the enzyme. These results also demonstrate the formation of a quaternary complex. The measured values of $1/T^*_{1,M}$ and $1/T^*_{2,M}$ are listed in Table 11.5 (Leigh, 1970) and the values of τ^*_c are calculated from the ratio of $T^*_{1,M}/T^*_{2,M}$ assuming no hyperfine contribution to $T^*_{2,M}$.

A comparison between Mn(II)–creatine distances in the abortive Mn–ADP–creatine–creatine kinase and active Mn–ATP–creatine–creatine kinase

TABLE 11.5

Mn–creatine distances in the quaternary Mn–ADP–creatine kinase–creatine complex†

	$1/T^*_{1,M}$ (s^{-1})	$1/T^*_{2,M}$ (s^{-1})	τ^*_c (ns)	r (Å)
CH$_2$	1350	2080	2·4 ⎫ 2·2	9·8
CH$_3$	1000	1700	2·0 ⎭	10·3

† Samples contained 10 mM Tris-Cl pH 8·0. Typical concentrations used were 65 mM creatine, 1·4 mM creatine kinase (sites) 1·2 mM ADP, 0·44 mM MnCl$_2$. The relaxation rates shown are an average of at least three experiments. Experimental uncertainties are ca. ±15% for the relaxation rate determinations. Measurements were carried out at 60 MHz.

complexes was made in the following way (Leigh, 1970). The equilibrium for the creatine kinase reaction may be displaced toward ATP by decreasing the pH. At pH 6·5 the calculated ratio of ATP–ADP is 57 with creatine–phosphocreatine = 11. The equilibrium mixture can be formed by adding phosphocreatine to the abortive complex. The results of such an experiment are shown in Fig. 11.15. Calculation of the Mn(II)–creatine distance in the active and abortive complexes showed little difference. Leigh (1970) ensured that the equilibrium had indeed been established and maintained during the course of the experiment (in spite of the ATPase activity of creatine kinase) by assaying enzymatically the ATP content of the n.m.r. sample.

There are some additional points that can be usefully made on this example that may be of general use. Firstly it is possible to observe the e.s.r. spectrum of Mn(II) in the various complexes. The K-band (equivalent to a hydrogen resonance frequency of 53·2 MHz) e.s.r. spectrum had a *derivative* line-width of ∼25 gauss (Reed and Cohn, 1972) and thus $\tau^*_S < 2·6 \times 10^{-9}$ sec, which is in good agreement with the value of τ^*_c used in this experiment. Secondly in very many experiments involving paramagnetic ions τ^*_c is often calculated from the proton relaxation enhancement data. In the quaternary complex $\varepsilon_q = 2$ (at 24·3 MHz) and even if it is *assumed* that $q^* = 1$ then a value of $\tau^*_c = 0·36 \times 10^{-9}$ s is calculated. Of course, we have already seen that this is not the case (see p. 225) and it is very likely that *no* water molecules are co-ordinated to the metal ion in the quaternary complex. Thus the value of τ_c^* calculated is obviously wrong, would lead to distances that are 33% shorter, and could change the entire conclusions on p. 322 regarding the model of the active complex. On the other hand the e.s.r. spectrum of the ternary complex is very similar to that of the quaternary complex (Reed and Cohn, 1972). In this case $\varepsilon_t = 20.5$ (at 24.3 MHz) and assuming $q^* = 3$ then τ_c^* (24.3 MHz) $= 1.25 \times 10^{-9}$ s. If this value of τ_c^* is identified with τ_s^*, then at 60 MHz (the frequency used in the experiments) the value of τ_s^* would be expected to be longer. This value of τ_c^* would lead to distances that are ca. 10% shorter. The use of proton relaxation enhancement (PRE) data to calculate correlation times in such experiments must be viewed critically in the light of the above. At high frequencies, the examples on PRE in Chapter 10 show that τ_M^* (for the water molecules) may become increasingly important in determining τ_c^*. Thus if PRE data *has* to be used at all, then it is advisable to calculate the value of τ_c^* when the contribution from τ_s^* is a maximum, i.e. at low frequencies. As τ_s^* generally increases with increasing frequency, the value of τ_c^* so obtained will represent a lower limit.

Appendix 11.1

Subroutine for calculation of binding constants in the presence of multiple equilibria. (Compiled by Dr. G. H. Reed)

```
C    SUBROUTINE FOR EQUIL CONC M,EM,MS,EMS,ES,S,E
C    ME,ET, AND ST ARE THE TOTAL CONCENTRATIONS OF METAL
C    ION, ENZYME AND SUBSTRATE, RESPECTIVELY
C    MD IS THE FREE METAL ION CONCENTRATION, THE OTHER SPECIES
C    ARE LOGICALLY DEFINED
C    EQUILIBRIUM CONSTANTS ARE DEFINED IN THE PAPER OF
C    REED, COHN, AND O'SULLIVAN (1970), J. Biol. Chem. 245, 6547.
C    WHEN THE SUBROUTINE IS CALLED, IT TAKES MT, ST, AND ET
C    ALONG WITH THE CURRENT VALUES FOR THE K'S, AND RETURNS
C    THE EQUILIBRIUM CONCENTRATION OF THE SPECIES LISTED IN
C    COMMON. NORMALLY CONVERGENCE IS RAPID. HOWEVER IF ANY
C    SPECIES GETS SMALL WITH RESPECT TO OTHERS, ROUND-OFF
C    ERRORS CAN OCCUR. IN SUCH A CASE IT IS BEST TO IGNORE
C    THE EQUILIBRIUM CONTAINING THIS SPECIES.
     REAL MD,MT,K1,KS,K2,KM,MS,MSX,MTX
     COMMON ST,ET,MT,K1,KS,KM,E,S,MS,EMS,MD,EM,ES
     F = 0.9*FT
     S = ST
     MD = MT
     EMX = 0
     ESX = 0
     MSX = 0
     EMSX = 0.1*ET
   2 ES = ESX
     EM = EMX
     MS = MSX
     EMS = EMSX
```

```
      EES = E+ES
      ESS = S+ES
      SUMK = EES+ESS+KS
      ESX = (SUMK-SQRT(SUMK*SUMK-4·0*EES*ESS))*0·5
      F = EES-ESX
      S = ESS-ESX
      SMM = MS+MD
      SSM = MS+S
      SUMK = SMM+SSM+K1
      MSX = (SUMK-SQRT(SUMK*SUMK-4·0*SMM*SSM))*0·5
      MD = SMM-MSX
      S = SSM-MSX
      SME = EM+MD
      SEE = E+EM
      SUMK = SME+SEE+KM
      EMX = (SUMK-SQRT(SUMK*SUMK-4·0*SEE*SME))*0·5
      E = SEE-EMX
      MD = SME-EMX
      SEEM = EMS+E
      SSEM = EMS+MS
      SUMK = SEEM+SSEM+K2
      EMSX = (SUMK-SQRT(SUMK*SUMK-4·0*SEEM*SSEM))*0·5
      E = SEEM-EMSX
      ETX = E+ESX+EMSX+EMX
      E = E+FT-ETX
      STX = S+ESX+MSX+EMSX
      S = S+ST-STX
      MTX = MD+MSX+EMSX+EMX
      MD = MD+MT-MTX
   32 IF(ABS((FSX-ES)/ESX).LT.0.1E-02) GO TO 33
      GO TO 2
   33 IF(ABS((MSX-MS)/MSX).LT.0.1E-02) GO TO 34
      GO TO 2
   34 IF(ABS((EMSX-EMS)/EMSX).LT.0.1E-02) GO TO 35
      GO TO 2
   35 IF(ABS((EMX-EM)/EMX).LT.0.1E-02) GO TO 36
      GO TO 2
   36 CONTINUE
      RETURN
      END
```

References

BENNICK, A., CAMPBELL, I. D., DWEK, R. A., RADDA, G. K., PRICE, N. C. and SALMON, A. J. (1971). *Nature (Lond.)*, **234**, 190.

BIRKETT, D. J., DWEK, R. A., RADDA, G. K., RICHARDS, R. E. and SALMON, A. J. (1971). *Eur. J. Biochem.* **20**, 494.

BLOEMBERGEN, N. and MORGAN, L. O. (1961). *J. Chem. Phys.* **34**, 842.

BROOKS, D. J., DWEK, R. A., and RADDA, G. K. (1972). Unpublished results.

COHN, M., LEIGH, Jr. J. S. and REED, G. H. (1971). *Cold Spring Harbor Symposia on Quantitative Biology*, **36**, 533.

DWEK, R. A., FERGUSON, S. J., RADDA, G. K., WILLIAMS, R. J. P. and XAVIER, A. V. (1973). To be published.

——, RADDA, G. K., RICHARDS, R. E. and SALMON, A. J. (1972). *Eur. J. Biochem.* **29**, 509.

DERANLEAU, D. A. (1969). *J. Am. chem. Soc.* **91**, 4044, 4050.

HENSON, R. (1972). *D.Phil. Thesis, Oxford.*

JONES, R., DWEK, R. A. and WALKER, I. O. W. (1972). *Eur. J. Biochem.* **28**, 74.

KASTENSCHMIDT, L. L., KASTENSCHMIDT, J. and HELMREICH, E. (1968a). *Biochemistry*, **7**, 4543.

——, ——, ——, (1968b). *Biochemistry*, **7**, 3590.

LEIGH, JR. J. S. (1971). *Ph.D. Dissertation, University of Pennsylvania.*

LUZ, Z. and SHULMAN, R. G. (1965). *J. chem. Phys.* **43**, 3570.

MILDVAN, A. S. and COHN, M. (1970). *Adv. Enzymol.* **33**, 1.

NAVON, G. (1970). *Chem. Phys. Lett.* **7**, 390.

REED, G. H., COHN, M. and O'SULLIVAN, W. J. (1970). *J. biol. Chem.* **245**, 6547.

——, DIEFENBACH, H. and COHN, M. (1972). *J. biol. Chem.* **247**, 3066.

—— and COHN, M. (1972). *J. biol. Chem.* **247**, 3073.

SALMON, A. J. (1972). *D.Phil. Thesis, Oxford.*

12

SPIN-LABEL PROBES

12.1. Introduction

THE term, spin label, is used to describe the stable free radicals that are used as reporter groups or probes. The term was first coined by Ohnishi and McConnell (1965), and Stone, Buckman, Nordio, and McConnell (1965). The most commonly used spin labels are molecules containing the nitroxide moiety namely,

TYPE (I)   and   TYPE (II)

which contain an unpaired electron localized mainly in a π-orbital on the nitrogen atom.

The nitroxide moiety is fairly unreactive under many conditions, for example it remains stable on heating up to ca. 80 °C and over the pH range 3–10. It can be reversibly reduced to the N-hydroxylamine by many mild reducing agents such as dithionite. The chemistry of the nitroxides is treated in considerable detail in the book by Rozantsev (1970) and the interested reader is referred to this work.

The stability of the spin labels thus makes them ideal probes and by a suitable choice of X in (I) and (II) it is possible to meet the specific requirements of many different biological problems. The spin labels may be either covalently attached (via the group X) to specific functional groups on the macromolecule under investigation or diffused into regions of interest, such as hydrophobic regions. Additionally, it is possible to make spin-labelled substrate or inhibitor analogues and thus many types of experiments can be done.

There are several good reviews on spin labels which concentrate mainly on the e.s.r. aspects (see e.g. Jost and Griffith, 1972 and references therein), but in this Chapter we shall be concerned mainly with those aspects of spin labels that are relevant to the types of n.m.r. studies already discussed. We shall deal mainly with labels in which X is an iodoacetamide group and mainly consider examples in which this spin label is covalently bound to a group on an enzyme, e.g. to a sulphydryl (SH) group.

The e.s.r. spectrum of the spin-label probe on the enzyme reflects both the mobility of the probe and the polarity of its environment. Under various conditions (e.g. variation of pH, or temperature or addition of ligands etc.) it is possible to cause changes in the e.s.r. spectrum. The spin-label probe can therefore be used to investigate conformational changes in its neighbourhood. The approach is more direct than that based on changes in the characteristic relaxation times (or enhancements) of the solvent water protons around a paramagnetic metal ion (i.e. the PRE method). This is because there are changes in the e.s.r. spectrum of the spin-label probe that can be interpreted unequivocally in terms of local conformational changes around the probe site.

Clearly a covalent probe such as the spin label has considerable advantage in studying conformational changes over the rather indirect PRE probes. Additionally of course the spin-label probe can 'report' on the effects of both diamagnetic and paramagnetic metal ions relatively easily. In the case of paramagnetic metal ions, in favourable cases, there is an apparent 'quenching' of the e.s.r. spectrum of the spin-label probe. This quenching can be related to the distance between the metal and the spin label probe.

While covalent spin-label probes are an attractive approach to studying macromolecular systems, there are problems which may best be solved by using spin-label probes on the ligands themselves. The spin-labelled ligands can be used to report changes in their environment in much the same way as the covalent probes. However there is the advantage that the spin-labelled ligand may be synthesized so that it can specifically be made to occupy a known ligand site. This approach may be useful too when the use of covalent spin-label probes results in labelling of an essential group. Because of the potential importance of this type of approach to solving biochemical problems we shall include some examples of non-covalent spin-label probes. The examples are limited to the effects of these probes on the nuclear relaxation rates of different ligand nuclei at other sites and on the enzyme itself. Few results as yet have been reported on the changes in the e.s.r. spectrum of such labels on addition of other ligands though again it is not difficult to envisage the types of problem where such an approach may be important.

In the following sections we attempt to illustrate some of the more general points about spin labels, concentrating mainly on their use as paramagnetic reference centres for distance determinations. However, because many readers will be unfamiliar with the use of spin labels, we start by describing the e.s.r. spectrum, in terms of the molecular motion of the spin label itself and illustrate how the changes in the e.sr. spectrum on addition of ligands may give information on enzyme conformation. Then we discuss their use in distance determinations from other nuclei, using n.m.r. spectroscopy and from another paramagnetic centre in the enzyme, using e.s.r. Finally, we consider briefly the effect of spin labels on the water proton relaxation rates and point out some of the difficulties that may arise in trying to interpret the effects quantitatively.

12.2. The motion of the spin label

12.2.1. *The effects of molecular motion on the e.s.r. spectra*

The e.s.r. spectrum of the nitroxide label consists of three lines, see Fig. 12.3 (assigned to the $m = 1$, $m = 0$, and $m = -1$, ^{14}N nuclear spin quantum states; the $m = 1$ state corresponds to the line at lowest field). Both

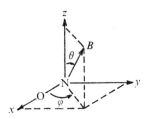

FIG. 12.1. The molecular co-ordinate system and angles used to define the direction of the external magnetic field (B). The angle θ is the angle from the direction of the applied magnetic field (B) to the molecular z-axis. The z-axis is parallel to the nitrogen 2p orbital associated with the unpaired electron. The angle ϕ is the angle between the x-axis and projection of B in the xy-plane.

the g-values and the hyperfine interactions are in general anisotropic, i.e. they depend on the orientation of the magnetic field the direction of which is specified in terms of a molecular co-ordinate system as shown in Fig. 12.1.

We start by considering the spectra obtained along each axis when the nitroxide radical is *oriented* in a rigid matrix (Fig. 12.2). Each three-line spectrum is characterized by one g-value and one value of the coupling constant, A. The g-value measured from the three-line spectra, g_{xx}, g_{yy}, and g_{zz} are all only slightly different in contrast to the coupling constants, for while A_{xx} and A_{yy} are approximately equal, they are both much smaller than A_{zz}.

If we now consider a completely random orientation of rigid nitroxide radicals in the magnetic field, then the spectrum in Fig. 12.3d is obtained. This essentially represents a sum of the spectra corresponding to all the possible orientations, of which Figs. 12.2a–c represent the extremes along each axis. The spectrum is called the rigid-glass spectrum and is observed for any randomly-orientated spin label assuming that there is no molecular motion, or whenever a small concentration of nitroxide radical is present in a rigid glass, polycrystalline sample, or powder.

The anisotropies in A and g can be averaged out by rapid isotropic tumbling motion. For example, Fig. 12.3 illustrates typical line-shapes which are observed on raising the temperature. Increasing the temperature causes the

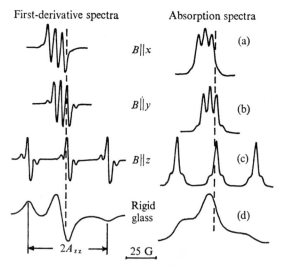

First-derivative spectra   Absorption spectra

$B\|x$ (a)

$B\|y$ (b)

$B\|z$ (c)

Rigid
glass   (d)

$2A_{zz}$ 25 G

FIG. 12.2. The 9·5 GHz e.s.r. spectra of representative spin labels in a rigid matrix. All e.s.r. spectra (except the bottom row) are of the nitroxide 2-doxylpropane (4′,4′-dimethyl-oxazolidine-N-oxyl derivative of acetone) oriented in cyclobutane-1,3-dione crystal. The host crystal orients the nitroxide molecules so that their x-, y-, and z-molecular axes are aligned in well-defined directions relative to the crystalline axes. The crystal was rotated until the laboratory magnetic field (B) was parallel to the $x(B \| x)$, $y(B \| y)$, and z molecular axis ($B \| z$). The rigid glass spectrum (bottom row) was obtained using a sample of randomly-oriented 12-doxylstearic acid in egg lecithin at -196 °C. The dashed lines mark the position of a 2,2-diphenyl-1-picrylhydrazyl reference sample ($g = 2\cdot0036$). (From Griffith, Libertini, and Birrell, 1971.)

solvent viscosity to decrease and the rotational motion to increase resulting ultimately in three sharp lines. Qualitative terms such as 'freely tumbling', weakly immobilized etc. are often used to describe the effects of the motion of the label on the e.s.r. spectrum and a rough guide is shown in Fig. 12.3. When complete averaging of the anisotropic effects occurs then the observed values of g and A are given by:

$$g = \tfrac{1}{3}(g_{xx}+g_{yy}+g_{zz}) \tag{12.1}$$

and

$$A = \tfrac{1}{3}(A_{xx}+A_{yy}+A_{zz}). \tag{12.2}$$

12.2.2. A simple general expression for the relaxation rate of each line in the e.s.r. spectrum of a fairly mobile nitroxide radical

In principle it is possible to describe the motion of the spin label in terms of the shape of the spectrum. While methods for computer simulation of the spectra are fairly well developed there are at present no rigorous quantitative theories that apply to the full range of mobilities of the label. However in the range of mobilities where the three e.s.r. lines do not overlap, (i.e. fairly

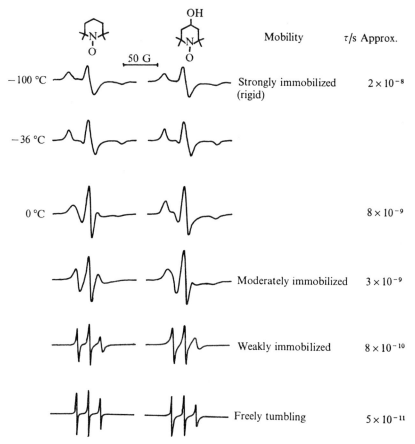

FIG. 12.3. The effect of viscosity on the 9·5 GHz e.s.r. spectra of two spin labels. Spectra were recorded using 0·5 mM spin label in glycerol. The approximate rotational correlation times and the description often used to describe such spectra are indicated. (Adapted from Jost, Waggoner, and Griffith, 1971.)

rapid tumbling) the treatment of Stone *et al.* (1965) can be used. We shall deal with this simple case here.

The transverse relaxation rate of each line $[T_2(m)]$ is given by

$$[T_2(m)]^{-1} = \tau \left[[3I(I+1)+5m^2]\frac{b^2}{40}+\frac{4}{45}(\Delta\gamma B_0)^2-\frac{4}{15}b\,\Delta\gamma B_0 m \right] \quad (12.3)$$

We may thus obtain eqn (12.4)

$$\frac{T_2(0)}{T_2(m)} = 1-\frac{4\tau b}{15}\,\Delta\gamma B_0 T_2(0)m+\frac{\tau}{8}\,b^2 T_2(0)m^2 \quad (12.4)$$

Putting in $M = \pm 1$ we have

$$\frac{T_2(0)}{T_2(-1)} - \frac{T_2(0)}{T_2(1)} = 8\tau b \,\Delta\gamma B_0 T_2(0)/15 \tag{12.5}$$

which is an expression which involves only the *linear terms*, in M. On the other hand

$$\frac{T_2(0)}{T_2(-1)} + \frac{T_2(0)}{T_2(1)} = 2 + 2\tau b^2 T_2(0)/8 \tag{12.6}$$

involves the addition of the *quadratic terms in* m. *The experimental values for* $T_2(0)/T_2(\pm 1)$ are obtained from the square root of the ratios of the heights of the experimental derivative curves.

12.2.3. *Definition of terms and units*

$\Delta\gamma$ is a measure of the anisotropy of the g-value

$$= -(\beta/\hbar)[g_{zz} - \tfrac{1}{2}(g_{xx} + g_{yy})]\dots \text{ s}^{-1}\,\text{G}^{-1} \tag{12.7}$$

where g_{zz} is the value of g along the z-axis. b is a measure of the anisotropy of the hyperfine interaction and is given by

$$b = \frac{4\pi}{3}(A_{zz} - A_{xx})\dots \text{ s}^{-1} \tag{12.8}$$

A_{zz} is the hyperfine coupling constant, in Hz, along the z-axis and B is that along the x- or y-axis, (since axial symmetry is assumed for the nitroxide label). In general, the values of A_{zz} and A_{xx} are *influenced by the polarity of the solvent*, their values usually decreasing as the solvent polarity is decreased.

T_2 is the line width and is obtained from eqn (12.9):

$$T_2^{-1} = \Delta\nu\pi\sqrt{3} \times 2{\cdot}8 \times 10^6 \text{ s}^{-1} \tag{12.9}$$

where $\Delta\nu$ is the peak to peak separation (G).

Stone *et al.* (1965) give the following condition for application of the equations for $T_2(M)^{-1}$: (1) there is axial symmetry in the anisotropic hyperfine interaction, i.e. $A_{xx} = A_{yy}$; (2) the tumbling motion is isotropic; (3) $(\pi a_0)^2 \tau^2 \ll 1$; (4) $b^2 \tau^2 \ll 1$; and (5) the tumbling motion is sufficiently slow that there are differences in line-width, i.e. $\omega_S^2 \tau^2 \gg 1$.

The use of eqns (12.5) and (12.6) allows the effects on the e.s.r. spectrum of a decrease in mobility of the label to be visualized (Fig. 12.3). As the label becomes more immobile, i.e. τ increases the high-field line ($m = -1$) is most affected, then the low-field line ($m = 1$) and finally the central line.

12.2.4. *Examples of the calculation and use of* τ

Phosphorylase b can be specifically labelled on one SH group in each subunit with a type I spin label, in which X is an iodoacetamide grouping, without any loss in enzymatic activity compared with the native enzyme (Campbell *et al.*, 1972).

The e.s.r. spectra at 9250 MHz of spin-labelled phosphorylase b is shown as the continuous line in Fig. 12.4 and is markedly different from that of the free label under similar conditions (Fig. 12.5). The changes in the e.s.r. spectrum arise from an increase in the tumbling (or correlation) time of the label when attached to the enzyme.

TABLE 12.1

Values of parameter used to calculate the rotational correlation time of the spin label in phosphorylase b

$$1 \text{ G} = 2 \cdot 8 \text{ MHz}$$
$$\hbar = 1 \cdot 054 \times 10^{-27} \text{ erg s.}$$
$$\beta = 0 \cdot 927 \times 10^{-20} \text{ erg G}^{-1}$$
$$g_{zz} = 2 \cdot 0027$$
$$g_{xx} = 2 \cdot 0089$$
$$g_{yy} = 2 \cdot 0061$$
$$A_{zz} = 87 \text{ MHz}$$
$$A_{xx}, A_{yy} = 14 \text{ MHz}$$
$$b = 3 \cdot 06 \times 10^{8} \text{ rad s}^{-1}$$
$$\Delta\gamma = 4 \cdot 22 \times 10^{4} \text{ s}^{-1} \text{ G}^{-1}$$
$$T_2(0)^{-1} = 46 \cdot 48 \text{ MHz}$$
$$T_2(0)/T_2(-1) = 4 \cdot 8$$
$$T_2(0)/T_2(1) = 1 \cdot 40$$

The value of the correlation time, τ, of the labelled enzyme calculated using the parameters in Table 12.1 is $2 \cdot 18 \times 10^{-9}$ s from the linear term [eqn (12.5)] and $2 \cdot 77 \times 10^{-9}$ s from the quadratic term [eqn (12.6)].

The discrepancy between these two calculated values of τ may arise from some anisotropic motion of the label group. In such a case the validity of eqn (12.3) may be questioned, since this *assumes an isotropic tumbling motion* (Stone *et al.*, 1965). However it *may* be possible that, if there is rapid motion about the bonds in the side chain of the iodoacetamide group, this would be sufficient to produce to an effective isotropic averaging (even though this motion may be fairly restricted). Nevertheless, the application of the above equations should provide a reasonable estimate of the value of τ. For comparison purposes, we note however that the value of τ for the free-spin label in pure water obtained using a similar analysis of the e.s.r. spectrum at 35 000 MHz is ca. $2 \cdot 5 \times 10^{-11}$ s (Stone *et al.*, 1965).

The value of the hyperfine interaction constants in this example are almost the same as that of the free label. This results, in part, from the short value of $\tau \approx 10^{-9}$ s. However, if we compare this with the value of $\approx 10^{-7}$ s for the tumbling time of the enzyme and note that the SH reactivity of the grouping

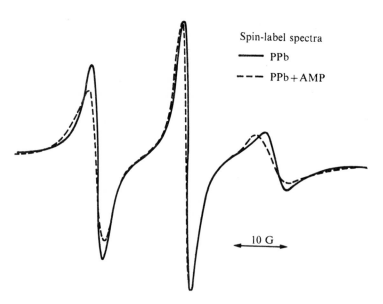

Spin-label spectra
——— PPb
– – – PPb+AMP

10 G

FIG. 12.4. The e.s.r. spectrum of spin-labelled phosphorylase b. (Solid curve) enzyme (100 μM in subunit) alone; (broken curve) enzyme with 800 μM AMP in 50 mM Tris–HCl buffer containing 100 mM KCl at pH 8·1. Temperature 25 °C. (From Campbell, Dwek, Price, and Radda 1972.)

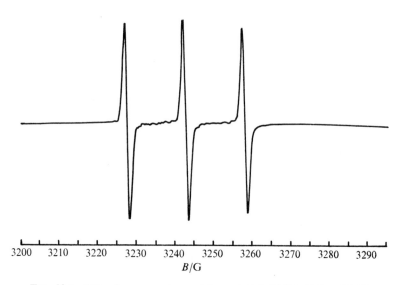

B/G

FIG. 12.5. X-Band e.s.r. spectrum of free spin label I (10^{-4} M) at 20 °C.

is similar to an unhindered SH group in model compounds, it may be concluded that the spin label is on the surface of the enzyme. The polarity of the environment of the spin label is thus almost unchanged on being bound to the enzyme.†

12.2.5. *Effects of ligands on the e.s.r. spectrum of labelled enzymes*

12.2.5.1. *Phosphorylase b.* The e.s.r. spectrum of spin-labelled phosphorylase b in the presence of saturating amounts of the effector AMP is shown in Fig. 12.4. The e.s.r. spectrum differs from that of the labelled enzyme in the lower heights of the low- and high-field peaks. We note from Fig. 12.3 that this type of behaviour corresponds to the label becoming more immobile. The value of τ as estimated from the eqns (12.5) and (12.6) in the previous section is now ca. $3 \cdot 6 \times 10^{-9}$ s. Thus the increase in τ provides direct evidence for an AMP induced conformational change in the enzyme.

The change in mobility of the spin label on phosphorylase b in response to the binding of the effector AMP follows very closely the appearance in enzyme activity (Fig. 12.6). Thus the spin label in this case is an ideal probe in that it responds to the addition of the ligand without measurably interfering with its action. The Hill coefficient n for both these titrations is $1 \cdot 34$ implying significant subunit interaction between the two protomers of phosphorylase b, both with respect to activity and the conformational

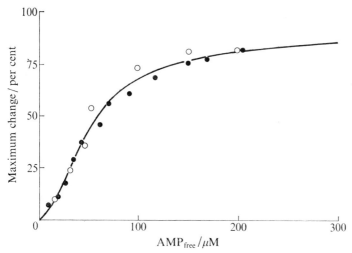

FIG. 12.6. Comparison of change in mobility and appearance of enzyme activity on addition of AMP to spin-labelled phosphorylase b. ($\bigcirc$) kinetic measurements, ($\bullet$) e.s.r. measurements. (From Campbell *et al.*, 1972.)

† Strictly speaking, under these conditions—one would have to obtain the spectrum under rigid lattice conditions to be able to demonstrate if there are any changes in the hyperfine splitting constants (and thus in the polarity of the environment) from those of the free label.

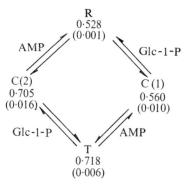

F IG. 12.7. Postulated scheme for the activation of phosphorylase b, as suggested by the spin–label probe. (The numbers represent the ratio of intensities of the low-field to centre lines in the e.s.r. spectrum; the numbers in parentheses are the standard errors.) (From Campbell *et al.* 1972.)

changes as detected by spin labels. Such experiments as these are important in establishing a relationship between the enzymatic and physico-chemical properties of the enzyme.

The changes in the e.s.r. spectrum of the label on addition of ligands can obviously be used to determine the enzyme–ligand binding parameters. There is however a practical difficulty with obtaining such parameters from changes in e.s.r. spectra. The absolute height (intensity) of the signal, for a given instrumental setting, depends very critically on the exact positioning of the sample tube (or cell) in the e.s.r. cavity. Thus it is necessary to be very careful in checking the reproducibility of the sample position each time. One method of overcoming this would be to compare the intensities of the signals with those from an internal standard. In the example of AMP binding to spin-labelled phosphorylase b, the centre peak of the e.s.r. spectrum (which does not change on addition of AMP) can be used as the 'internal standard'.

In this case the change in the *ratio* of (say) the low field to centre peak, on addition of AMP can be used to monitor the binding. Obviously this method is only possible when one feature of the spectrum remains constant. If the AMP caused larger changes in the mobility of the spin label, then the centre peak would change and an alternative method would have to be used.

Using the ratio of the intensities of the low-field to middle-field lines as a characteristic parameter of the e.s.r. spectrum, of spin-labelled phosphorylase b, the effects of glucose-1-phosphate and AMP (at saturating ligand concentrations) is summarized in Fig. 12.7. The implication of these observations is that it is necessary to postulate different conformations of the enzyme in the unliganded and liganded states and that glucose-1-phosphate, AMP, and the two together induce different transitions. These four conformational states were also postulated on the basis of the PRE results (see Chapter 11). It is important to rely on a combination of techniques in postulating the existence of different conformational states. The spin-label probe is sensitive to relatively small changes in the protein structure so that it is possible that the differences may not be detectable in the overall shape and size of the enzyme.

The second method of activating phosphorylase b is by covalent modification. The activation of phosphorylase b by phosphorylase kinase, Mg(II), and ATP can be followed by observing the e.s.r. spectrum of the label during the conversion (Fig. 12.8). The label again becomes more immobilized (as in

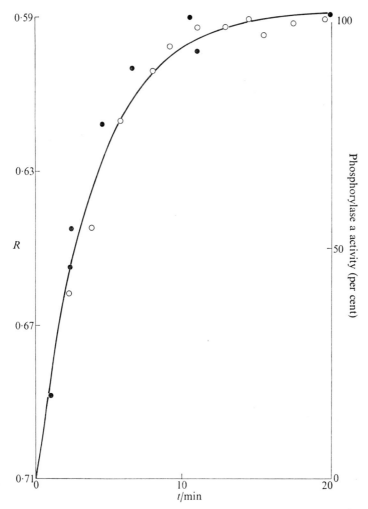

FIG. 12.8(a). Conversion of phosphorylase b to a monitored by e.s.r. (O) using a spin label probe and by activity (●) measurements. R is the ratio of low-field to centre-line intensities in the e.s.r. spectrum.

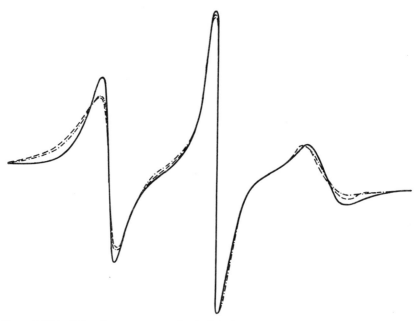

FIG. 12.8(b). X-band e.s.r. spectra of spin–labelled phosphorylase at 22 °C and pH 8·5, illustrating the near-equivalence (as detected by e.s.r.) of phosphorylase a and phosphorylase b+AMP., (———) phosphorylase b (50 μM), (– – – –) phosphorylase b+AMP (1 mM), or phosphorylase a (47 μM)+AMP (220 μM), (— · — · —) phosphorylase a (48 μM). (From Dwek, Griffiths, Radda, and Strauss, 1972.)

the activation by AMP), as phosphorylase a is formed (Dwek *et al.*, 1972). The e.s.r. spectrum of pure spin-labelled phosphorylase a indicates a mobility of the label close to that of the active b form (b+AMP). As from the PRE data this suggests that phosphorylation of the enzyme locks the conformation in the same form as that obtained by addition of saturating AMP concentrations to the unphosphorylated enzyme. This provides a further check that the changes detected by the spin-label probe are conformational changes in this particular system.

12.2.5.2. *Creatine kinase.* The importance of using a variety of spin-label reagents in enzyme modification studies is well illustrated by the reaction of spin labels of type (I) and (II) (where X = iodoacetamide) with creatine kinase. With a type (I) spin label the reaction is relatively non-specific and about three times slower than with a type (II) label which is specific for only one SH group per subunit (Taylor, Leigh, and Cohn, 1969). However the labelling of this SH group gives a substantial decrease in activity (of between 75–100%) compared with the native enzyme; the exact amount depends on the source of creatine kinase (Leigh, 1971). The enzyme however is not assumed to be grossly distorted by spin labelling since nucleotides and metal nucleotide complexes still form enzyme–substrate complexes with the modified enzyme, and have approximately the same affinity as with the native enzyme (Taylor, 1969). The e.s.r. spectrum of creatine kinase with a type (II) label attached is shown in Fig. 12.9. The spectrum of the label on the enzyme corresponds to that of a highly-immobilized species ($\tau \approx 5 \times 10^{-8}$ s) with a separation of the outer peaks of ca. 58 G at 22°C. This value may be compared to ca. 31 G for the free radical in water and ca. 64 G for the solid powder spectrum.

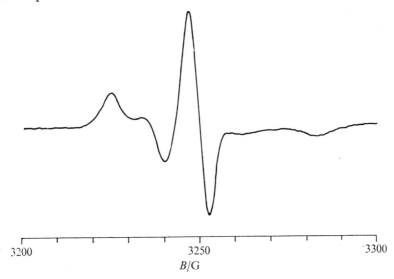

| 3200 | | | | | 3250 | | | | 3300 |

B/G

FIG. 12.9. X-band e.s.r. spectrum of spin–labelled creatine kinase. The five-membered ring N-(1-oxyl-2,2,5,5-tetramethyl-3-pyrrolidinyl)iodoacetamide was used for labelling. Protein concentration 6·2 mg ml^{-1}. (From Leigh, 1971.)

Titration of spin-labelled creatine kinase with alkaline earth ion–ADP complexes gives a reduction in intensity of ca. 20% in the centre peak while the low- and high-field peaks shift outwards (Fig. 12.10). (However the effects on the e.s.r. spectrum are not very different for the Mg(II), Ca(II), Sr(II), and Ba(II) complexes in spite of large differences in the activation of the enzymatic reaction.) That the effects of metal–ADP complexes result from a decrease in mobility of the label is substantiated by decreasing the temperature of the spin-labelled enzyme from 25 to 3 °C, when similar spectra are obtained.

The small changes observed in the e.s.r. spectrum of labelled creatine kinase on addition of ligands may result because the label is very immobile. The SH group may be in a rigid region of the enzyme, probably a cleft, and the value of τ will be almost that of the tumbling time of the whole enzyme molecule. In such cases, the spin label may be sensitive only to gross conformational changes. At the other extreme is spin labelled phosphorylase b for

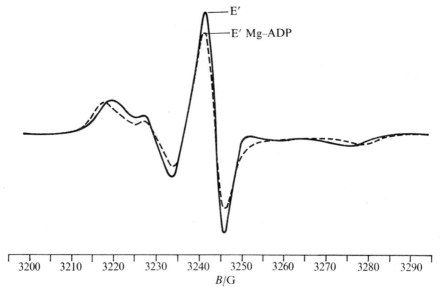

FIG. 12.10. The effect of Mg–ADP on the e.s.r. spectrum of spin–labelled creatine kinase (E') (0·14 mM active sites). (From Taylor, Leigh, and Cohn, 1971.)

which the label is sensitive to local changes, but again the observed effects are relatively small unless there are large changes in τ. We may then suggest that conditions of 'intermediate mobility' may be most favourable for the detection of small conformational changes by e.s.r. Such appears to be the case for phosphofructokinase with a type (I) label (Jones, Dwek, and Walker, 1972).

Sufficient has been said to indicate the potential of this type of probe, particularly when used in conjunction with other techniques, such as PRE. We have dwelt on the shape of the e.s.r. spectrum and the calculation of τ because we shall see that it is useful to have some idea of the mobility of the label and the electron spin relaxation time in quantifying the effects of the spin label on the ligand nuclear relaxation rates.

12.3. Calculation of distances between selected nuclei and a spin label centre by observation of the changes in the relaxation rates of the nuclei

The theory is very similar to that already encountered for the case of paramagnetic ions. The effects of the spin label on the relaxation rates of a particular nucleus related to the distance between the spin label and the nucleus by the Solomon–Bloembergen equations. For convenience we restate the assumptions implicit in the usual forms of these equations, and discuss the validity of these for relaxation by spin labels.

(1) The tumbling motion is isotropic—*not valid*
(2) The g-value is isotropic   —*not valid*
(3) The e.s.r. spectrum consists of a single Lorentzian line with only one $\tau_{1,S}$ which equals $\tau_{2,S}$ —*not valid*
(4) There is no zero field splitting .—*valid* since this does not occur with organic radicals for which $S = \frac{1}{2}$.
(5) The nuclear spin splitting of the electron resonance line $\ll \omega_S$
        —*valid* since these splittings are ca.
15–100 MHz and ω_S is usually ca. 9000 MHz in most experiments.

Assumption (1) is not, in general, fulfilled. If the label is rigidly held so that it tumbles at the same rate as the macromolecule, then its motion may be considered to be almost isotropic. However, if the spin label is relatively mobile, it is possible that there could be a partial averaging out of the dipolar interactions. Assumption (2) is not likely to lead to large errors because $g_{\parallel}$ and $g_{\perp}$ for the radicals are almost the same and thus an average g-value can be used. The presence of a multiplet (assumption (3)) can be dealt with using the treatment outlined in Chapter 9, i.e. measuring the line-width (or intensity) of each resonance.

Turning now to the use of such probes, we note that, in certain cases, it may be possible to observe the effect of the spin label on the nuclear relaxation rates of the macromolecule itself. However the more usual case will be the observation of the changes in relaxation rates of ligand nuclei. Under conditions of rapid exchange, it is then possible to calculate the distances of the ligand nuclei from the spin label on the macromolecule.

A field which will become increasingly important in the future is the use of spin-labelled substrate molecules. One problem in which these techniques would be useful is probing the stereochemical relationship between various sites on macromolecules and, in particular, on allosteric enzymes. The use of such probes is easier experimentally than the use of paramagnetic ion probes, as noted in the introduction to this Chapter, making this type of approach to biochemical problems attractive.

12.4. Some examples of distance calculations involving spin labels

12.4.1. *The proximity of the nucleoside monophosphate and triphosphate binding sites on* E-coli *deoxyribonucleic acid (DNA) polymerase (Krugh, 1971)*

DNA polymerase plays a role in the reproduction and repair of DNA. It catalyses the formation of a phosphodiester bond between the α-phosphate of a deoxyribonucleoside 5'-triphosphate and the 3'-hydroxyl terminus of a DNA chain. In the presence of Mg(II) the enzyme has two nucleotide binding sites. One site binds deoxyribonucleoside triphosphates and several nucleoside analogues but the triphosphate group appears essential. The other site binds a variety of analogues of the mono- and di-phosphates of deoxyribonucleoside but there is a specificity for nucleotides with a 3'-hydroxyl group in the ribo configuration and a 5'-orthophosphate linkage. This specificity suggests that it is this site that binds the primer terminus of a DNA chain. If this is the case, then the mono- and tri-phosphate binding sites should be adjacent. In an elegant series of experiments Krugh (1971) showed that this was indeed the case.

These experiments involved binding a spin label analogue of ATP (Fig. 12.11), called Tempo-ATP, to the triphosphate site when the effects of the

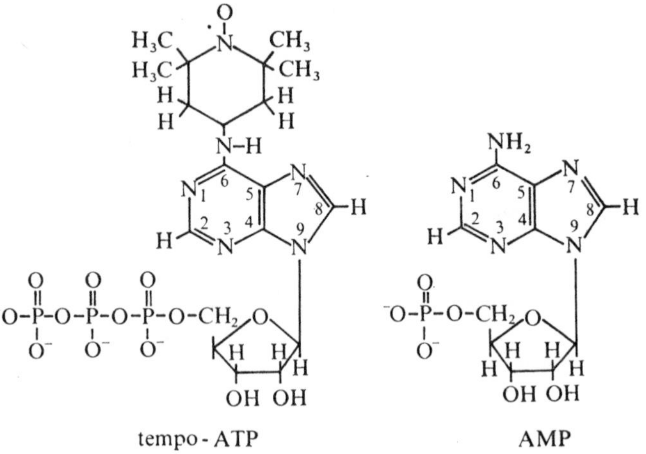

tempo - ATP   AMP

FIG. 12.11. Structures of tempo-ATP and AMP.

unpaired electron on the nuclear relaxation rates of AMP bound in the monophosphate site could then be monitored. To ensure that the observed changes in relaxation rates are a result of interaction when both substrates are bound to the enzyme molecule, the Tempo-ATP was displaced using an excess of thymidine-5'-triphosphate (TTP), which competes for the ATP site. This effectively allows the contribution only from the bound Tempo-ATP to be calculated. Tables 12.2 and 12.3 list the results of Krugh's experiments for the H(2) hydrogen of AMP. The calculation of the relaxation rates is reasonably straightforward from inspection of Table 12.2. The AMP molecule can exist in two sites bound to the macromolecule or 'free' in solution. The relaxation rate in each site consists of both paramagnetic (P) and diamagnetic (D) contributions, i.e. the observed relaxation rates are given by:†

$$\frac{1}{T_{i,\text{obs}}} = \left(\frac{P_{\text{M}}q}{T_{i,\text{M}}}\right)_{\text{P}} + \left(\frac{P_{\text{A}}q}{T_{i,\text{A}}}\right)_{\text{P}} + \left(\frac{P_{\text{M}}q}{T_{i,\text{M}}}\right)_{\text{D}} + \left(\frac{P_{\text{A}}q}{T_{i,\text{A}}}\right)_{\text{D}} \qquad (12.10)$$

$$i = 1, 2$$

† The asterisks are again omitted throughout this chapter, since there is never likely to be any ambiguity.

TABLE 12.2

Nuclear magnetic resonance relaxation measurements of the C(2) hydrogen of AMP (Krugh, 1971 © 1971 Amer. Chem. Soc.)

Soln.	AMP (M)	DNA polymerase (μM)	Tempo ATP (μM)	TTP (mM)	$1/T_1$ (s^{-1})	$1/T_{1,P}$† (s^{-1})	$1/T_{1,P}$ (s^{-1})	Temp (°C)	Frequency (MHz)
1	0·038	52·3	0	0	0·243		0·89	35	100
2	0·036	50·4	126	0	0·472	0·099	3·51	35	100
3	0·030	48·5	122	11	0·373		1·32	35	100
4a	0·137	43·1	83	0	0·410	0·033	1·45	35	100
4b	0·137	43·1	83	0	0·467	ca. 0		27	100
5a	0·132	41·5	80	11	0·377		1·02	25	100
5b	0·132	41·5	80	11	0·470			27	100
6a	0·096	45·4	133	0	0·505	0·132	1·82	35	60
6b	0·096	45·5	133	0	0·407	0·022		35	100
6c	0·096	45·4	133	0	0·357			40	100
7a	0·092	43·6	128	15	0·373		0·52	35	60
7b	0·092	43·6	128	15	0·385			35	100

† The concentration of Mg^{2+} is 5 mM and the enzyme solutions are in a D_2O–5 mM phosphate buffer (uncorrected pH = 7·4). The experimental uncertainty is usually 0·01–0·015 s^{-1}. Solutions 1, 2, 3, 6, and 7 were time averaged and only one value of $1/T_1$ was measured. For solutions 6 and 7 the data showed more scatter than usual and the uncertainty is probably ±0·02 s^{-1}.

where $P_A q$ is the fraction of nuclei free in solution. The term $(P_A q/T_{i,A})_P$ may be thought of as representing 'outer sphere' relaxation. In these examples the relaxation rate observed on the addition of TTP effectively consists of the contributions from the last three terms in eqn (12.10). To obtain $(1/T_{i,M})_p$ it is necessary to know $P_M q$. Since the enzyme is saturated with AMP then:

$$P_M = \frac{[\text{Tempo ATP}]_{\text{bound}}}{[\text{AMP}]_{\text{total}}}.$$

TABLE 12.3

Separation of the nucleotide binding sites (Krugh, 1971)

Enzyme soln	$1/T_{1,P}$ (s^{-1})	$T†_{1,M}$ (sec)	$\left(\dfrac{(1/T_{1,\rho})_P}{(1/T_{1,P})}\right)$‡	τ (s)	r (Å)§	Enzyme¶ Act. (%)
1, 2, 3	0·099	$7·9 \times 10^{-3} (\pm 2 \times 10^{-3})$	22	9×10^{-9}	7·4	70
4a, 5a	0·033	$4·4 \times 10^{-3} (\pm 3 \times 10^{-3})$	13	7×10^{-9}	7·0	100
6a, 7a	0·132	$2·1 \times 10^{-3} (\pm 1 \times 10^{-3})$	10	$9·6 \times 10^{-3}$	6·9	10
		Average value $r = 7·1 \pm 0·6$ Å (estimated error)				

† The uncertainties in $T_{1,M}$ are estimates.
‡ From line-width measurements these values are 20, 4, and 9 respectively.
§ Separation between unpaired electron of Tempo-ATP and C(2) hydrogen of AMP.
¶ Tested at the completion of the experiments.

The data in Table 12.2 are included because the changes in relaxation rates are fairly typical of those obtained with many spin labels. The data from solutions 4 and 5 indicate that the effects of the spin label on relaxation rates may decrease with decreasing temperature. This would mean that fast exchange conditions are not applicable at least at the lower temperature. As Krugh (1971) points out, however, the effects of the spin label on the spin–lattice relaxation rates are small in these experiments and variable-temperature studies may often be inconclusive. Other arguments may then be necessary to establish the relative magnitudes of $T_{1,M}$ and τ_M. At present we shall assume that fast-exchange conditions do apply. The amount of bound Tempo-ATP is calculated from its binding constant (ca. 10 μM). Other binding studies indicate that DNA polymerase has a single binding site for 3'-hydroxyl-ribonucleotides and thus $q = 1$.

The values of $1/T_{1,\rho}$ in Table 12.3 refer to the magnitude of the spin–lattice relaxation rates in the rotating frame. Effectively, this technique is equivalent to measuring T_1 at very low fields. The technique is described in Chapter 14 but the important result to note here is that $1/T_{1,\rho}$ tends towards the value of $1/T_2$. (A simple explanation of this is seen by inspection of Fig. 2.12. As the frequency is reduced, the value of τ_c, at which T_1 and T_2 diverge, becomes longer. Thus at very low values of the frequency $T_1 \rightarrow T_2$.) An alternative method, but often less accurate is to measure the line-widths and hence calculate T_2 from the usual relationship $1/T_2 = \pi \, \Delta\nu$.

Table 12.3 lists the values of τ_c calculated from the ratio of $T_{1,P}/T_{2,P}$ ($\approx T_{1,P}/(T_{1,\rho})_P$) *assuming* fast chemical exchange, and also the resulting values of r. The values of r are approximately those expected from models of a DNA chain if Tempo-ATP is placed in the position of the next nucleotide residue.

The value of τ_c probably corresponds to the value of $\tau_{1,s}$, a measure of which could be obtained from the e.s.r. spectrum, although this was not done in this instance. This could be important if fast exchange conditions do *not* apply since an upper limit for the distance could still be obtained. In this instance it has been assumed that $T_{1,M} \gg \tau_M$ and also $T_{2,M} \gg \tau_M$. The large difference in $1/T_{1,P}$ and $1/T_{2,P}$ suggests that this is probably true for $T_{1,M}$.

12.4.2. *Distances between bound saccharides, histidine-15 and tryptophan-123 on lysozyme (Wien, Morrisett, and McConnell, 1972)*

This is an important example, as are all cases, when there is the possibility of comparing the distances obtained by magnetic resonance methods with X-ray crystallographic data.

The experiments involved using spin labels covalently bound to the enzyme and also covalently bound to the inhibitors. The first series of experiments involved lysozyme spin labelled with the bromamide spin label, at the N(3) atom of histidine-15, and the effects of the label on various inhibitors were observed. In the other series of experiments the changes in line-width of the C(2) hydrogen of histidine-15 were monitored on addition of spin-labelled inhibitors. The various spin labels used are listed in Fig. 12.12.

In these experiments, the fraction of inhibitor molecules bound to the enzyme is comparable if not greater than the fraction 'free' in the solution. The usual equations for $1/T_{1,P}$ and $1/T_{2,P}$ derived for the case of only a small fraction bound are not applicable. Some details of the relevant equations for this example are given in the Appendix to this chapter and the conditions for the validity of fast chemical exchange are discussed. For the present we assume that this is so. The problem of calculating distances then follows the normal procedure of evaluating $1/T_{2,M}$ and calculating r from the Solomon–Bloembergen equations using the required correlation time. The relevant correlation time may be τ_R, τ_M, $\tau_{1,S}$, or a combination. τ_R has been estimated as 10^{-8} s (Dubin, Clark, and Benedek, 1971) from the depolarization of a laser beam in lysozyme solutions. (The value estimated from Stokes' Law (p. 242 is ca. 0.5×10^{-8} s.) From analysis of the spin-labelled spectra $\tau_{1,S} > 1.4 \times 10^{-7}$ s, but at high concentrations of spin labels ($>10^{-3}$ M) there is the possibility of spin exchange between different spin-label molecules. This would tend to shorten $\tau_{1,S}$ but Wien et al. (1971) consider that this is only important in the experiments involving Tempo-acetamide for which $\tau_{1,S} > 2.2 \times 10^{-8}$ s. The value of τ_M for the lysozyme inhibitors used here may be estimated as $>10^{-5}$ s. This assumes that the 'on' rate constant for these labelled inhibitors are the same as for NAG and di-NAG and that any differences in binding constant reflect the 'off' rate constant which may be equated with $1/\tau_M$ (see e.g. p. 232). Thus the value of τ_R is the relevant correlation time to use in calculating distances.

Figure 12.13 shows the effect of binding N-acetyl glucosamine (NAG) to lysozyme covalently labelled on the N(3) of histidine-15 with (V). The two methyl peaks corresponding to the α- and β-anomers of NAG are broadened differentially; the β-peak experiences the larger broadening. The calculated distances between the free electron on the nitroxide attached to histidine-15 and the acetamido methyl group of α- and β-NAG are 15 Å and 13 Å respectively. This suggests that the α- and β-NAG bind in different configurations and is in accord with the X-ray crystallographic results and the results obtained from chemical shift measurements of the bound inhibitor resonances (see p. 126). The X-ray data indicate that α- and β-NAG bind in subsite C and using a model based on this data, the distance of the α- and β-acetamido methyl groups from the middle of the imidazole ring of histidine-15 is 18 Å.

The e.s.r. spectrum of the spin label corresponds to that of a fairly mobile label ($\tau_R \approx 3 \times 10^{-9}$ s). Model studies (Wien et al., 1973) indicate that the nitroxyl ring can assume two extreme orientations: one with the ring sticking out into the solvent and the other with the ring pointed inwards and occupying

21

NAG —— NAG —— CH_2 —— Tempyro

(I)
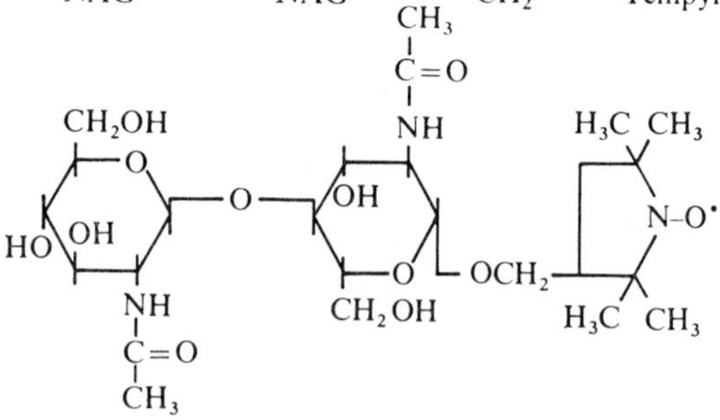

(II)
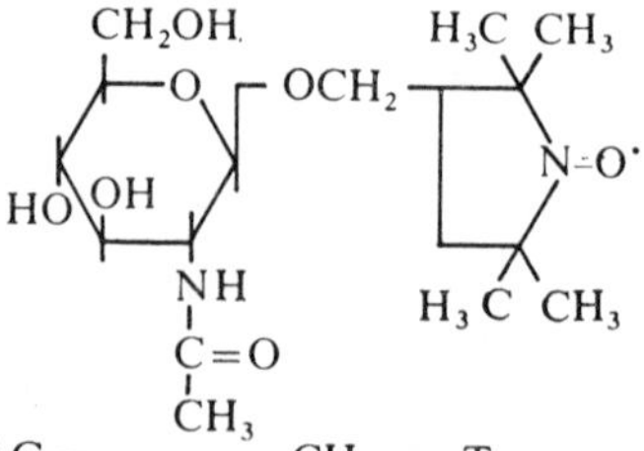

NAG —— CH_2 —— Tempryo

(III)
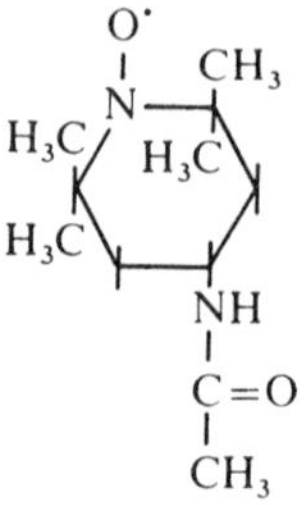

Tempoacetamide

(IV)
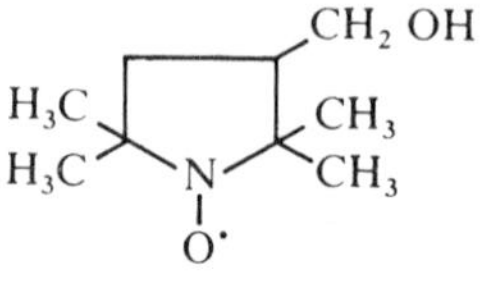

Tempyrocarbinol

(V)
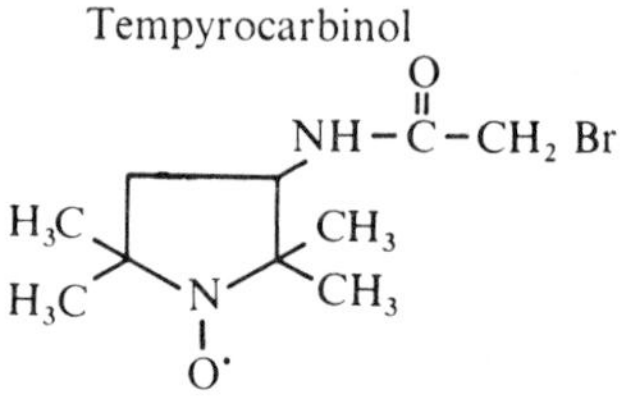

Tempyrobromacetamide

FIG. 12.12. Some spin labels used in the lysozyme example. (I), (II), and (III) are spin labelled *inhibitor* analogues while (V) was attached *covalently* to N(3) of His-15 in lysozyme. (IV) does not bind to the enzyme and was used as a control in experiments with (III). Tempyro ≡ *N*-(1-oxyl-2,2,5,5-tetramethyl-3′-pyrrolidinyl); tempo ≡ *N*-(1-oxyl-2,2,6,6-tetramethyl-4-piperidinyl); NAG ≡ *N*-acetylglucosamine.

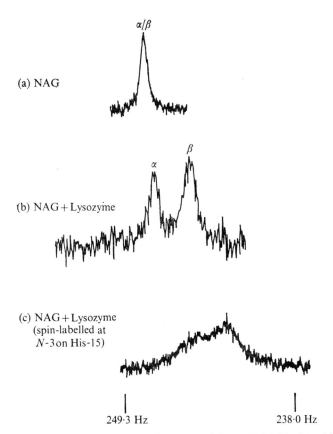

FIG. 12.13. Methyl resonance spectrum of 20 mM NAG at pH 4·5 (a) alone (single scan) (b) plus 2 mM lysozyme (3 CAT scans) (c) plus 2 mM 3-spin-labelled His-15-lysozyme (single scan). Vertical lines indicate chemical shift in Hz from external TMS. (From Wien *et al.*, 1972. © 1972 *Amer. Chem. Soc.*)

a hydrophobic pocket. Of course all intermediate orientations are possible. The effects of the motion of the spin label might be that r takes on a distribution of values. Because the relaxation effects depend on $1/r^6$ the calculated value of r will be heavily weighted towards that value corresponding to the distance of closest approach. If the nitroxide radical is oriented inwards, the free electron bearing nitrogen is ca. 7 Å (the length of the Tempyroacetamide group) closer to the methyl groups than the imidazole ring. This would mean that the distance, as measured on the model, from the free electron to the methyl group of NAG in subsite C, is ca. 11 Å.

In another series of experiments the effects of various spin labelled inhibitors on the C(2) hydrogen of histidine-15 in lysozyme were monitored. The distance from the unpaired electron on the spin-labelled Tempo-acetamide (III) to the histidine was calculated to be 17 Å. To compare this with the

X-ray model it is necessary to know where in lysozyme this pseudo saccharide is bound. The 6 Å crystallographic study on the lysozyme–tempoacetamide complex by Berliner (1971) provides the answer. The results are somewhat unexpected for by analogy with NAG, the spin label would be expected to bind selectively to the C-site on lysozyme. However while tempoacetamide does appear to bind to subsite C, the occupancy of this site is low. The label also appears to have a second weak binding site at subsite A, and, more importantly for the studies here, a third *strong* binding site located near tryptophan-123. This third site is the closest to histidine-15. The calculated distance would then refer to the distance from the unpaired electron on the label to the C(2) hydrogen histidine-15 when the label is in this strong binding site. Measurements from the lysozyme model give the distance from the centre of the tryptophan-123 indole ring to the C(2) hydrogen as 18 Å. In this experiment a high concentration of label was needed (ca. 7 mM) to observe a reasonable effect. In such a situation, the concentration of the free (unbound) spin label is sufficient to result in a broadening of the histidine resonance. Some idea of the magnitude of this effect was estimated by measurement of the histidine C(2) hydrogen resonance in the presence of Tempyrocarbinol(IV) which does not bind to lysozyme. (see Table 12.4.)

The other type of experiments that can be done is to measure the distances between various sites on the enzyme. In this instance the distances between the strong Tempoacetamide binding site and subsites B and C could be estimated by measuring the line width of the acetamido methyl signals of di-NAG which occupies these sites. No observable changes in line-width were observed and thus the distance between either methyl group and the electron on the nitroxide group of tempoacetamide is >15 Å. Measurements on the model indicate that the distances from the reducing and non-reducing groups to the centre of the indole ring of Trp-123 are 18 Å and 28 Å, respectively.

The mode of binding of NAG–NAG–CH$_2$–Tempyro(I) and NAG–CH$_2$–Tempyro(II) to lysozyme is unknown at present. By comparison with the known mode of binding of NAG and di-NAG etc. (I) would be expected to occupy subsites B, C, and D while (II) should occupy C and D. In both cases Wien *et al.* (1972) suggest that the tempyro ring is in site D. The quantitative analysis of the broadening of the histidine C(2) hydrogen resonances caused by labels (I) and (II) suggest that the distances from the nitroxyl nitrogen to the C(2) hydrogens are 19 Å for (II) and 20 Å for (I). This can be compared to the distance measured from the model of 21 Å between the middle of the imidazole ring of histidine-15 and the nitrogen on the label in subsite D.

A summary of all the experimental results is presented in Table 12.4 and Fig. 12.14. The distances obtained from the n.m.r. experiments denoted M are compared with those measured from a model of lysozyme based on the X-ray data. (These distances are denoted X.) In general there is good agreement between the two sets of measurements though it is better in those

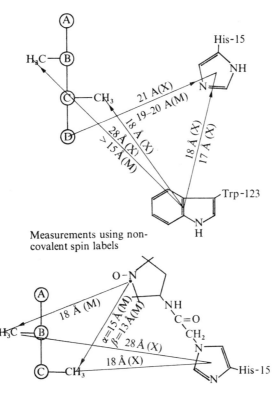

Measurements using non-
covalent spin labels

Measurements using covalently spin-labelled lysozyme

FIG. 12.14. Diagram comparing distances between groups on lysozyme as determined by
measurements on a model based on X-ray diffraction data (X) and determined from
^{1}H n.m.r. line broadening experiments (M). Lines point *from* the site of the spin label *to* the
relaxed hydrogen(s) being observed. In the upper diagram, the paramagnetic effect arises
from either tempo-acetamide bound near tryptophan-123 or the tempyro moiety held in
subsite D. The hydrogen resonance lines arise from the C(2) hydrogen line of His-15 in one
case, and from the acetamide methyl groups of di-NAG (or NAG) bound in subsites B and
C (or C alone) in the other cases. For the lower diagram, the paramagnetic effect originates
at the Tempyro acetamido moiety attached to histidine-15. The observed resonances are
those of the acetamido methyl groups of di-NAG (or NAG) bound in subsites B and C (or
C alone). The probable distance between the methylene (attached to N(3) of His-15) and
the nitroxyl nitrogen is of the order of 7 Å. (From Wien *et al.*, 1972. © 1972 Amer.
Chem. Soc.)

experiments employing spin-labelled inhibitors than in those where the
labelled enzyme is used. This is probably due to the uncertainty of the
average position of the chemically attached nitroxyl ring.

In this particular example it was possible to obtain τ_R by an independent
method. Interestingly, if τ_R had been estimated by Stokes' Law, it would
have been about a factor of two smaller. The use of this value of τ_R would
give distances which are 12% shorter. The more usual way of obtaining a

TABLE 12.4

Experimental parameters used in distance calculations involving lysozyme

Mixture components[a]	Resonance line	Observed line-width[b]	Broadening due to uncomplexed nitroxide[b]	Initial line-width[b]	Net broadening	K_a	P_M	Calc. r (Å)[e]	Model r (Å)[f]
3-SLHis-15-lysozyme (2 mM)	Acetamido methyl								
+α-NAG (12 mM)		2·4	0·4[g]	1·2[d]	0·8	50[i]	0·05	15	(18)[j]
+β-NAG (8 mM)		3·6	0·4[g]	1·2[d]	2·0	50[i]	0·05	13	(18)[j]
3-SLHis-15-lysozyme (2 mM)	Non-reducing acetamido methyl								
+di-NAG (2 mM)		8·4	0	5·3[d]	3·1	5000[i]	0·725	18	(28)[k]
lysozyme (7 mM)	C(2) hydrogen of His-15	8·0	2·0[h]	4·0	2·0	33[e]	0·286	17	18
+TEMPO-acetamide (14 mM)	Non-reducing acetamido methyl								
lysozyme (2 mM)		2·6	≤0·2	2·4[d]	≤0·2	5000[i]	0·016	≤15	28
+TEMPO-acetamide (8 mM)	methyl								
+di-NAG (25 mM)									
Lysozyme (7 mM)	C(2) hydrogen of His-15	5·0	0	4·0	1·0	185[e]	0·393	20	21
+NAG-CH₂-TEMPYRO (6 mM)									
Lysozyme (7 mM)	C(2) hydrogen of His-15	4·8	0	4·0	0·8	150[e]	0·20	19	21
+NAG-NAG-CH₂-TEMPYRO (3 mM)									

a Components were dissolved in D_2O and the solution adjusted to pH 4·5.

b In Hz.

c As M^{-1} and determined by e.s.r.

d This is the line-width of the saccharide in the absence of the nitroxide. For example, in experiments employing 3-spin-labelled His-15-lysozyme, the native enzyme was substituted.

e These values were calculated using eqn (12.10) and eqn (9.2) and the assumptions discussed in the results section.

f These values were determined by measurements on a Kendrew model of lysozyme built according to the coordinates of Phillips (1967).

g These values were estimated from linewidth measurements of NAG in the presence of spin-labelled Ser-195-chymotrypsin to which the saccharide does not bind.

h This value was estimated by substituting TEMPYRO-carbinol(IV) for TEMPO-acetamide. The alcohol does not bind to lysozyme.

i Values reported by Chipman and Sharon (1969)

j This distance (18·1 Å) is the model distance between C(2) hydrogen of His-15 and the acetamido methyl of β-NAG in subsite C, and cannot be exactly equal to the calculated distance (13·8 or 11·4 Å). See text.

k This distance (28·4 Å) is the model distance between the C(2) hydrogen of His-15 and the non-reducing acetamido methyl of di-NAG in subsite B, and cannot be exactly equal to the calculated distance of 18 Å. See text.

value of τ_c is to measure the ratio of $T_{1,P}/T_{2,P}$. However, in general, the value of τ_c will be such that $\omega_I\tau_c > 1$. In this region $1/T_{1,P}$ may be quite small (as we saw in the previous example) and difficult to measure accurately. However these values of τ_c suggest that $T_{1,P}/T_{2,P} \approx 20$ and thus the changes in line-width caused by the spin label will be relatively easier to monitor. So that if the value of τ_c can be estimated independently it will obviously be quicker and easier to calculate distances from line-width measurements.

12.4.3. *The distance between a covalent spin label on phosphofructokinase and a Mg–ATP site* (Jones, Dwek and Walker 1973).

Phosphofructokinase (PFK) is a key regulatory enzyme of glycolysis. This enzyme catalyses the transfer of the terminal phosphate group from ATP to fructose-6-phosphate and has an absolute requirement for a divalent cation. The metal ion is thought to bind to the enzyme in the form of a complex with ATP. The enzyme isolated from mammalian skeletal muscle has a complex subunit structure probably based upon the association of identical protomers each of mol. wt. 90 000. This example is included because it represents the type of problems that NMR might be able to solve. The crystal structure of the enzyme is unknown and there is even some controversy as to the number of subunits. There is some evidence to suggest that there are at least three distinct sites at which various nucleotides can bind in competition (Mathias and Kemp, 1972). These sites are catalytic, inhibitory and activating, and n.m.r. could well provide the stereochemical relationship between them.

PFK has one extremely reactive SH group to which a spin label can be covalently bound and using a type (I) label, 4-(2-iodoacetamido)-2,2,6,6-tetramethyl piperidino-oxyl, the enzyme retains at least 60% of its original activity at pH 8·2. The e.s.r. spectrum of labelled PFK is shown in Fig. 12.15, and is characteristic of a 'partially' immobilized spin label ($\tau_R > 4 \times 10^{-9}$ s). The rotational correlation time of the molecule is ca. 10^{-7} s (Jones *et al.*, 1973a) and thus the spin-label spectrum would appear to be dominated by the motion of the label relative to the enzyme.

An unusual feature of spin-labelled phosphofructokinase is that the e.s.r. spectra show evidence of two components under all the conditions investigated. The splitting becomes most marked in the low-field line in the presence of metal–ATP complexes (Jones *et al.*, 1972) and raises the question of whether two sites are labelled. However, the labelling procedure is highly specific for a single thiol group and identical spectral shapes are obtained in all labelling experiments, so Jones *et al.* favoured an alternative explanation. The relative intensity of the two components is altered by environmental changes, and the intermediate forms have isosbestic points (Jones *et al.*, 1972). Figure 12.15 illustrates this effect as the pH is decreased in Tris–phosphate buffer. The spectral change, centred on pH 6·4, can be directly correlated with a dissociation of the enzyme from the tetramer to the dimer

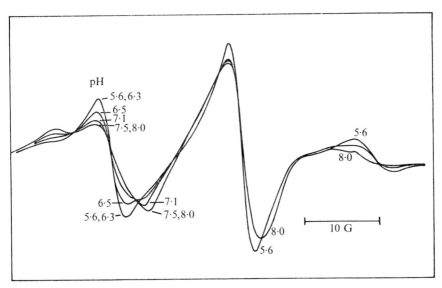

FIG. 12.15. The effect of pH on the e.s.r. spectrum of spin–labelled phosphofructokinase in 0·3 M Tris–phosphate buffer, 1 mM EDTA. The pH value was adjusted by additions of phosphoric acid. Note the isosbestic points, indicating an equilibrium between two conformational states. (From Jones *et al.* 1973a.)

(mol. wt. 180 000) observed in the ultracentrifuge. These changes may suggest that although the label is bound at a unique site it can 'flip' between two environments which may differ in polarity and rotational mobility. Qualitatively it can be seen that dissociation might be expected to favour the more polar and more mobile state of the label, particularly if the labelling site is near the subunit interface.

The addition of Mg–ATP to labelled PFK causes marked changes in the e.s.r. spectrum of the label (Fig. 12.16). Figure 12.17 shows the absorption spectra of the unliganded and liganded forms of the enzyme and the values of $\tau_{2,S}$ for each resonance line in this figure are given in Table 12.5. The values calculated from the absorption curves agree reasonably well with those obtained from the peak to peak separation of the corresponding derivative curves (Fig. 12.16). However, in view of the complexity of the spectra, the values should be considered to be very approximate. Computer simulation of the spectra would be required to obtain accurate values of $\tau_{2,S}$ but here we indicate a less sophisticated approach that is accessible to a larger number of workers.

We have dwelt on the estimation of $\tau_{2,S}$ because of its importance in the calculation of distances of nuclei from the spin label. As we noted in the lysozyme example, an independent estimate of τ_c is most helpful in experiments where only line-widths are to be measured. Again, we note that the

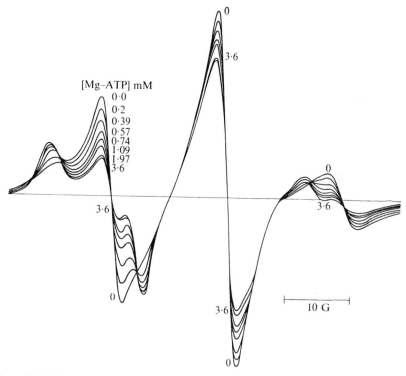

FIG. 12.16. E.s.r. spectra of spin-labelled phosphofructokinase (120 μM) in tris–HCl buffer, pH 7·5, titrated with Mg–ATP. The concentration of Mg–ATP is given at the curves, which are uncorrected for a dilution of 5–10%. Note the isosbestic points, which indicate the presence of two conformational states. (From Jones *et al.* 1972.)

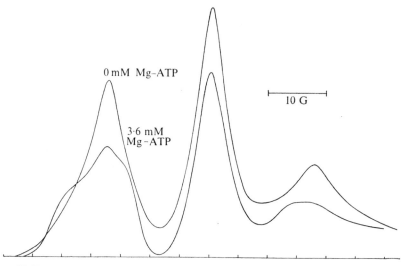

FIG. 12.17. E.s.r. absorption spectra of spin–labelled phosphofructokinase in the presence and absence of Mg–ATP obtained by integration of the corresponding two derivative spectra in Fig. 12.16.

TABLE 12.5

Comparison of the values of $\tau_{2,S}$ from the absorption spectrum and the approximate peak to peak separation in the derivative spectrum

Ligand		Value of $\tau_{2,S}$ for each resonance $\times 10^9$ (s)			Mean value of $\tau_{2,S} \times 10^9$ (s)
		low field	centre	high field	
—	Absorption	14·4	17·8	9·1	13·8
	Derivative	20·0	19·7	14·6	18·1
Mg–ATP	Absorption	8·7	17·8	6·5	11·0
	Derivative	4·5	16·5	7·8	9·6

changes in $1/T_1$ due to the spin label on the hydrogens of the Mg–ATP complex are small (Table 12.6). The variations in the value of τ_c calculated from the ratio of $T_{1,M}/T_{2,M}$ probably reflects this. The calculation of τ_c by this method assumes fast exchange. In this example this was verified by measuring the changes in line-width as a function of temperature, i.e. the line-width increased with decreasing temperature. That τ_c is indeed dominated by τ_S follows from the estimated values of $\tau_R \approx 10^{-7}$ s and $\tau_M > 5 \times 10^{-6}$ s.†
However the mean value of τ_c agrees rather well with that obtained from e.s.r. measurements. Two points however should be noted: (1) the value of τ_c obtained from the ratio of $T_{1,M}/T_{2,M}$ is $\tau_{1,S}$ and $\tau_{1,S} > \tau_{2,S}$; (2) the value of $\tau_{2,S}$ is obtained from the e.s.r. spectrum at X-band. This corresponds to a hydrogen resonance frequency of 14 MHz. At 90 MHz (the n.m.r. frequency) the value of $\tau_{2,S}$ may have increased. The τ_c from n.m.r. data should thus be longer than that from the e.s.r. spectrum. The agreement between the two methods in this case may therefore be fortuitous, in part, and probably reflects the experimental uncertainties in the ratio of $T_{1,M}/T_{2,M}$ and in the approximate method used for obtaining $\tau_{2,S}$.

TABLE 12.6

Longitudinal and transverse relaxation rates (s^{-1}) of ATP hydrogens and their distances from the spin-label unpaired electron at 23 °C and 90 MHz in spin-labelled phosphofructokinase
(37·5 μM/90 000 mol. wt), [MgATP] = 25 mM, $P_M = 0.0015$, pH 7·5
(From Jones *et al.*, 1973a)

	Labelled $1/T_{1,obs}$	Native $1/T_{1,obs}$	$1/T_{1,M}$	Labelled $1/T_{2,obs}$	Native $1/T_{2,obs}$	$1/T_{2,M}$	$\dfrac{T_{1,M}}{T_{2,M}}$	$10^9\tau_c$ (s)	r (Å)
H(2)	1·00	0·81	125	14·7	8·7	3950	31	13	7·2
H(8)	2·15	1·73	276	11·3	8·2	2100	8	6	7·5
H(1')	1·43	1·15	184	14·7	9·7	3290	18	10	7·0

† Estimated from the relationship $K_D = (k_{on}/k_{off}) = k_{on}\tau$. with $k_{on} < 10^{10}$ s^{-1} and $K_D \approx 500$ μM.

The question of interest now is to which site do the distances refer. ATP is both a substrate and an allosteric inhibitor and it seems that it does not bind to the activator site (Mathias and Kemp, 1972). We thus have a choice of two sites. Similar experiments to those above with the Mg–ITP–spin-labelled-PFK complex resulted in very small changes in the relaxation rates of the ITP hydrogens (Jones et al., 1973b). ITP is a good substrate and a much poorer inhibitor than ATP. Thus it would appear that the spin label is near to the inhibitory site. Clearly this is not conclusive but by using suitably designed competition experiments with different nucleotides it should be possible to sort this out.

12.4.4. Distances between the nitroxide moiety and selected hydrogens of substrates in the complex Mg–ADP–creatine-spin-labelled creatine kinase (Leigh, 1971; Cohn, Leigh and Reid, 1971a)

We saw earlier that the determination of the Mn(II)–substrate distances in the abortive Mn–ADP–creatine–creatine kinase complex was straightforward only for the creatine CH_2 and CH_3 hydrogens. The main difficulties arise from the presence of Mn(II)–substrate complexes. It will be recalled that if there is a substantial concentration of metal ion–substrate complex any effects from the smaller concentration of the ternary complex (metal–substrate–enzyme) may be difficult to detect. With a covalent spin-label probe these difficulties are largely absent mainly because there is no equivalent spin-label substrate complex. Additionally, with these spin-label probes, the diamagnetic control for the experiments is achieved quite simply by using a mild reducing agent.

The distances from the nitroxide moiety to various protons on ADP and creatine are listed in Table 12.7. The correlation times used in the calculation

TABLE 12.7

Distances (in Å) between nitroxide moiety covalently bound to creatine-kinase and various nuclei on substrates (Leigh, 1971; Cohn et al., 1971a)

	ADP hydrogens			Creatine hydrogens	
	H(2)	H(8)	H(1′)	CH_2	CH_3
Mg–ADP–creatine	7·9	7·3	7·9	9·5	9·3

Correlation times used in the calculation of distances

Method	$10^8 \tau_c$ (s)
$T_1/T_{2,M}$	1·2
τ_R	2·6 ⎫ 1·3
$\tau_{2,S}$	2·45 ⎭

of these distances are also shown. τ_R was estimated as in Section 10.10.3, using a partial specific volume of 0.744 cm^3 g^{-1}, while τ_{2S} was estimated from the e.s.r. spectrum. The value of τ_c obtained by combining these values i.e. $(1/\tau_R + 1/\tau_{2,S})$ is in good agreement with that from the ratio of $T_{1,M}/T_{2,M}$, although for similar reasons to those discussed in the previous example this may be fortuitous.

12.5. Calculation of the distance between a spin-label probe and a second paramagnetic centre

In this section we shall consider how the rigid lattice line shape of an electron spin resonance signal is influenced by dipole–dipole interaction with a second spin. Under certain conditions there is no observable broadening of the e.s.r. signal, only an apparent decrease (or 'quenching') in the spectrum amplitude. The analysis of this diminution in amplitude allows the calculation of the distance between the two spins (Leigh, 1970). We shall be concerned here only with the effects of a paramagnetic ion on the e.s.r. spectrum of a spin label when attached to a macromolecule.

The theory of relaxation effects of dipole–dipole interactions of small molecules, undergoing rapid rotation and characterized by short correlation times are quite well understood. Theoretical treatments have been considered for the other extreme of rigid molecules and long correlation times. However, as yet, Leigh's theory (1970) is the only one for the intermediate case, often encountered with macromolecules, where the rotational correlation times are long but the correlation times, characterizing the dipolar interaction between the two spins, are short. The development of the theory followed the observation that the addition of Mn–ADP to a solution of spin-labelled creatine kinase caused a dramatic decrease (ca. 95%) of the amplitude of the spin-label e.s.r. spectrum without appreciably broadening the spectrum.

Leigh's theory assumes that the relative geometry of the two spins does not change appreciably during the life-times of the spin states. The e.s.r. line-width of the observed spin (in these examples that of the spin label), is then given by:

$$\delta B = C(1-3\cos^2\theta'_R)^2 + \delta B_0 \tag{12.11}$$

where δB_0 is the natural e.s.r. line-width of the spin label in the absence of the paramagnetic ion, θ'_R is the angle between the applied magnetic field and the vector joining the two unlike spins and

$$C = \frac{g\beta^3\mu^2}{\hbar r^6}.\tau, \tag{12.12}$$

r is the distance between the two unlike spins, μ is the magnetic moment of the paramagnetic ion (in Bohr magnetons), g is $\frac{1}{3}(2g_\perp + g_\parallel)$ and τ is the correlation time for the modulation of the dipolar interaction. Since the theory is concerned with relatively rigid centres Leigh (1970) has identified the value of τ in eqn (12.12) as that of the electron spin relaxation time of the spin creating the relaxing field (i.e. the paramagnetic ion in this case). The value of C obviously depends on the geometry of the system and, in principle, C can be obtained from a comparison of the observed and simulated spectra. For the parameters likely to be applicable to nitroxide radicals, Leigh (1970) has calculated the value of C as a function of the relative amplitude of the centre peak (Figs. 12.18 and 12.19).

The physical basis for the decrease in signal amplitude can be seen from inspection of the equation above. The two interacting spins are rigid and with a random orientation of molecules in the magnetic field all values of θ'_R are possible. For most values of θ'_R, the term $C(3 \cos^2\theta'_R - 1) \gg \delta B_0$ and the resulting spectra will be too broad to observe. However, there will be certain orientations of the molecules for which θ'_R has a value such that

$$(3 \cos^2\theta'_R - 1) = 0.$$

In such cases the unperturbed spectrum will be observed, but it will be diminished, since only a small number of spins will have values of θ'_R which allow the above condition to be satisfied.

The theory assumes that the spins are relatively rigid ($\tau_R \gg \tau_{1,S}$) and further, that $\omega_S \tau_{1,S} \gg 1$. If the spin label is rapidly rotating then the validity

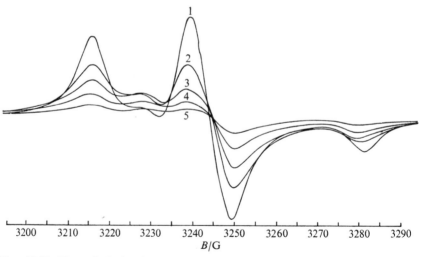

FIG. 12.18. Theoretical nitroxide e.s.r. spectra. The parameters used were $g_{xx} = 2\cdot0089$, $g_{yy} = 2\cdot0061$, $g_{zz} = 2\cdot0027$, $A_{xx} = 1\cdot25 \times 10^8$ rad s^{-1}, $A_{yy} = 0\cdot986 \times 10^8$ rad s^{-1}, $A_{zz} = 5\cdot63 \times 10^8$ rad s^{-1}, $\delta B_0 = 4\cdot59$. The value of C in curves 1–5 is 0, 3, 10, 30, and 100, respectively. (From Leigh, 1970.)

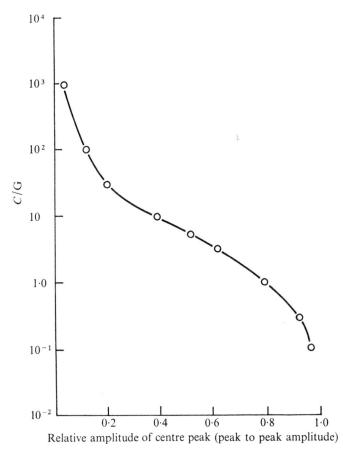

FIG. 12.19. Variation of the relative amplitude of peak height of a 'nitroxide' radical with the dipolar interaction coefficient C. (Supplied by Dr. J. S. Leigh.)

of this theory is difficult to assess. However the effect of the rapid motion will probably tend to reduce the mean dipolar interaction and, if this is the case, the value of the distance will be an upper limit.

In order to indicate the sensitivity of the value of r to the values of τ and of the relative amplitude of the centre e.s.r. peak, r is plotted as a function of these two variables in Fig. 12.20. The value of r is quite sensitive to changes in the correlation time and the relative signal amplitude. For example, if the relative amplitude is 80%, a change in the correlation time from 3×10^{-10} to 3×10^{-9} s changes r from 16 to 24 Å. Because of this sensitivity and often the difficulty of evaluating the correlation time accurately some care is necessary in interpreting distances obtained by this method.

Below we shall consider an example of the application of this theory to

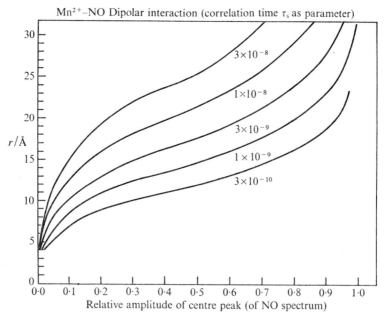

FIG. 12.20. The distance between Mn(II) and the nitroxide spin is plotted as a function of the amplitude of the centre peak of the nitroxide e.s.r. spectrum. The series of curves as indicated represent the function at different values of the correlation time, τ_s (the longitudinal electron spin relaxation time of Mn(II) in the enzyme complex). (From Cohn Diefenbach and Taylor, 1971*b*.)

determine the distance between Mn(II) and a spin-label probe in the Mn–ADP– spin-labelled creatine kinase system. By combining this result with the distances, from the spin label and from the Mn(II) ion, to selected nuclei on ADP and creatine, a map of the substrates at the active site can be drawn. However because of the importance of the determination of distances between the two probes in such triangulation procedures, it will be necessary in the future to check the application of the theory on suitable model systems.

12.6. Calculation of the Mn(II) to the nitroxide spin-label distance in the Mn(II)–ADP–spin-labelled–creatine kinase complex

The e.s.r. spectrum of spin-labelled creatine kinase corresponds approximately to that expected for a rigid lattice spectrum. The addition of saturating concentrations of Mn(II)–ADP results in ca. 95 per cent reduction in signal amplitude. For comparison purposes Fig. 12.21 shows the results with the Co(II)–ADP and Ni(II)–ADP complexes. The decrease in signal height is much smaller and could reflect variations in the metal–spin label distance or more probably the much smaller values of $\tau_{1,S}$ for Co(II) and Ni(II) complexes.

The value of $\tau_{1,S}$ for the Mn(II) complex was estimated from the enhancement of the water relaxation rate in the Mn(II)–ADP–spin-labelled creatine kinase complex ε_t. At 6 °C the value of ε_t is ca. 25 (at 24·3 MHz). Assuming $q^* = 3$, then $\tau_{1,S} \approx 1\cdot5\times10^{-9}$ s (Cohn et al., 1971b). The e.s.r. frequency used in these experiments corresponds to a hydrogen resonance frequency of ca. 14 MHz. Strictly speaking, the value of $\tau_{1,S}$ at this frequency should be used. It is possible, however, that the above value is correct since the value of $\tau_{1,S}$ is calculated on the assumption of a negligible contribution from τ_M to τ_c (see p. 245 and 282). If τ_M does contribute to τ_c, $\tau_{1,S}$ (at 24·3 MHz) would be longer than that calculated. However its value at 14 MHz would be expected to be shorter thereby, perhaps resulting in some self compensation in using the above value.

A value of C for this system was obtained from theoretical predictions of the e.s.r. spectrum expected for nitroxide spin labels in the presence of a second spin. Knowing $\tau_{1,S}$, r was thus calculated as ca. 7·5 Å.

In this particular case an elegant control can be carried out to show that the effects of the spin label on the e.s.r. spectrum arise from the proximity of the Mn(II)–ADP complex to the spin label and not merely from its presence in the solution. It is known that ADP and Mn(II)–ADP compete for the same site and the titration curve in Fig. 12.22 shows that at high ADP levels the amplitude of the central peak begins to rise towards its initial value as ADP displaces Mn–ADP from the binding sites on the enzyme.

Cohn et al. (1971b) also consider how the interaction between the two paramagnetic species would be different if the spin label was free in solution

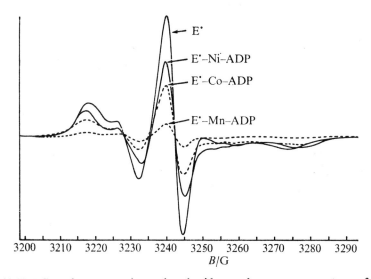

FIG. 12.21. Effect of paramagnetic metal nucleotide complexes on e.s.r. spectrum of spin-labelled creatine kinase. The solutions contained enzyme, 9 mg ml⁻¹ in 0·05 M N-ethyl-morpholine-Cl, pH 8·0, (E·), no metal nucleotide; (E·–Ni–ADP), 1 mM NiCl₂ and 1 mM ADP; (E·–CoADP), 3 mM CoCl₂ and 3 mM ADP; (E·–Mn ADP), 2·5 mM MnCl and 2·5 mM ADP. T = 22 °C. (From Taylor et al. 1969.)

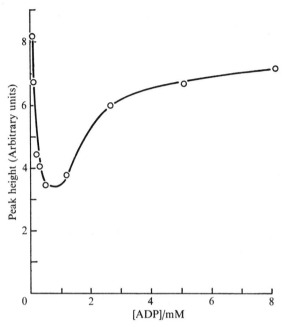

FIG. 12.22. Titration of the spin-labelled creatine kinase at constant Mn(II) with ADP. The amplitude of the centre e.s.r. line is plotted versus ADP concentration. The solutions contained enzyme, 4·2 mg ml⁻¹; MnCl₂, 1 mM; 0·05 M N-ethylmorpholine-Cl, pH 8·0. $T = 22$ °C. (From Taylor *et al.* 1969.)

rather than covalently bound to the enzyme. In contrast to the covalent spin label, for which Mn–ADP only causes a decrease in signal amplitude, for the free spin label there is a change in shape (i.e. a broadening) [Fig. 12.23]. This broadening will also cause the signal amplitude to decrease. However double integration of the derivative curves in Fig. 12.23 gives areas under the respective absorption curves within 1%, indicating that the total number of spins is unchanged (Cohn *et al.*, 1971b). This is not the case for the spectra of the spin-labelled enzyme spectra in Fig. 12.23 resulting from the addition of Mn–ADP. For example, there is an apparent loss of 90% of the total spin in spectrum 3 compared to spectrum 1. Of course there is no loss of spin in this system, for as we have seen for most of the spins, the dipolar interaction is very large compared to the unperturbed line-width and the resulting signals are too broad to observe.

12.7. The topography of the substrates at the active site of creatine kinase

We are now in a position to combine the results from two paramagnetic reference points, Mn(II) and the spin label, in the Mn–ADP–creatine-creatine kinase complex. The model constructed of the topography of the

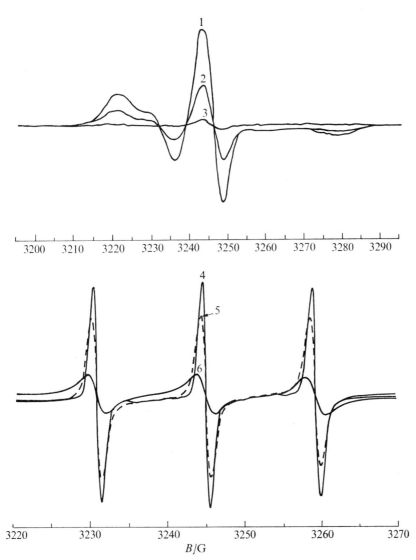

FIG. 12.23. Comparison of the effect of Mn–ADP on the e.s.r. difference spectra of the spin label (a) covalently bound to creatine kinase, (b) free in solution. Series (a) E·, 4 mg ml⁻¹ in 50 mM N-ethylmorpholine-Cl, pH 7·9 $T = 20$ °C; Curve (1) no additions, (2) 2·5 mM Mn ADP, (3) 7·7 mM Mn–ADP. Series (b); N(1-oxyl-2,2,5,5-tetramethyl-3-pyrrolidinyl)-iodoacetamide, 0·1 mM in 50 mM potassium–HEPES, pH 7·9, $T = 20$ °C; curve (4) no additions; (5) 5·5 mM Mn–ADP; (6) 55·0 mM Mn–ADP. (From Cohn *et al.* 1971*b*.)

TABLE 12.8

Internuclear distances in Å in the 'active' complex of creatine kinase (Leigh, 1971; Cohn *et al.*, 1971a)

| | ADP | | | | | Creatine | | |
	H(2)	H(8)	H(1′)	P_α	P_β	CH$_2$	CH$_3$	Mn(II)
Mn(II)	6·0	3·4	—	3·0	3·0	9·8	10·3	—
NO	7·9	7·3	7·9	—	—	9·3	9·5	8·0

substrates at the active site involves the calculated distances given in Table 12.8 and is shown in Fig. 12.24. The Mn(II)—^{31}P distances for the α- and β-group were assumed to be the same as in the Mn–ADP complex.

We noted earlier that in the active complex (with ATP replacing ADP) the Mn(II)–creatine distances were unchanged. This means that Mn(II) is not co-ordinated to the γ-phosphate of ATP in the complex with creatine (q.v. Section 9.12) (Leigh, 1971; Cohn *et al.*, 1971a). Attempts to 'place' the metal ion on the β,γ-phosphates and maintain contact between the creatine nitrogen and the transferable phosphate lead to Mn–creatine hydrogen distances of less than 8 Å which is not really compatible with the experimental results (Table 12.8). Thus a significant mechanistic implication may be that the metal ion is *not* co-ordinated to the phosphoryl group to be transferred. Another feature is that the distances of the methyl and methylene hydrogens of creatine to Mn(II) are too great (>9 Å) for the direct co-ordination of

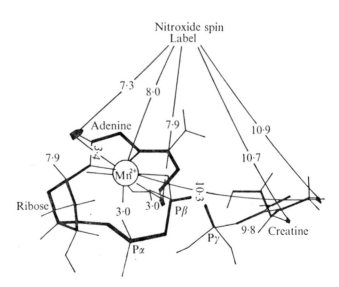

FIG. 12.24. Molecular model of topography of substrates at the active site of creatine kinase. (From Cohn *et al.* 1971a; Leigh 1971.)

creatine to Mn(II). This means that the metal ion does not activate the guanidino substrate.

In this particular instance there are many assumptions in building this map of the substrate at the active site: in particular the labelling of the essential SH group which results in the enzyme becoming inactive. This may be due to a conformational change involving the SH group on the enzyme so that conclusions based on the use of the nitroxide probe could have little relationship to those in the active enzyme. However, the procedures involved for the creatine kinase example do show the type of information and the conclusions that can be obtained.

12.8. The effects of spin-label probes on the proton nuclear relaxation rates of the solvent water

Although several studies of relaxation rates of solvent water in systems containing spin labels have been reported, no detailed analyses in terms of molecular motion of the water hydrogens have been reported.

In many solvents containing organic free radicals, the relaxation rates of the solvent hydrogens have been interpreted satisfactorily in terms of a dipolar interaction between the electron and hydrogen spins modulated by the random relative translational motion of the molecules (Dwek, Richards, and Taylor 1969). It would seem reasonable to use such an approach in interpreting the relaxation rates of the solvent protons in solutions containing spin-labelled enzymes (or macromolecules). The relevant correlation time is τ_D, that of random translational diffusion and the equations for the relaxation rates are similar to those for the case of outer sphere relaxation (see p. 190). In general, the values of τ_D are $\sim 10^{-10}$–10^{-11} s. Thus it is reasonable to assume that $\omega_I \tau_D \ll 1$. We may then write eqns (9.17) and (9.18) as:

$$\frac{1}{T_{1,P}} = \frac{1}{T_{1,\text{obs}}} - \frac{1}{T_{1,(0)}} = \frac{N_S \gamma_I^2 \gamma_S^2 \hbar^2 \pi}{50 \mathscr{D} d} \left[\tfrac{28}{3} f_t(\omega_S \tau_D) + 4\right] \qquad (12.13)$$

$$\frac{1}{T_{2,P}} = \frac{1}{T_{2,\text{obs}}} - \frac{1}{T_{1,(0)}} = \frac{N_S \gamma_I^2 \gamma_S^2 \hbar^2 \pi}{50 \mathscr{D} d} \left[\tfrac{26}{3} f_t(\omega_S \tau_D) + \tfrac{14}{3}\right] \qquad (12.14)$$

The explicit form of $f_t(\omega_S \tau_D)$ has been given previously (p. 190). We note simply that when $\omega_S \tau_D \ll 1$, $f_t(\omega_S \tau_D) = 1$. Thus for the ω_S dispersion, the ratio of $T_{1,P}$ at high $[f_t(\omega_S \tau_D) \approx 0]$ to low $[f_t(\omega_S \tau_D) \approx 1]$ frequencies is 10:3. In the above equations, d is the distance of closest approach between the solvent hydrogens and the free radical electron, D is the relative diffusion coefficient, and N_S is the number of electron spins per ml of solution. τ_D is given by

$$\tau_D = d^2 / \mathscr{D} \qquad (12.15)$$

An analysis of relaxation rates measured at different frequencies can yield information on d, τ_D, and $\mathscr{D}$ (Dwek et al., 1969).

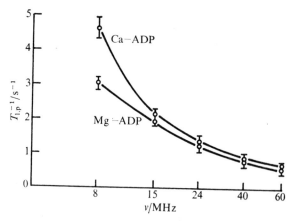

FIG. 12.25. Frequency dependence of spin–lattice relaxation rates of solvent water protons for Mg–ADP (25 mM) and Ca–ADP (8 mM) complexes of spin–labelled creatine kinase (0·21 mM active sites) at pH = 8·0 and $T = 0$ °C. (From Taylor, McLaughlin, and Cohn 1971.)

Let us now consider the results of the water proton relaxation rates in the Mg–ADP– and Ca–ADP–spin-labelled creatine kinase complexes (Cohn *et al.*, 1971b; Taylor *et al.*, 1971). The values of the hydrogen spin–lattice relaxation rates, $1/T_{1,P}$, as a function of frequency are plotted in Fig. 12.25. The functional form of $f_t(\omega_S \tau_D)$ is shown in Fig. 12.26, and we note that at high frequencies the value of $1/T_{1,P}$ becomes constant. This behaviour is approximately that observed for the spin-labelled complexes. This suggests

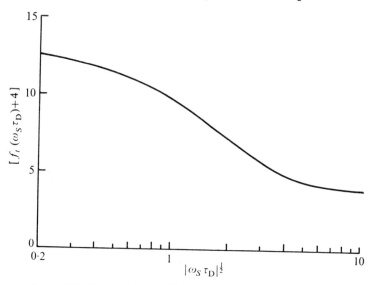

FIG. 12.26. The variation of $[f_t(\omega_S \tau_D)+4]$ with $(\omega_S \tau_D)^{\frac{1}{2}}$ in eqn (12.13).

that the value of $\tau_D \approx 2 \times 10^{-11}$ s. However the ratio of the values of the relaxation times at high to low frequencies is generally greater than the expected value of 10/3. One possible explanation is that there could be a contribution to the observed proton relaxation rates from the hydration spheres of adjacent amino acid residues or even that of the metal nucleotide complex. The relevant correlation times for such relaxation would then be either the rotational correlation time (τ_R) or the electron spin–lattice relaxation time ($\tau_{1,S}$). τ_R can be estimated from Stokes' Law as ca. 3×10^{-8} s and from the e.s.r. spectrum the value of τ_S is ca. 4×10^{-8} s. This means that at high frequencies $\omega_I \tau_{1,S}$ or $\omega_I \tau_R \gg 1$ and correspondingly less, of course, at low frequencies. The result would then be that relaxation by these processes would only make a significant contribution at the low frequencies, and thus increase the expected ratio of the relaxation times at high and low frequencies. The usual Solomon–Bloembergen equations would apply then to the relaxation rates of this 'bound water', in contrast to the previous case. However even with measurements over a larger frequency range than that shown in Fig. 12.25, the analysis would be difficult and ambiguous. What does seem reasonably clear is that relaxation by translational diffusion is often important. In general though as a result of the short value of τ_D, the enhancements of water relaxation rates in the system *will be small*, even at low frequencies and low temperatures where the theory predicts the relaxation rates will be largest.

Appendix 12.1

The conditions for 'fast exchange' in the example involving the lysozyme system

The usual equation for $1/T_{2,P}$ applies only when the population of the 'bound' site is very much less than that of the bulk site. It can be shown, however, that in practice it is a reasonable approximation to use the Swift–Connick equation for $1/T_{2,P}$ up to a bound fraction of ca 0·3.

A more general equation for the value of $1/T_{2,P}$ could be derived from the complete expression for the line-shape function (McConnell, 1958). An analysis of this expression shows that it can be divided into three terms. Two of these correspond to the usual Swift–Connick line shape functions for *each* site. The third term involves both sites and only contributes when the exchange between them is fast. For example, in the slow-exchange region, where the A and M resonances are resolved, the line-widths of each resonance are given by equations of the form

$$\frac{1}{T_{2,P}} = \frac{1}{T_{2,M}} + \frac{P_M}{\tau_M} \qquad (12.16)$$

where P_M is the fraction bound in the M site. This equation corresponds to eqn (9.28) (neglecting outer sphere terms). Following the treatment in Chapter 9, we can obtain a limiting expression for the region of exchange narrowing as

$$\frac{1}{T_{2,P}} = \frac{P_M \cdot q}{T_{2,M}} + \frac{P_A \cdot q}{T_A} + \tau(P_A q)^2 (P_M q)^2 \cdot (\Delta\omega_A - \Delta\omega_M)^2 \qquad (12.17)$$

where $\tau = \tau_A + \tau_M$ and $\Delta\omega_A$ and $\Delta\omega_M$ are the corresponding shifts in the two sites. Under conditions of rapid exchange, as expected,

$$\frac{1}{T_{2,P}} = \frac{P_M q}{T_{2,M}} + \frac{P_A q}{T_{2,A}} \qquad (12.18)$$

The conditions for *fast exchange* are thus: $\tau_M \ll T_{2,M}$ and $\tau_A \ll T_{2,A}$ which are equivalent to

$$\frac{1}{\tau} \gg \frac{P_A \cdot q}{T_{2,A}}, \frac{P_M \cdot q}{T_{2,M}} .$$

and

$$\frac{P_A \cdot q}{T_{2,A}} + \frac{P_M \cdot q}{T_{2,M}} \gg \tau(P_A q)^2 (P_M q)^2 (\Delta\omega_A - \Delta\omega_M)^2 \qquad (12.19)$$

The value of τ_M for the inhibitors used in the work of Wien *et al.* (1973) is estimated as $\tau_M > 10^{-5}$ s. It can be shown that the values of $T_{2,M}$ and $T_{2,A}$ calculated in almost all of these experiments are such that the above conditions generally hold and *conditions of fast exchange are valid.* As a corollary we may note, with hindsight, that the shift between the sites due to the presence of the spin label ($\Delta\omega_A - \Delta\omega_M$) is expected to be small (ca. 1–2 Hz) because of the large distances involved (Wien *et al.*, 1973).

References

BERLINER, L. (1971). *J. molec. Biol.* **61**, 189.

CAMPBELL, I. D., DWEK, R. A., PRICE, N. C., and RADDA, G. K. (1972). *Eur. J. Biochem. 30*, 339.

CHIPMAN, D. M. and SHARON, N. (1969). *Science,* **165**, 454.

COHN, M., LEIGH, JR. J. S., and REED, G. H. (1971a). *Cold Spring Harbor Symposia on Quantitative Biology,* **36**, 533.

——, DIEFENBACH, H., and TAYLOR, J. S. (1971b). *J. biol. Chem.* **246**, 6037.

DUBIN, S. B., CLARK, N. A., and BENEDECK, G. B. (1971). *J. chem. Phys.* **54**, 5158.

DWEK, R. A., GRIFFITHS, J. R., RADDA, G. K., and STRAUSS, U. (1972). *FEBS Lett.* **28**, 161.

——, RICHARDS, R. E., and TAYLOR, D. (1968). *A. Rev. N.M.R.* **2**, 394.

GRIFFITH, O. H., LIBERTINI, L. J., and BIRRELL, G. B. (1971). *J. phys. Chem.* **75**, 3417.

JONES, R., DWEK, R. A., and WALKER, I. O. (1972). *FEBS Lett.,* **26**, 92.

——, ——, ——, (1973). *Eur. J. Biochem. 34*, 28.

——, ——, ——, (1973b) unpublished results.

JOST, P. and GRIFFITH, O. H. (1972). In *Methods in pharmacology,* (ed. C. Chignell) Vol. 2, 223, Appleton-Century-Crofts, New York.

——, WAGGONER, A. S. and GRIFFITH, O. H. (1971). In *Structure and function of biological membranes,* (ed. L. Rothfield) p. 83, Academic Press, New York.

KRUGH, T. R. (1971). *Biochemistry* **10**, 2594.

LEIGH, JR. J. S. (1970). *J. chem. Phys.* **52**, 2608.

——, (1971). *Ph.D. Dissertation*, University of Pennsylvania, Philadelphia.

MATHIAS, M. M. and KEMP, R. G. (1972). *Biochemistry* **11**, 578.

MCCONNELL, H. M. (1958). *J. chem. Phys.* **28**, 430.

OHNISHI, S. and MCCONNELL, H. M. (1965). *J. Amer. chem. Soc.* **87**, 2293.

PHILLIPS, D. C. (1967). *Proc. natn. Acad. Sci.* **57**, 484.

ROZANTSEV, E. G. (1970). *Free nitroxyl radicals*, Plenum Press, New York and London.

STONE, T. J., BUCKMAN, T., NORDIO, P. L., and MCCONNELL, H. M. (1965). *Proc. natn. Acad. Sci. U.S.* **54**, 1010.

TAYLOR, J. S., LEIGH, JR. J. S., and COHN, M. (1969). *Proc. natn. Acad. Sci. U.S.* **64**, 219.

——, (1969). *Ph.D. Dissertation*, University of Pennsylvania, Philadelphia.

——, MCLAUGHLIN, A., and COHN, M. (1971). *J. biol. Chem.* **246**, 6029.

WIEN, R. W., MORRISETT, J. D., and MCCONNELL, H. M. (1972). *Biochemistry* **11**, 3707.

13

SOME EXAMPLES OF THE USE OF NUCLEI WITH QUADRUPOLE MOMENTS

13.1. Introduction

FOR nuclei with spin quantum numbers greater than $\frac{1}{2}$, the distribution of positive charge over the nucleus may be aspherical and this situation may be described in terms of a nuclear electric quadrupole moment. If the electron distribution about the nucleus has less than cubic symmetry, the resulting electric field gradient can interact with the nuclear electric quadrupole moment. Any fluctuations of the electric field gradient coupled to the quadrupole moment can provide an *additional* mechanism for nuclear relaxation. The relaxation rates are given by

$$\frac{1}{T_1} = \frac{1}{T_2} = \frac{3}{40} \frac{2I+3}{I^2(2I-1)}\left(1+\frac{\tilde{\eta}^2}{3}\right)\left(\frac{e^2Qq}{\hbar}\right)^2\tau_c = K\left(\frac{eQq}{\hbar}\right)^2\tau_c \qquad (13.1)$$

$\omega_I \tau_c \ll 1$

where (eq) is the electric field gradient to which the electric quadrupole moment of the nucleus (eQ) is coupled. $\tilde{\eta}$ is an asymmetry parameter which relates to the field gradient symmetry around the nucleus and τ_c is the correlation time characterizing the fluctuations of the electric field gradient. The relevant correlation time may be rotational or diffusional in a liquid. In a molecule such as carbon tetrachloride, the quadrupole coupling arises from the electric fields produced by the C—Cl bond, so that *rotation* of the molecule will cause fluctuations in this interaction, which lead to the nuclear relaxation. On the other hand, the electron distribution about the bromine nucleus in the bromide ion in solution is almost symmetrical, but the symmetry is momentarily reduced by collisions with solvent molecules or with other ions in the solution. In this case the transient quadrupole coupling is associated mainly with a diffusional correlation time.

The electric field gradients produced by chemical bonds are often very strong and the electric quadrupole interaction may dominate completely the relaxation time of the nuclei. The quadrupolar relaxation is often so strong that nuclear resonances become very broad and difficult to detect. For example, the ^{35}Cl chloride ion line-width from an aqueous solution of NaCl is ca. 15 Hz, whereas that from pure liquid CCl_4 is ca. 10 KHz.

The sensitivity to its environment of the relaxation times of a nucleus with a quadrupole moment gives the impetus to the study of such nuclei in biological systems. The binding of ions or ligands (possessing quadrupole moments) to macromolecules can produce large changes in the relaxation times of the ion as a result of changes in electric field gradients and correlation times. In most liquids, typical values of the correlation times are 10^{-12} s, but when an ion binds to a macromolecule, these values may be reduced by three orders of magnitude or more, corresponding to the slower rotational motions of the macromolecule. This change in τ_0 may often be dominant in such systems in causing short relaxation times. A change in both (eq) and τ_0 can lead to very dramatic results.

TABLE 13.1

Nuclear properties of some naturally-occurring n.m.r.-active nuclei with quadrupole moments compared with those of hydrogen

Isotope	Natural abundance %	Resonance frequency MHz in field of 10^4 G	Sensitivity at const. field relative to H = 1·000	Nuclear spin in units of $\hbar$	Nuclear electric quadrupole moment (barns)
^{1}H	99·985	42·5759	1·000	$\frac{1}{2}$	0
^{2}D	0·015	6·5357	$9\cdot65 \times 10^{-3}$	1	$2\cdot77 \times 10^{-3}$
^{17}O	0·037	5·772	$2\cdot91 \times 10^{-2}$	$\frac{5}{2}$	$-4\cdot0 \times 10^{-3}$
^{14}N	99·635	3·076	$1\cdot01 \times 10^{-3}$	1	2×10^{-2}
^{23}Na	100	11·262	$9\cdot27 \times 10^{-2}$	$\frac{3}{2}$	0·1
^{25}Mg	10·05	2·606	$2\cdot68 \times 10^{-3}$	$\frac{5}{2}$	0·22
^{39}K	93·08	1·987	$5\cdot08 \times 10^{-4}$	$\frac{3}{2}$	0·07
^{41}K	6·91	1·092	$8\cdot39 \times 10^{-5}$	$\frac{3}{2}$	—
^{43}Ca	0·13	2·865	$6\cdot39 \times 10^{-2}$	$\frac{7}{2}$	—
^{85}Rb	72·8	4·111	$1\cdot05 \times 10^{-2}$	$\frac{5}{2}$	0·31
^{87}Rb	27·2	13·932	$1\cdot75 \times 10^{-2}$	$\frac{3}{2}$	0·15
^{133}Cs	100	5·585	$4\cdot74 \times 10^{-2}$	$\frac{7}{2}$	$-0\cdot3 \times 10^{-2}$
^{35}Cl	75·53	4·1717	$4\cdot70 \times 10^{-3}$	$\frac{3}{2}$	$-7\cdot89 \times 10^{-2}$
^{37}Cl	24·47	3·472	$2\cdot71 \times 10^{-3}$	$\frac{3}{2}$	$-6\cdot21 \times 10^{-2}$
^{79}Br	50·54	10·667	$7\cdot86 \times 10^{-2}$	$\frac{3}{2}$	0·33
^{81}Br	49·46	11·498	$9\cdot85 \times 10^{-2}$	$\frac{3}{2}$	0·28
^{127}I	100	8·5183	$9\cdot34 \times 10^{-2}$	$\frac{5}{2}$	$-0\cdot69$

The concentration of macromolecules in solution is generally quite small. The inherent lack of sensitivity of quadrupolar nuclei makes them difficult to detect by n.m.r. Practical concentrations with present instrumentation are ca. 10 mM for ^{14}N and those for other nuclei can be gauged from Table 13.1. Thus, at present, detection of resonance of nuclei bound to macromolecules (usually in the μM concentration range) is not possible in the majority of systems. As before, fast chemical-exchange conditions provide the biochemical amplifier. The theory of exchange in such systems is similar to that described previously. However there may be slight complications if the exchange lifetime, τ_M, is comparable with τ_R, the rotational correlation time. Marshall (1970) has shown that for a quadrupolar nucleus undergoing fast exchange between two environments the value of τ_0 in eqn (13.1) depends both on τ_R and τ_M, i.e.

$$\frac{1}{\tau_c} = \frac{1}{\tau_R} + \frac{1}{\tau_M} \qquad (13.2)$$

$$\tau_M \ll \tau_R$$

We return to this point on pp. 345, 349. In the fast-exchange limit, if $\tau_M \gg \tau_R$, then, as expected, the relaxation rates of the nuclei are given by the mean of the relaxation rates in each environment weighted by their respective populations. The relaxation rates in the bound site are given by eqn (13.1). Because the relaxation rates in the bound site may be so large, the addition of a macromolecule with 10 μM binding sites for an ion can give rise to detectable effects on the relaxation.

What of the drawbacks? Inspection of Table 13.1 shows that the sensitivity of the n.m.r. detection method for these nuclei is considerably less than for hydrogen. The relatively high concentrations of these nuclei then needed for observation in macromolecular systems may lead to non-specific binding and other non-specific effects. Even if it is possible to arrange the system such that the relaxation effects of one particular site dominates, as in the 'halide probes' (see below), there must always be some doubt as to the effect of the large concentrations of ions on the structure of the macromolecule. It may also be difficult to obtain quantitative results in terms of changes in τ_c and (eqQ) because, for instance, the use of eqn (13.1) to obtain information on (eqQ) and τ_c is only valid if the 'white spectrum' approximation ($\omega_I \tau_c \ll 1$) is valid. In the non-white region (i.e. as the motion of the 'lattice' becomes more rigid) the equation can be easily modified for nuclei with $I = 1$ but for nuclei with $I > 1$ the theory becomes rather complex. In this case the relaxation times are no longer simple exponential functions and measurements of relaxation rates could be difficult to interpret. This point is discussed further for the case of ^{35}Cl ($I = \frac{3}{2}$) binding to haemoglobin (see pages 344, 345). However, as usual, the observation of a *change* in relaxation time can still be used to obtain information, such as binding parameters, without any knowledge of the details of the molecular motion causing the change.

Although it is too early to claim success involving the study of the quadrupolar nuclear resonances in biology, there are signs that their study could be useful in some systems particularly those of membranes which are involved in the transport of ions. At present, however, we shall confine ourselves to a discussion of some of the problems that have been encountered and the difficulties in their interpretation. Undoubtedly the most studied of the quadrupolar nuclei are those of ^{23}Na and ^{35}Cl and most of our discussion will be concerned with these.

13.2. Sodium-23

There are many references to the use of ^{23}Na n.m.r. in biological tissues, but undoubtedly the most definitive and important papers are those

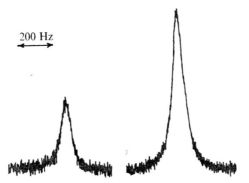

FIG. 13.1. Absorption spectrum of ^{23}Na obtained from sodium linoleate in water (left) and from a reference sample of NaOH in water and glycerol (right). The integrated intensity of the visible experimental signal is 39% of that for the reference signal, although both samples contained the same concentration and quantity of ^{23}Na nuclei. (From Shporer and Civan, 1972.)

by Shporer and Civan (1972) and Edzes, Rupprecht, and Berendsen (1972), and these serve as a convenient introduction to the literature.

The experiments by Shporer and Civan were on the ^{23}Na spectrum of liquid crystals of sodium linoleate in water. Figure 13.1 shows a representative absorption spectrum of ^{23}Na in a sample of sodium linoleate in water compared with that of a control sample of NaOH in glycerol. The integrated intensity of the ^{23}Na resonance is only 39% that of the reference signal although both samples contain the same concentration of ^{23}Na nuclei. Figure 13.2 shows a spectrum of the same sample, but it is displayed in the more sensitive derivative mode. In addition to the central line, two satellite signals are now distinguishable (no satellite signals were visible in the control system).

The ^{23}Na nucleus possesses a spin quantum number $I = \frac{3}{2}$. In a magnetic field the nucleus may exist in one of four different energy states characterized by the magnetic quantum numbers $m = \frac{3}{2}, \frac{1}{2}, -\frac{1}{2}$, or $-\frac{3}{2}$.

The interaction of the *nuclear moment* alone with the magnetic field gives rise to three permitted transitions of *equal* energy from $m = \frac{3}{2}, \frac{1}{2}$, and $-\frac{1}{2}$ to $m' = \frac{1}{2}, -\frac{1}{2}$, and $-\frac{3}{2}$ (see e.g. Fig. 2.1). In an asymmetric environment, however, because of the interaction of the quadrupole moment, the transitions are no longer of equal energy, and the quadrupole splitting of the ^{23}Na resonance results in a central line with two satellites on either side. The distance or frequency shift ν_m (in Hz) between the outer and central resonance is given by:

$$\nu_m = \tfrac{1}{2}\nu_a(3\cos^2\varepsilon - 1)(m - \tfrac{1}{2}) \tag{13.3}$$

where ε is the angle between the electric field gradient and the magnetic field

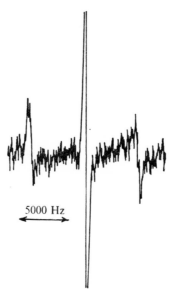

FIG. 13.2. Derivative of the ^{23}Na signal obtained from sodium linoleate in water. In addition to the central signal, satellite lines are easily distinguishable. (From Shporer and Civan, 1972.)

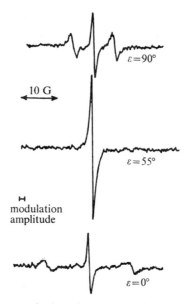

FIG. 13.3. ^{23}Na n.m.r. spectra of oriented NaDNA at different angles, ε, between fibre axis and magnetic field. Resonance frequency 16 MHz, magnetic field 14·2 kG. The NaDNA sample contained about 6 g NaCl and 65 g H_2O (84% R.H.) per 100 g dry weight. (From Edzes *et al.*, 1972.)

B_0, ν_a is the frequency of the central signal and is given by:

$$\nu_a = \left(\frac{e^2qQ}{h}\right)\frac{3}{(4I^2-2I)} \tag{13.4}$$

Under conditions of rapid tumbling, the three spectral lines converge to a single line characterized by a single relaxation time. In many biological systems nuclei may be in very rigid environments and the three resonance lines may be observable. For example, Fig. 13.3 shows the spectrum of ^{23}Na in orientated DNA fibres. We note that as ε is altered the shift between the central and outer components changes. (At $55°$, the so called 'magic angle', $(3\cos^2\varepsilon-1) = 0$ and so no splitting is observed.) Similar spectra have also been observed for ^{7}Li bound to orientated DNA fibres.

Theoretical considerations predict that the central line accounts for only 40% of the total integrated signal intensity, and in many cases this will be the only signal observed. In non-orientated or isotropic samples all values of ε are equally probable. As the value of ε changes the position of the satellite peaks will change so that the net result over all ε will be a very broad line which is difficult to detect. The larger the quadrupolar coupling constant, the larger the maximum separation of the satellite peaks and the broader the 'satellite' resonances in isotropic samples. Thus many workers have noted that the intensity of the ^{23}Na signal obtained from almost all tissues studied has been close to 40% of that anticipated on the basis of the total ^{23}Na content. The interpretation has usually been that the binding of 60% of the ^{23}Na to the tissue gives rise to a ^{23}Na n.m.r. signal that is undetectable while the other 40% is 'free' and gives an observable ^{23}Na n.m.r. signal. An *alternative* explanation in the light of the above would seem to be that because of quadrupolar interactions *all the sodium nuclei* contribute to an easily detectable signal with 40% of the total intensity; the remaining 60% of the intensity is 'smeared' out by quadrupolar effects and is very difficult to detect.

There still remains an unsolved problem. If the apparent loss of signal of bound ^{23}Na nuclei in tissue is due to a quadrupolar interaction what is the physical basis for the operation of this interaction? Two suggestions have been considered: (1) an immobilization of all the ^{23}Na nuclei and (2) an ordering effect induced by the liquid crystalline structure. The relatively small quadrupolar coupling constants in Na–linoleate (22·2 KHz) and in Na–DNA (ca. 22 KHz) would seem to argue against the former possibility. The ordering effect may reflect either or both of two possible phenomena: (1) The sodium nuclei are in anisotropic domains which produce ordering of the electric field gradients, or (2) Rapid exchange between free and an unspecified fraction of bound nuclei, for which the three transitions are then weighted averages of those of *all* the sodium nuclei.

Most of the sodium in fresh tissues exists as extracellular sodium. To imply that there are anisotropic domains suggests that there is structuring of the water and studies of the water resonance could then give important information. The alternative suggestion of fast chemical exchange is more difficult to understand and we now consider this more fully.

The situation in the tissue may perhaps best be visualized as an exchange process between free and bound sodium nuclei. The *bound* signal from the

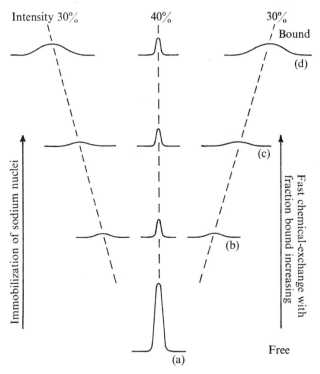

FIG. 13.4. Illustration of change in line shape as nuclei become immobilized. The same shape can obtain from fast chemical exchange between the free and bound sites, if the bound site is highly immobilized.

sodium nuclei will consist of three lines, while in the *free* sodium these three lines will be degenerate. Figure 13.4 shows how the resonance positions will alter as the sodium nuclei become immobilized. Similar spectra will also be observed if there is fast chemical exchange between the free and bound sodium nuclei, the observed spectrum will be a weighted average of the spectra in the two environments. Even if a small fraction is bound, as in Fig. 13.4(b), we note that the apparent intensity of the weighted average spectrum will drop quite quickly to 40% as the other two resonance transitions are too broad to detect.

The explanations of Shporer and Civan (1972) for the 'loss of intensity' of the resonances only apply if the quadrupolar interaction energy is less than the Zeeman energy. In such a case, the spectrum of the bound nuclei has the characteristics of the solid state. However, when the quadrupolar interaction is of the same order as the Zeeman energy, theory predicts that no nuclear resonance will be observed for the bound site. Thus, although the explanation of Shporer and Civan is attractive and seems intuitively correct, it is still too early to rule out other possibilities completely. Whatever the explanation, measurements of the relaxation rates under conditions of fast exchange (rather than intensities) will still allow information on the bound site to be obtained. It is also probable that eqn (13.1) does not apply in such systems (since $\omega_I \tau_c < 1$) and a more complete analysis is necessary (Bull, 1973). This is discussed further for the case of the relaxation of ^{35}Cl by binding to haemoglobin (see page 343).

A system in which the complications arising from quadrupolar interaction are probably absent is that of sodium binding to ionophores such as valinomycin (Haynes, Pressman, and Kowalsky, 1971). Ionophores facilitate the passive diffusion of alkali cations across biological and artificial membranes. They are thus good model compounds for membrane studies. The binding of ^{23}Na to ionophores results in large chemical shifts (ca. 10 p.p.m.) and in a broadening of the ^{23}Na resonance. The relaxation rates of the bound ^{23}Na nuclei are increased by about a factor of ten in the majority of cases studied so far. By calculating a value of τ_c, the rotational correlation time, from the Stokes–Einstein equation (see p. 242), values of the ^{23}Na quadrupole coupling constant were obtained in the range 0·47–1·64 MHz, for various ionophores. These values are somewhat low for ^{23}Na compounds and probably reflect the symmetrical array of oxygen atoms around the ^{23}Na atom. This field of work with ^{23}Na and other metal ions could well produce some interesting developments in our understanding of membrane systems.

13.3. Potassium-39

Potassium-39 is present in relatively high concentrations in most living cells. Although measurements of ^{39}K resonances in electrolytes were reported in 1966 (Deverell and Richards) few measurements in biochemical systems have been carried out. Some preliminary measurements of ^{39}K resonances have been reported on 'intracellular potassium' but these results must be regarded in a somewhat critical light in view of the conclusions mentioned above regarding quadrupolar nuclei within tissues. Other preliminary work on ^{39}K resonances involves the measurement of ^{39}K line-width in the pyruvate kinase system (Bryant, 1970).

Pyruvate kinase (PK) catalyses the final step in the sequence of enzymic reactions which converts glucose to pyruvate in the 'glycolytic pathway'. The step involving PK is

$$H^+ + CH_2{=}C{-}CO_2^- + ADP^{3-} \xrightleftharpoons{(Mg^{2+}, K^+)} CH_3{-}C{-}CO_2^- + ATP^{4-}$$

with OPO_3^{2-} on the left carbon and O (double bond) on the right carbon.

While the requirement of the enzyme for divalent ions is shared by all phosphotransferases this is not so for the monovalent ion. The role of the monovalent ion is not well defined. In PK it is thought that the monovalent ions promote a conformational change in the enzyme which influences the binding constant for the divalent ion and which is necessary for the reaction to take place. Some other work on this enzyme was discussed in Chapter 10.

When pyruvate kinase is added to a 2 M KCl solution at pH = 7, the line-width of the ^{39}K resonance is increased (Bryant, 1970). Control experiments with bovine serum albumin which causes much smaller effects on the line-width suggest that non-specific effects such as viscosity are small. Thus the broadening of the ^{39}K resonance in the presence of PK is a direct result of interaction with the enzyme. It would be worthwhile to undertake a more complete study of the line-widths under a variety of conditions.

The few measurements that have been done in biochemical systems are no doubt a result of the low n.m.r. sensitivity of the nucleus. Relatively high concentrations of ^{39}K are required to observe the resonance. In general other probes such as ^{205}Tl and ^{133}Cs are often used. In the PK system, for example, the ^{39}K ion is replaceable by ^{205}Tl, ^{133}Cs, NH_4^+, and ^{87}Rb. We turn next to the use of ^{133}Cs n.m.r. in the same reaction.

13.4. Caesium-133 (Barker, Radda, and Richards, 1970)

The addition of pyruvate kinase to solutions containing Mg^{2+} and CsCl, gives chemical shifts of the CsCl resonance (relative to saturated aqueous CsCl). The interpretation of chemical shifts in electrolyte solutions is not straightforward since ion–ion, ion–solvent, and ion–ligand interactions will all contribute to the shift. Any modification of these factors caused by the presence of the protein will lead to a chemical shift. The effects of changes in ion–ion interactions in the presence of the enzyme can be eliminated by extrapolating the observed shift to zero CsCl concentration, where ion–ion interactions, of course, disappear. The ^{133}Cs chemical shifts in the presence and absence of PK are shown in Fig. 13.5. The difference in the extrapolated shifts at infinite dilution is ca. 100 Hz. Assuming a dissociation constant for the Cs–PK complex of 25 mM and assuming conditions of rapid exchange the shift of the ^{133}Cs *bound* to the enzyme (relative to that in the presence of buffer) is ca. 3×10^4 Hz.

The spin–lattice relaxation times of ^{133}Cs complexes are generally quite long compared with other quadrupolar nuclei. This arises because of the small nuclear quadrupole moment and because in the majority of complexes, the bonding of ^{133}Cs is ionic; thus the electric field gradients will be smaller than those for nuclei like ^{25}Mg or ^{35}Cl (say) which may experience some degree of covalent bonding. Interestingly, the values of ^{133}Cs relaxation times are different in Tris or triethanolamine buffers, being higher by a factor of two in Tris. The value in triethanolamine is similar to that in water but the value in

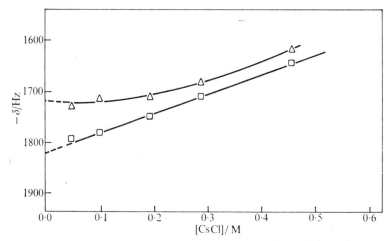

FIG. 13.5. Shifts of Cs resonances in 0·5 ml samples of a solution of pyruvate kinase (10 mg per ml) in triethanolamine chloride buffer ($\triangle$) and of the buffer alone ($\square$) at 14 ± 2 °C at 27·97 MHz. The shifts were measured at 50 kG on a broad line n.m.r. spectrometer, relative to aqueous, saturated CsCl. The pyruvate kinase causes a shift to low field. (From Barker *et al.* 1970.)

Tris probably reflects the tighter binding of ^{133}Cs to it. Competition with potassium ions confirms this idea of binding. Thus, as usual in biochemical systems, care is required in choosing a suitable buffer.

Some preliminary investigations have indicated that there are small changes in the ^{133}Cs relaxation rates on addition of the enzyme PK. Substitution of Mn^{2+} for Mg^{2+} in the enzyme can lead to further changes in the ^{133}Cs relaxation rates that are related to the distance between the two metal binding sites. Inspection of Table 10.9 in Chapter 10, however, indicates that the dipolar effects of a paramagnetic ion are likely to be small for ^{133}Cs because of the small gyromagnetic ratios. High concentrations of Mn^{2+} (relative to the enzyme) which may then be required to obtain measurable effects may result in non-specific binding of the Mn^{2+} to the enzyme. The change in relaxation rates may then be due to any of the non-specific Mn^{2+} sites nearer to the Cs^+, rather than the strong Mn^{2+} binding site and suitable controls have then to be devised.

13.5. Nitrogen-14

Recent instrumental developments have encouraged an exploration of the possibilities of studying ^{14}N resonances in biochemical systems (Richards and Thomas, 1973). For reference, the ^{14}N shifts and line-widths of the amino acids and some other selected molecules are given in Appendix 13.1. As expected, when the nitrogen is in an asymmetric environment, the resonance lines are often too broad to detect. Thus, for example, the amide resonance in dipeptides and higher peptides is not usually observable. In some cases, e.g. the nucleotides and their derivatives, only very broad resonances can be observed precluding the measurement of the shifts of the individual nitrogens.

As with many biochemical experiments involving nuclear resonances it is important to establish their sensitivity to pH. A typical example for the changes in line-width of ^{14}N resonances as a function of pH is shown in Fig. 13.6 for the resonances of histidine. The ^{14}N resonances show large changes at their own respective pK_a values. The largest change is for ^{14}N resonance of the α-amino group which must reflect the change in symmetry, from approximately tetrahedral before deprotonation to a distorted structure incorporating the lone pair on deprotonation. In addition the imidazole nitrogens are also sensitive to the titration of the α-amino group while the titration of the carboxyl group affects all the resonances. The changes in chemical shift of the ^{14}N resonances are not so dramatic (Fig. 13.7). The apparent lack of shift in the α-amino nitrogen is consistent with the shift change of <10 p.p.m. between non-protonated and protonated primary amines (Richards and Thomas, 1973).

The very broad lines expected for ^{14}N resonances in macromolecules will make their detection almost impossible. Thus in macromolecular systems, the method of approach will be to study resonances of ^{14}N in the ligand. We noted in the introduction that the binding of a small ligand to a macro-molecule can lead to dramatic changes in the relaxation rates of any quadrupolar nuclei of the ligand. This dominant effect may arise from the

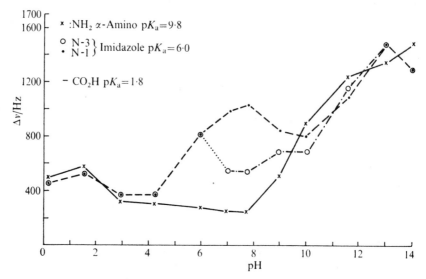

FIG. 13.6. The variation of ^{14}N line-widths of histidine (200 mM) with pH. (From Richards and Thomas, 1973.)

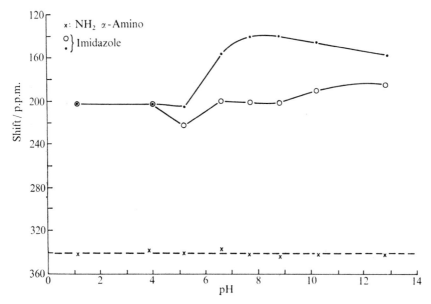

FIG. 13.7. The variation of ^{14}N shifts of histidine (200 mM) with pH. (From Richards and Thomas, 1973.)

large increase in the rotational correlation time, τ_c, for the ligand on the macromolecule (since $1/T_2 \propto \tau_c$). As usual to obtain information on the bound ligand, conditions of fast exchange are necessary. Richards and Thomas (1973) chose the binding of imidazole to bovine serum albumin (BSA) to illustrate how large these effects could be.

BSA undergoes a rapid, reversible conformational change at around pH 4·0, the binding of positive ions increases and the disruptive repulsion expands the molecule (Schlessinger, 1958). The effects of pH on the relaxation times of imidazole in the absence and presence of BSA are shown in Fig. 13.8. As most of the imidazole is not bound to BSA, the curves are all very similar. However it is clear that there is a definite enhancement in the relaxation rate. If it is assumed that the BSA is saturated with imidazole, then on average the value of $1/T_2$ for imidazole when bound to BSA is ca. 10^3 times greater than when it is free in solution. The rotational correlation time of 'free' imidazole is ca. $0·4 \times 10^{-11}$ s and the corresponding value for BSA would be ca. 10^{-8} s. Thus the large change in $1/T_2$ on binding arises primarily from the change in correlation time.

Between pH 3·0 and 4·5 it is clear that there is some discontinuity in the imidazole titration curve in the presence of BSA. This probably reflects the conformational change in the BSA molecule as the positively-charged imidazole binds.

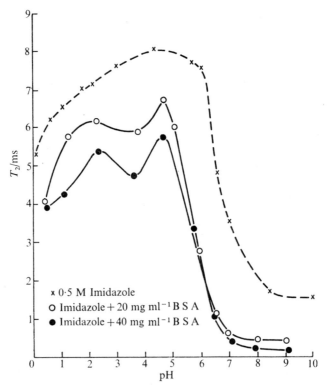

F IG. 13.8. The effect of pH on the spin–spin relaxation time T_2 of 0·5 M Imidazole+BSA. (From Richards and Thomas 1973.)

13.6. Halide-ion probes in the absence of metal ions

NMR studies on the binding of halide ions to proteins is again based on relaxation and exchange of the quadrupolar halide nuclei at suitable sites. Inspection of eqn (13.1) shows that everything else being equal the magnitude of the relaxation rates and thus the sensitivity of the method depends on the square of the quadrupole moment. The order for the halides is thus:

$$^{127}I > {}^{79}Br > {}^{81}Br > {}^{35}Cl > {}^{37}Cl.$$

The ion most frequently encountered in biological systems is the least sensitive, namely, chloride.

The binding of halides to proteins may give interactions of the type $Cl^-...R^+$, where R^+ may be groups such as NH_3^+ in the basic side-chain groups of the protein. If the field gradient and correlation times are sufficiently large in this site, then there will be a significant contribution from this site to the observed halide resonances under conditions of exchange averaging. However even under conditions of slow exchange it is possible that the relaxation rates of the halide ions may be altered, since the protein may cause changes in the solvent structure. (It is known that quadrupolar relaxation of halide ions is sensitive to changes in solvent structure.) Further ion–ion interactions are also known to be important so that changes in ionic strength would be expected to change the relaxation rates too.

The type of measurements that can be done is illustrated by the binding of ^{81}Br to various proteins in aqueous solutions. For example, on addition of 0·35 mM lysozyme to a solution of 0·5 M KBr at pH $= 7·5$, the ^{81}Br$^-$ line-width is increased by almost a factor of three (Zeppezauer, Lindman, Forsén, and Lindquist, 1969). That the effect does not merely reflect an increase in the viscosity of the solution is confirmed by measuring the ^{85}Rb relaxation times in solutions of RbBr. If the effect was due to alterations of solvent structure or simple viscosity effects, the ^{85}Rb relaxation times should be similarly affected. In fact no change was observed. (^{85}Rb was used in preference to ^{39}K because of its higher n.m.r. sensitivity.) Fast-exchange behaviour was verified in two ways: (1) increase of temperature caused a decrease in the relaxation rates of the ^{81}Br$^-$ ion, consistent with the behaviour expected for fast chemical exchange and (2) the ratio of the relaxation times of ^{79}Br$^-$ to ^{81}Br$^-$ theoretically is 1·545 based on the values of the quadrupolar coupling constant in eqn (13.1). If there is a contribution from chemical exchange this ratio would be reduced (see eqn (9.26)).

The changes in relaxation time of the Br$^-$ ion on addition of an enzyme is indicative of direct evidence for bromide binding, albeit non-specific; and any quantitative interpretation is difficult. The addition of other ligands however may cause further changes in the relaxation times of the halides and binding data and perhaps information on protein conformation may then be obtained from such measurements.

By far the most elegant and detailed study of the binding of anions to macromolecules is that of chloride binding to haemoglobin (Chiancone, Norne, Forsén, Antonini, and Wyman, 1972; Bull, Andrasko, Chianconce, and Forsén, 1973). Among the many factors which influence the oxygen equilibrium in haemoglobin is the presence of salts. At low salt concentrations (<1 M) the heterotropic effects on the equilibrium appear to be due to anion binding, while at high salt concentrations the picture is a little less clear. For example, in the presence of sodium the oxygen affinity continuously decreases with increasing salt concentration while in the presence of lithium it reaches a minimum at <1 M after which it sharply increases (Antonini and Brunori, 1971). This is an ideal problem in which the study of the nuclear resonances of the ions themselves can give information on their interaction with haemoglobin. The results may then allow a separation of the cation and anion contributions to the equilibrium.

The dependence of the excess line-width of ^{35}Cl$^-$ on NaCl concentration at constant protein concentration at pH $\approx 7·5$ is shown in Fig. 13.9. We note

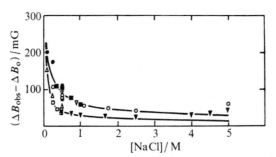

FIG. 13.9. Excess line-width as a function of NaCl concentration at pH 7·45–7·50. Haemoglobin derivative: (○), (■), (●) HbO₂; (▽) Hb⁺; (□), (▼), (△) Hb. The results are normalized to a protein concentration of 1·5%, the actual measurements being carried out in concentrations of 4% (▼), 2·5% (○, □), 1·2% (◑), (▽), and 0·6% (●, △). Full lines are theoretical ones and are calculated with the binding constants given in Table 13.2. (From Chicancone *et al.* 1972.)

immediately that there is a difference between oxy- and deoxy-haemoglobin further if there is only one type of a binding site for each form then the excess line-width should drop to zero as the concentration of chloride is increased. (This follows simply because the fraction of Cl⁻ 'bound' is decreasing.) However, in Fig. 31.9 it is apparent that at very high concentrations of chloride the line-widths are still finite. Chiancone *et al.* (1972) interpreted this as a result of the presence of two different binding sites for Cl⁻. If the Cl⁻ dissociation constants are quite different, then the binding process is described by two widely-separated hyperbolae in accordance with the equation:

$$\Delta B_{obs} - \Delta B_0 = (P_M q)_A \Delta B_A + (P_M q)_B \Delta B_B = [Hb_t]\left(\frac{q_A \Delta B_A}{K_A + [Cl]} + \frac{q_B \Delta B_B}{K_B + [Cl]}\right)$$

$$(13.5)$$

In this example ΔB_{obs} and ΔB_0 are the observed line-widths in gauss in the presence and absence of the protein. A and B refer to the two sites, $(p_M q)_A$ and $(p_M q)_B$ are the fraction of chloride nuclei bound at each site where the characteristic line width is ΔB_A and ΔB_B. q_A and q_B are the number of binding sites. K_A and K_B are the dissociation constants for the Cl⁻ ions from these sites and $[Hb_t]$ represents the total haemoglobin concentration. The curve in Fig. 13.9 is calculated from the parameters given in Table 13.2.

In the absence of other information it is of course not possible to separate the products $q \Delta B$ into their two components. A further complication that has been neglected in this analysis is any effects due to the dissociation of the haemoglobin into its subunits which may be occurring under these conditions.

In contrast to the case of haemoglobin, the dependence of the excess line width of chloride concentration in *myoglobin* can be accounted for on the

TABLE 13.2

Numerical values of constants, corresponding to the calculated curves shown in Fig. 13.9

	HbO$_2$	Hb
K_A(M)	0·1	0·01
$q_A \Delta B_A$(G)	139	54
K_B(M)	10	10
$q_B \Delta B_B$(G)	1300–1500	650–850

basis of only one type of binding constant ($K_D = 2$ M). Further there is no difference in the line-width between oxy- and deoxy-myoglobin. In the case of myoglobin the oxygen affinity is known to be unaffected by the presence of salts. This would seem to suggest that in haemoglobin the high affinity chloride sites are related to oxygen binding, and is consistent with the binding of chloride ions to these sites being stronger in deoxy- than in oxy-haemoglobin.

It may be possible to draw on these results in discussing the well-known effect of organic phosphates, such as 2,3-diphosphoglycerate and ATP, on the oxygen equilibrium of haemoglobin. Perhaps there is a similar differential binding between the oxy- and deoxy-forms. Then too the fact that the effect is largely suppressed in the presence of high salt concentrations suggests that there may be a direct competition between these ions and chloride ions (Benesch and Benesch, 1969). Figure 13.10 shows the effects of ATP concentration on the ^{35}Cl$^-$ line-width. At pH 7·5 there was little or no effect of ATP on the ^{35}Cl$^-$ line-width in 0·2 M NaCl in the presence of either oxy- or deoxy-haemoglobin and this may be a consequence of the relative binding constants of ATP and Cl$^-$. However at pH values between 6·0 and 6·7, the effect of ATP on the ^{35}Cl line-width is marked and can be accounted for on the basis of competition for the high affinity chloride sites. Chiancone *et al.* (1972) mention that this raises an interesting point. Perutz (1970) had suggested that in deoxyhaemoglobin the ATP occupies the central cavity of the ligand axis of the molecule, which is uniquely suited to accommodate it. On oxygenation this cavity is reduced in size and ATP is expelled. In the case of the chloride ions binding at the same site, no such explanation based on steric effects is really tenable. One possible explanation is of course that the ATP and Cl$^-$ sites are different and that the binding of ATP causes a conformational change that directly affects the Cl$^-$ binding. This ambiguity always exists in 'competition experiments'. Perhaps, by extending the studies to other phosphates which compete with ATP, this problem may be solved. Another possibility based on the observation that ATP binds both to the oxy- and deoxy-forms (see Fig. 13.10) is that there may be a weaker binding site for ATP for which the chloride competes.

The above analysis of binding has been done without any discussion as to the nature of the relaxation mechanism responsible for the line-width

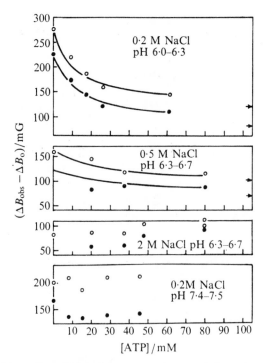

FIG. 13.10. Excess line-width in the presence of oxy- and deoxy-haemoglobin as a function of ATP added. Haemoglobin derivatives: (○) HbO_2; (●) Hb. Lines are theoretical ones and were calculated with dissociation constants as follows; for HbO_2 $K_{Cl} = 0.05$ M $K_{ATP} = 0.0025$ M; for Hb, $K_{Cl} = 0.0025$ M $K_{ATP} = 0.00017$ M. Arrows are the asymptotic values at infinite ATP concentration. (From Chiancone et al., 1972.)

ΔB. To obtain details of molecular motion it is necessary to have explicit equations for the linewidth. Equation (13.1) is unlikely to apply strictly to this case, since by assuming a value of τ_c, equal to the rotational correlation limit of the protein ($\sim 2 \times 10^{-8}$ s), it can be estimated that $\omega_I \tau_c \sim 1$. Hubbard (1970) has shown that for the non-white ($\omega_I \tau_c \gtrsim 1$) region, for the case for which $I = \frac{3}{2}$, the relaxation rates are in fact the sum of four exponential terms. However, under conditions where the population of one site is very small, this is reduced to two exponential terms (Bull, 1973). If it is assumed that the experimentally-observed relaxation rates do not deviate significantly from an exponential function then the two remaining exponential terms can be combined. The following equations are then obtained:

$$\frac{1}{T_{1,obs}} = \frac{1}{T_{1,A}} + \frac{P_M q}{(1-P_M)}\left(\frac{0.8}{T'_{1,M}+\tau_M} + \frac{0.2}{T''_{1,M}+\tau_M}\right) \tag{13.6}$$

where

$$\frac{1}{T'_{1,M}} = \frac{1}{10}\left(\frac{e^2 qQ}{\hbar}\right)^2\left(\frac{\tau_c}{1+4\omega_I^2\tau_c^2}\right) \tag{13.7}$$

and

$$\frac{1}{T''_{1,M}} = \frac{1}{10}\left(\frac{e^2qQ}{\hbar}\right)^2\left(\frac{\tau_c}{1+\omega_I^2\tau_c^2}\right) \tag{13.8}$$

where q represents the number of binding sites and $1/T_{1,A}$ is the spin–lattice relaxation rate at site A (for which $\omega_I\tau_c \ll 1$). Equation (13.6) reduces to the usual Luz–Meiboom equation (eqn (9.25)) when $\omega_I\tau_c \ll 1$. The expression for $1/T_2$ is expected to be similar to the Swift–Connick equation (eqn (9.26)). However by making the reasonable assumption that the relaxation rate of a quadrupolar nucleus when bound to a macromolecule is very efficient, such that $(1/T'_{2,M}+1/T''_{2,M}) > \Delta\omega_M^2$, some simplification results, i.e.

$$\frac{1}{T_{2,obs}} = \frac{1}{T_{2,A}}+\frac{P_Mq}{(1-P_M)}\left(\frac{0.6}{T'_{2,M}+\tau_M}+\frac{0.4}{T''_{2,M}+\tau_M}\right) \tag{13.9}$$

where

$$\frac{1}{T'_{2,M}} = \frac{1}{20}\left(\frac{e^2qQ}{\hbar}\right)^2\left(\tau_c+\frac{\tau_c}{1+\omega_I^2\tau_c^2}\right) \tag{13.10}$$

and

$$\frac{1}{T''_{2,M}} = \frac{1}{20}\left(\frac{e^2qQ}{\hbar}\right)^2\left(\frac{\tau_c}{1+4\omega_I^2\tau_c^2}+\frac{\tau_c}{1+\omega_I^2\tau_c^2}\right) \tag{13.11}$$

where $1/T_{2,A}$ is the transverse relaxation rate at the A site. Equations (13.6) and (13.9) reduce to eqn (13.1) when $\omega_I\tau_c \ll 1$.

In a subsequent series of experiments on the binding of chloride to haemoglobin Bull et al. (1973) measured the temperature and frequency dependence of T_1 and by comparison with the T_2 values attempted to calculate τ_M/q. Unfortunately, neither T_1 nor T_2 is completely dominated by τ_M so that only limits of τ_M/q could be obtained; the upper limit was 10^{-6} and the lower $\sim 10^{-8}$ s. The lower limit is obtained by use of the following equation (Marshall, 1970) (see p. 330)

$$\frac{1}{\tau_c} = \frac{1}{\tau_R}+\frac{1}{\tau_M}$$

with τ_R ca. 10^{-8} s.

13.7. Halide-ion probes in the presence of metal ions

As a means of circumventing the relatively large effects of non-specific binding of halide ions to macromolecules, Stengle and Baldeschwieler (1966) made use of the fact that the relaxation rates of halide ions are markedly increased when the halide nucleus is in a covalent bond. For instance that the ^{35}Cl line-width in 2 M NaCl, is doubled on addition of 10^{-3} M $HgCl_2$ because of exchange of the type:

$$2Cl^- + HgCl_2 \rightleftharpoons HgCl_4^{2-}$$

In the $HgCl_4^{2-}$ ion the chloride is covalently bonded to the mercury and because of the large electric field gradients in the bond is relaxed very effectively. If the Hg is bonded to a macromolecule, the longer rotational correlation time will make the relaxation rate even larger. For example, addition of ca. 10^{-5} M haemoglobin and ca. 10^{-5} M $HgCl_2$ to a solution containing ca. 0·5 M Cl^- ions, increases the Cl^- relaxation rate by an order of magnitude.

The experimental method consists of labelling specific SH groups in the protein with Hg, adding halide ions to the solution, and studying the line-widths of the halide ion resonances. The solvated halide ion is essentially symmetric, unlike the halide ions bound to the Hg, and from changes in line width, the accessibility and concentration of specific sites on the protein can be determined. Stengle and Baldeschwieler also suggested that by labelling a substrate in this manner, substrate binding to enzymes might be followed.

Of course, information on the bound site can only be obtained if conditions of rapid chemical exchange are valid. If the value of the quadrupolar coupling constant of the nucleus in the site can be estimated, then the value of τ_0 can be calculated. The change in relaxation rates may also be used as a convenient monitoring observable in experiments involving changes in pH, temperature, and the addition of other ligands. Of course, if structure–function relationships are to be obtained it is essential that the probe does not alter the characteristics of the macromolecule. For example, the binding of p-Hg–benzoate to Cys-93 in the β-chain of haemoglobin changes the Hill coefficient from $n = 2\cdot8$ to $n = 1$ and also alters the spin state of the haeme iron, so this would not be a good probe. Hg itself may not do this but this has to be checked before any interpretation of ligand induced changes etc. is made using this probe.

The halide ion probe technique can also be extended to systems containing metal ions other than Hg. Obvious systems are then metalloenzymes such as carboxypeptidase and carbonic anhydrase which contain zinc, or enzymes like pyruvate kinase, which may be activated by zinc. The efficiency of using other metal ions will depend on the type of bond formed between the metal and exchanging halide, for this will determine the magnitude of the electric field gradients.

Some examples of n.m.r. studies of halide ions in biological systems are listed in Table 13.3. We shall only consider three examples here: (1) When the halide ion binds to a Hg label which is bound to a protein; (2) when halide ions bind to a Hg label which is on a substrate; and (3) the binding of halide ions to a metalloenzyme.

Human haemoglobin contains two reactive and four unreactive SH groups. The reactive SH groups are those which show a chemical reactivity comparable with simple thiols. The unreactive SH groups have a very low

TABLE 13.3

Some biological applications of halogen n.m.r.

N.m.r. nucleus	Metal ion	Biomolecules studied	Reference
^{35}Cl	Hg	haemoglobin	a, g, o
^{35}Cl	Hg	hapten-antibody interactions	b
^{35}Cl	Hg, Cd, Zn	bovine mercaptalbumin	c
^{35}Cl	Zn	carbonic anhydrase	d
^{35}Cl	Zn	Zn–nucleotide–diphosphate complexes	e
^{35}Cl	Zn	Zn–ADP complexes	f
^{35}Cl	Hg	ferric peroxidase, myoglobin, haemoglobin	g
^{35}Cl	Hg	synthetic poly-L-glutamate	h
^{35}Cl	Hg	α-chymotrypsin	i
^{35}Cl	Zn	pyruvate kinase	j
^{35}Cl	Zn	liver alcohol dehydrogenase	k, l
^{35}Cl ^{81}Br	Zn Cu	alkaline phosphatase	m
^{81}Br		lysozyme, α-chymotrypsin, seralbumin, ceruloplasmin, alcohol dehydrogenase, carbonic anhydrase	n

a. T. R. STENGLE and J. D. BALDESCHWIELER, (1966), *Proc. natn. Acad. Sci. U.S.*, **55**, 1020; (1967) *J. Am. Chem. Soc.*, **89**, 3045.

b. R. P. HAUGLAND, L. STRYER, T. R. STENGLE, and J. D. BALDESCHWIELER, *Biochemistry*, 1967, **6**, 498.

c. R. G. BRYANT, *J. Amer. chem. Soc.*, 1969, **91**, 976.

d. R. L. WARD, *Biochemistry*, 1969, **8**, 1879, 1970, **9**, 2447.

e. J. A. HAPPE and R. L. WARD, *J. Amer. chem. Soc.*, 1969, **91**, 4096.

f. R. L. WARD and J. A. HAPPE, *Biochem. Biophys. Res. Comm.*, 1967, **28**, 785.

g. W. D. ELLIS, H. B. DUNFORD, and J. S. MARTIN, *Canad. J. Biochem.*, 1969, **47**, 157.

h. R. G. BRYANT, *J. Amer. chem. Soc.*, 1967, **89**, 2496.

i. A. G. MARSHALL, *Biochemistry*, 1968, 7, 2450.

j. G. L. COTTAM and R. L. WARD, *Arch. Biochem. Biophys.*, 1969, **132**, 308.

k. B. LINDMAN, M. ZEPPEZAUER, and A. AKESON, *Biochem. biophys. Acta*, 1972, **257**, 173.

l. R. L. WARD and J. A. HAPPE, *Biochem. Biophys. Res. Comm.* 1971, **45**, 1444.

m. H. CSOPAK, B. LINDMAN and H. LILJA, *FEBS. Letts.* 1970, **9**, 189.

n. M. ZEPPEZAUER, B. LINDMAN, S. FORSÉN, and J. LINDQVIST, *Biochem. Biophys. Res. Comm.*, 1969, **37**, 137.

o. CHIANCONE, E., NORNE, J. E., FORSÉN, S., ANTONINI, E., WYMAN, J., 1972, *J. mol. Biol.* **70**, 675.

reactivity in native haemoglobin but become indistinguishable from the reactive groups when the protein is denatured. Much attention has been focused on the function of the unreactive SH groups, and Cecil and Thomas (1965) have suggested that they are involved in an intrachain structure. The halide-ion probe technique is a further method for studying SH groups.

The binding of $HgCl_2$ to human oxyhaemoglobin can be monitored by dissolving the protein in 0·5 M NaCl and observing the ^{35}Cl resonance. The line-width of the ^{35}Cl resonance changes as protein—S—Hg—Cl groups are formed and the results are conveniently shown as a titration in Fig. 13.11. When all available SH groups have reacted, further addition of $HgCl_2$ has little effect on the spectrum. The end point at a ratio of $(HgCl_2)$/protein equal to two shows that the Hg binds to two reactive SH groups per molecule.

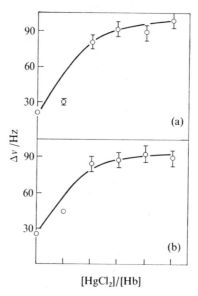

[HgCl₂]/[Hb]

F IG. 13.11. Titration of human oxyhaemoglobin versus $HgCl_2$ at various NaCl concentrations. All solutions were adjusted to pH 7 with 0·05 M phosphate buffer: (a), NaCl = 1·0 M, Hb = $2·6 \times 10^{-5}$ M; (b), NaCl = 4·0 M, Hb = $5·1 \times 10^{-5}$ M. (From Stengle and Baldeschwieler 1967 © 1967 Amer. Chem. Soc.)

In 4·0 M NaCl, the haemoglobin molecule dissociates into halves and no further SH groups have reacted (Fig. 13.11b). This is consistent with the suggestion by Cecil and Thomas (1965) that the unreactive SH groups are involved in intrachain binding rather than inter-chain binding.

The advantage of studying SH groups by this method is that it is possible, in principle, to obtain information on the mobility of the SH groups. To calculate τ_c, the correlation time, it is necessary to know the quadrupolar coupling constant. As a rough approximation, this may be taken to be the same as in $HgCl_2$ (ca. 40 MHz). Thus if $1/T_2^*$ is the ^{35}Cl relaxation time in the presence of the enzyme and $1/T_{2,A}$ that in the absence of enzyme and $\Delta\nu^*$ and $\Delta\nu$ the corresponding line-widths then:

$$\frac{1}{T_2^*} - \frac{1}{T_{2,A}} = \pi(\Delta\nu^* - \Delta\nu) = (P_M q)K(e^2 qQ)\tau_c \qquad (13.12)$$

where $(P_M q)$ is the fraction of chloride ions bound per labelled molecule and K is a constant incorporating the factors in eqn (13.1). (For a spin of $I = \frac{3}{2}$, K is $2\pi/5$, if the asymmetry factor is ignored.) Thus the value of τ_c in haemoglobin is ca. 10^{-9} s. The overall tumbling time of the molecule can be estimated as ca. 10^{-7} s so the value of τ_c probably corresponds to motions within the protein chains. Further, the value of τ_c calculated from the data in Fig. 13.11, using the same assumptions as above, indicates that τ_c increases by a factor of

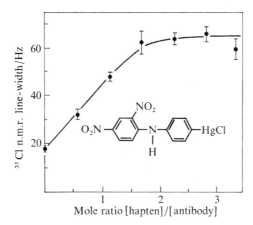

FIG. 13.12. ^{35}Cl n.m.r. titration of antidinitrophenyl antibody showing observed ^{35}Cl n.m.r. line-widths as a function of the DNP–mercurial: antibody mole ratio. (From R. P. Haughland Stryer, Stengle, and Baldeschwieler 1967. © 1967 Amer. Chem. Soc.)

four (approximately) as the haemoglobin molecule dissociates, perhaps indicative of a conformational change involving the SH group.

We mentioned in the introduction (p. 330) that if fast chemical exchange occurs, τ_c can contain contributions from τ_R and τ_M. If, in this example, τ_M does make a significant contribution to τ_c, then this will give a value of τ_c that is much shorter than τ_R. The 'apparent' changes in τ_c may then reflect changes in τ_M. At present there is insufficient information for the exchange life-times of halides in mercury halide bonds to know if this is the case. It is included only as a cautionary warning.

An example of the mercury label on a ligand rather than the protein is that of a mercury-labelled hapten (Haughland et al., 1967). The interaction of this with antidinitrophenyl antibody can be monitored by use of ^{35}Cl n.m.r. The line-width of the ^{35}Cl resonance in a 1 M aqueous NaCl solution was changed marginally on addition of the antibody, perhaps as a result of non-specific binding. Addition of the labelled hapten leads to a broadening of the resonance, and a titration curve showing the observed line-width as a function of hapten–antibody concentration is shown in Fig. 13.12. The authors claim that measurable effects are observed with antibody concentrations as low as 6×10^{-7} (antibody) in 1 M NaCl. That the effects are so large probably means that the mercury atom is accessible to the solvent and therefore the bulk Cl^- ions. This idea can be extended by substituting the metal label at various sites on a substrate and it may be possible to map the accessibility of various parts of a bound substrate.

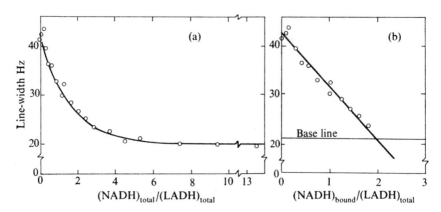

FIG. 13.13. The observed ^{35}Cl n.m.r. line-width as a function of the ratio of NADH to LADH for 13 mM LADH (pH 7·5) in 0·5 M NaCl; (a) total [NADH] (b) bound [NADH] determined spectrophotometrically. The residual ^{35}Cl n.m.r. line-width, after saturation of the enzyme with NADH, may arise from non-specific binding of chloride ions to the enzyme, or from viscosity effects. (From Ward and Happe, 1971.)

The final example we shall consider is that of chloride ion binding to the metalloenzyme, liver alcohol dehydrogenase (LADH). This enzyme has a molecular weight of 80 000, and two structurally and chemically identical subunits each of which contains two zinc atoms. It can be shown spectroscopically that the addition of o-phenananthroline, a strong chelator of zinc ions, interacts with only one zinc ion in each subunit. Further o-phenanthroline has been shown to be kinetically competitive with the reduced coenzyme NADH. These and other studies have given rise to the idea that two zinc ions are involved in the binding of NADH, while the other two zinc ions are 'buried' within the enzyme and play a role in maintaining the three-dimensional structure of the enzyme. However in contrast to this, recent evidence (Iweibo and Weiner 1972) suggests that no zinc atoms are involved in coenzyme binding. Iweibo and Weiner (1972) showed that in the absence of zinc the binding of NADH to the enzyme shows negative co-operativity, while in the presence of zinc the binding shows no co-operativity. In both cases NADH had the same affinity for the tight binding sites. Iweibo and Weiner suggested that one type of zinc site is involved in the formation of the nascent tertiary structure of the enzyme whilst the second is involved in maintaining the catalytically active form of the enzyme, rather than being involved in NADH binding.

Chloride ion is a competitive inhibitor with respect to NADH. The addition of horse LADH to a 0·5 M solution of NaCl results in a broadening of the ^{35}Cl resonance line-width (Ward and Happe 1971) which decreases upon addition of NADH (Fig. 13.13a). By determining the amount of NADH bound per mole of LADH spectrophotometrically, Ward and Happe (1971) showed that the titration in Fig. 13.13a could be interpreted in terms of two moles of NADH bound per mole of enzyme (Fig. 13.13b). They suggested that the ^{35}Cl probe interacts with the zinc ions located at the active site of LADH and that NADH prevents chloride ions from sampling this site. In a further series of experiments Ward and Happe showed that orthophenanthroline produced no change in ^{35}Cl line width, but that o-phenanthroline and NADH compete.

One speculative suggestion to explain the above observation could be that o-phenathroline and chloride ions bind at different zinc ions and that neither has a *direct* effect of NADH binding but rather an effect mediated through a conformational change in the enzyme. If this is so, the net result of any changes at the other metal site would have to be small to be consistent with the experimental observation. Changes in electric field gradients around the zinc, because of changes in the symmetry of the zinc site, would have to be offset by changes in the relative amount of chloride 'bound' to the zinc.

13.8. Reflections on the halide-ion probe methods

The high halide ion concentrations used in all the experiments referred to in this section are no longer necessary in order to observe the halide ion resonance. Advances in instrumentation have improved the sensitivity of spectrometers by an order of magnitude. The use of lower concentrations may largely only diminish effects of non-specific binding. The main use of the halide probe method will then probably be in the field of metalloenzymes, rather than as a technique for studying SH groups of proteins. Biochemistry is rich in techniques, chemical and spectral, for studying reactivities, mobilities, and conformational changes of SH groups. In metalloenzymes, the information obtained will be mainly empirical rather like the empirical uses of proton relaxation enhancement or those spectral techniques in which a change in any parameter is observed, i.e. the determination of binding parameters and, perhaps the detection of conformational changes.

References

ANTONINI, E. and BRUNORI, M. (1971). *Haemoglobin and myoglobin in their reactions with ligands*, North-Holland Publ. Co., Amsterdam.

BARKER, R. W., RADDA, G. K., and RICHARDS, R. E. (1970). unpublished results.

BENESCH, R. E. and BENESCH, R. (1969). *Fötsvarsmedicin*, **5**, 154.

BRYANT, R. G. (1970). *Biochem. biophys. Res. Commun.* **40**, 1162.

BULL, T. E., ANDRASKO, J., CHIANCONE, E., and FORSÉN, S. (1973). *J. molec. Biol.* **73**, 251.

——, (1973). *J. magn. Res.* **8**, 344.

CECIL, R. and THOMAS, M. A. W. (1965). *Nature (Lond.)*, **206**, 1317.

CHIANCONE, E., NORNE, J. E., FORSÉN, S., ANTONINI, E., and WYMAN, J. (1972). *J. molec. Biol.* **70**, 675.

DEVERELL, C. and RICHARDS, R. E. (1966). *Molec. Phys.* **10**, 55.

EDZES, H. T., RUPPRECHT, A., and BERENDSEN, H. J. C. (1972). *Biochem. biophys. Res. Commun.* **46**, 790.

HAYNES, D. H., PRESSMAN, B. C., and KOWALSKY, A. (1971). *Biochemistry* **10**, 852.

HAUGHLAND, R. P., STRYER, L., STENGLE, T. R. and BALDESCHWIELER, J. D. (1967). *Biochemistry*, **6**, 498.

HUBBARD, P. S. (1970). *J. chem. Phys.* **53**, 985.

IWEIBO, I. and WEINER, H. (1972), *Biochemistry* **11**, 1003.

MARSHALL, A. G. (1970). *J. chem. Phys.* **52**, 2527.

PERUTZ, M. (1970), *Nature (Lond.)*, **228**, 734.

RICHARDS, R. E. and THOMAS, N. A. (1973), to be published.

SCHLESSINGER, B. S. (1958). *J. phys. Chem.* **62**, 916.

SHPORER, M. and CIVAN, M. M. (1972). *Biophys. J.* **12**, 115.

STENGLE, T. R. and BALDESCHWIELER, J. D. (1966). *Proc. natn. Acad. Sci.* **55**, 1020.

——, ——, (1967). *J. Am. Chem. Soc.* **89**, 3045.

WARD R. L. and HAPPE, J. A. (1971), *Biochem. biophys. Res. Commum.* **45**, 1444.

ZEPPEZAUER, M., LINDMAN, B., FORSÉN, S., and LINDQVIST, J. (1969). *Biochim. biophys. Res. Commun.* **37**, 137.

Appendix 13.1.

Nitrogen-14 shifts and line-widths of some biological molecules (Richards and Thomas, 1973)

(The shifts are relative to the nitrate resonance of NH_4NO_3 (4·5 M) in HCl (3M))

Compound	Structure	Concentration (mM)	pH	Shift (p.p.m.)	$\Delta\nu$ (Hz)
Glycine	NH_2—CH_2—CO_2H	204	7·27	346	115
Alanine	CH_3 / NH_2—CH—CO_2H	200	7·42	332	117
Valine	CH_3—CH—CH_3 / H_2N—CH—CO_2H	192·3	7·42	339	144
L-leucine	CH_3 / CH—CH_2—CH / NH_2, CO_2H / CH_3	205	7·47	336	207
Isoleucine	CH_3 NH_2 / CH_3—CH_2—CH—CH—CO_2H	199	7·47	340	213
Serine	CH_2—OH / NH_2—CH—CO_2H	205	7·14	340	137

APPENDIX 13.1 *(Continued)*

Compound	Structure	Concentration (mM)	pH	Shift (p.p.m.)	$\Delta\nu$ (Hz)
Threonine	CH_3—CH—OH / NH_2—CH—CO_2H	205	7·05	342	137
Aspartic acid	CH_2—CO_2H / CH—CO_2H / NH_2	204	7·29	336	137
Asparagine	CH_2—$CONH_2$ / CH—CO_2H / NH_2	196	7·26	—CH—NH_2 339, $\ominus CONH_2$ 268	160, 583
Glutamic acid	CH_2—CO_2H / CH_2 / CH—CO_2H / NH_2	202	7·50	333	187
L-Glutamine	CH_2—$CONH_2$ / CH_2 / CH—CO_2H / NH_2	207	7·35	—CH—NH_2 334, $\ominus CONH_2$ 264	172, 500

Compound	Structure		pH		
Lysine	$(CH_2)_4-NH_2$ $NH_2-CH-CO_2H$	202	7·38	α-NH_2 335 ε-NH_2 342	225 192
Hydroxylysine	CH_2-NH_2 $CH-OH$ $[CH_2]_2$ $CH-CO_2H$ NH_2	165	7·20	α-NH_2 336 ε-NH_2 348	340 350
L-Histidine	CH_2-CH NH_2 CO_2H (imidazole ring)	220	7·09	α-NH_2 335 N-imid N-3 197 N-1 147	205 745 1000
Arginine	$NH=C-NH_2$ NH $[CH_2]_3$ $CH-CO_2H$ NH_2	208	7·45	α-NH_2 335 Others 312	289 ca. 1000
DL-Phenylalanine	CH_2-CH NH_2 CO_2H (phenyl ring)	191	7·2 9·4	346 323	260 1530

APPENDIX 13.1 (*Continued*)

Compound	Structure	Concentration (mM)	pH	Shift (p.p.m.)	$\Delta \nu$ (Hz)
Tyrosine	HO—C₆H₄—CH₂—CH(NH₂)—CO₂H	178	10.56	312	1870
Tryptophan	indolyl—CH₂—CH(NH₂)—CO₂H	55·5 (max)	7·15	α-NH$_2$ 343	488
				NH	
				293	ca. 1000
L-Cysteine	CH₂—SH / CH—CO₂H / NH₂	184	7·32	335	174
L-Cystine	(NH₂—CH—CO₂H / CH₂—S—)₂	196	10·58	333	950
L-Cysteic acid	NH₂—CH—CO₂H / CH₂—SO₃H	191	7·20	336	222
L-Methionine	[CH₂]₂—SCH₃ / CH—CO₂H / NH₂	211	7·10	335	195

Compound	Structure				
L-Ethionine	$[CH_2]_2$—SCH_2CH_3 / CH—CO_2H / NH_2	164	10·80	341	2200
L-Proline		216	7·35	322	137
L-Hydroxyproline		210	7·50	323	216
DL-Ornithine	$[CH_2]_3$—NH_2 / NH_2—CH—CO_2H	220	7·15	α-NH_2 339 / δ-NH_2 343	212 / 150
Betaine (*NNN*-Trimethylglycine)	$(CH_3)_3\overset{\oplus}{N}$—$CH_2$—$CO_2H$ CH_3 $Cl^{\ominus}$	208	7·10	328·8	4·0
DL-α-Amino butyric acid	CH_3—CH_2—CH / NH_2 / CO_2H	211	7·2	336	141
DL-$\alpha\gamma$-Diamino butyric acid	NH_2 / CH_2—CH_2—CH / NH_2 / CO_2H	209	7·42	α-NH_2 336 / γ-NH_2 343	259 / 163

APPENDIX 13.1   (*Continued*)

Compound	Structure	Concentration (mM)	pH	Shift (p.p.m.)	$\Delta\nu$ (Hz)
Imidazole	imidazole ring (N—N—H)	200	7·08	196	91
Taurine	NH_2—CH_2—CH_2—SO_3H	205	7·20	343	82
L-Histamine	CH_2—CH_2—NH_2 on imidazole ring (N—N—H)	200	7·12	α-NH_2 345 Imid N-3 196 N-1 144	186 392 547
3-Nitrotyrosine	HO—(ring, NO_2)—CH_2—CH(NH_2)—CO_2H	160	10·59	NH_2 ca. 337 NO_2 41	1740 456
3,5-Dinitrotyrosine	HO—(ring, NO_2, NO_2)—CH_2—CH(NH_2)—CO_2H	156	10·78	NH_2 337 NO_2	1810 516
Creatine	NH_2—C(=NH)—N(CH_3)—CH_2—CO_2H	110	5·8	ca. 335	1360

	Structure			$>NH$ / $>N-CH_3$	
Creatinine	CH_3-N-C (with NH, NH, $CH_2-C=O$)	190	7·10	$=N-H$ 334 / 226	800 / 920
Phosphocreatine	$NH=C$ with $NHPO_3H_2$, $NCH_3CH_2CO_2H$	105	7·20	ca. 310	1515
Urea	NH_2-C-NH_2, $=O$	200	6·96	300	221
Glucosamine	CH_2OH ... OH OH NH_2 HO	200	7·21	327	532
Guanidine	$(H_2N)_2C=NH$	200	7·11	301	341
Hexamethylenetetramine		ca. 200		330	3400
Glycyl-glycine	NH_2-CH_2-CO- / HO_2C-CH_2-NH	208	7·06	$\ominus NH_2$ 350 / $\ominus NHCO$ 262	210 / 1020

APPENDIX 13.1 (*Continued*)

Compound	Structure	Concentration (mM)	pH	Shift (p.p.m.)	$\Delta\nu$ (Hz)
Glycyl-L-tyrosine	NH_2-CH_2-CO / $HO-\text{(ring)}-CH_2-CH-NH$ / CO_2H	135	7·1	348	635
Glycyl-L-histidine	NH_2-CH_2-CO / $HO_2C-CH-NH$ / CH_2 / imidazole ring	105	7·2	$-NH_2$ 352 / Imid 187	465 / 1440
Glycyl-methionyl-glycine	NH_2-CH_2-CO / $[CH_2]_2-CH-NH$ / SCH_3 $CO-NH$ / HO_2C-CH_2	118	7·24	348	670
Penicillamine	$(CH_3)_2-CH-SH$ / NH_2-CH / CO_2H	202	3·83	336	253

Benzyl penicillin (Na salt)	150	7.21	ca. 320	ca. 2 kHz
Choline chloride $CH_3\overset{+}{N}-CH_2-CH_2-OH$ Cl^-	100	7.1	326.8	5.8
Choline phosphate $(CH_3)_3\overset{+}{N}-CH_2-CH_2$ O $HO-P=O$	70	7.0	326.5	5.0
Adenine 9-β-D-ribofuranosyladenine	70	3.0	185	1020
Adenosine	70	2.2	ca. 239	2600
AMP Adenosine-5′-monophosphate	50	4.2	ca. 245	5430
Guanidine	50	Alkali	182	1220
Guanosine 9-β-D-ribofuranosylguanine	50	0.5	195	1710

14

A GLOSSARY OF SOME
INSTRUMENTAL TECHNIQUES

14.1. Introduction

THROUGHOUT the text several instrumentational techniques have been mentioned. It is the intention of this section to indicate the meaning of these in simple terms. The enthusiast will find a reading list at the end of this section from which more details may be obtained.

14.2. Continuous-wave techniques

The most usual spectrometer operates in the continuous wave (c.w.) mode, where the small radio-frequency (r.f.) field B_1 is applied continuously to the sample by means of an r.f. coil. To trace out a spectrum, either the frequency of B_1 is varied slowly at constant B_0, or B_0 is varied slowly at constant frequency of B_1. The resonance condition of each line is thus obtained in turn. From the spectrum the parameters δ, J, and T_2 may be obtained; T_2 is related to the line-width at half-height $\Delta\nu$ (in Hz) by

$$\frac{1}{T_2} = \pi\,\Delta\nu$$

14.3. Multiscan averaging techniques

Inherent in most electronic circuitry is a certain noise background. In many cases the n.m.r. signal is too weak to be easily detected above the noise background of the instrument. The most widely used technique to overcome this problem is that of multiscan averaging. This exploits the fact that noise, by definition, is a random function of time, so that if one re-examines the n.m.r. signal a second time, the noise component being random will have changed and to a certain degree will tend to cancel the noise component observed in the first scan. The technique requires a storage device of many separate storage channels, typically a small computer with 1000–4000 separate data storage locations. This storage device is known as a *c*omputer of *a*verage *t*ransients or CAT. The n.m.r. spectrum is converted to a digital form and stored in the computer (CAT).

A second spectrum is then recorded, under identical conditions to the first and the data in corresponding channels is added. True n.m.r. signal components thus grow linearly with the number of scans, but the noise level being random only grows as the square root of the number of scans taken. This results in an improvement in signal-to-noise ratio, proportional to the square root of the number of scans.

14.4. Double irradiation techniques

The assignment of resonances in complex molecules is often far from trivial. We have seen that a variety of chemical perturbations can be helpful in this context. One way however that requires only an instrumental perturbation is to introduce a *second* radio-frequency excitation B_2.

14.4.1. *Spin decoupling*

The spin multiplet structure on a given resonance line, A, due to spin–spin coupling to a neighbouring nucleus B can be made to coalesce to a single line if B is irradiated with B_2 sufficiently strongly. If nucleus B then changes its alignment in the main magnetic field B_0, at a rate fast compared with the spin–spin coupling constant J_{AB} then the hyperfine structure disappears. (Nucleus B is changing direction at a rate too fast for A to detect its orientation.) The technique is known as '*spin decoupling*' and is used to prove that A and B are spin–spin coupled, or to simplify the A spectrum or to locate the B resonance if it is obscured by other overlying resonances.

When A and B are the same nuclear species (e.g. both hydrogens) the technique is called *homonuclear decoupling*. When A and B are different nuclear species (e.g. ^{13}C and ^1H) the technique is called *heteronuclear decoupling*.

14.4.2. *Noise decoupling*

Optimum decoupling of B from A occurs when the irradiating frequency B_2 is exactly centred on the B spin multiplet. Simultaneous decoupling of A from several different neighbour nuclei B, C, D etc. would therefore seem to require quite complex equipment for *multiple* irradiation. In practice for *heteronuclear decoupling*, this can be achieved quite simply by means of a noise decoupler which 'washes out' all spin–spin couplings to the second nuclear species.

The irradiation source (noise source) B_2 contains frequency components over a wide chemical shift range, so that it 'visits' all the possible chemical shift values B, C, D etc., in turn. This is of great practical importance in carbon-13 spectroscopy where, by noise irradiation of the hydrogen spectrum, all hydrogen–carbon splittings can be removed from the carbon-13 spectrum, leading to great simplification.

14.4.3. *The nuclear Overhauser effect*

This is a more subtle double irradiation effect and can lead to structural information as well as providing an increase in signal-to-noise ratio, which is particularly important in ^{13}C spectroscopy.

Nuclei at resonance absorb energy from the irradiating source B_1. To have continuous resonance absorption the nuclei must give up this energy to the 'lattice'. As we have seen (p. 16) the lattice provides a fluctuating component at the Larmor frequency which causes the nuclei to relax back to the equilibrium situation. Sometimes the agent most responsible for spin–lattice relaxation of nucleus A is a near-neighbour of nucleus B. (The internuclear distance r_{AB} is important since the interaction between two magnetic dipoles is proportional to r_{AB}^{-6}.) When this dipolar interaction is the *dominant* mechanism for transfer of energy from A to the lattice, it is possible by irradiation of the B resonance to increase the intensity of the A resonance to $(1 + \varepsilon)$. ε is called the nuclear Overhauser enhancement. and can have values up to a *maximum* value given by $\varepsilon = \frac{1}{2}(\gamma_A/\gamma_B)$, where γ_A and γ_B are the magnetogyric ratios of the two nuclei involved. For like nuclei, $\varepsilon = \frac{1}{2}$, so a 50% increase in intensity is possible. We saw that this effect was helpful in distinguishing two possibilities for the structure of the Ln(III)–cyclic AMP complex (p. 73). In one instance the proximity of the two hydrogens resulted in an observable Overhauser enhancement.

In the observation of ^{13}C spectra, irradiation of directly-bonded hydrogens can result in a threefold improvement in sensitivity ($\varepsilon = 2$). This is because those hydrogens are usually the main source of the ^{13}C spin–lattice relaxation, having relatively strong dipoles at small internuclear distances. Because ^{13}C spectra in natural abundance (ca. 1%) are inherently weak double irradiation can be very important and generally this is done as a matter of routine, using noise decoupling.

14.4.4. *Internuclear double resonance (INDOR)*

This is used as an aid to assignment and involves monitoring a given resonance, A, while sweeping the irradiation frequency B_2. Any correlation between the resonance irradiated by B_2 and resonance A will result in a change in the B_1 intensity. Such a correlation would arise from coupled resonances (i.e. spin decoupling) or resonances corresponding to nuclei causing Overhauser effects.

14.5 Continuous-wave methods for measurements of T_1 and T_2

14.5.1. *Progressive saturation method for* T_1 *and* T_2

A technique which is often referred to in the literature for the determination of T_1 is that of progressive saturation. It must be emphasized that it is in general not a very precise method. The technique is based on the theoretical line shape for an n.m.r. absorption line, the equation for which is

$$L = \frac{M_0(\gamma B_1)T_2}{1+(\omega-\omega_I)^2 T_2^2+(\gamma B_1)^2 T_1 T_2}. \tag{14.1}$$

M_0 is the equilibrium magnetization and is proportional to the concentration of observed species. (γB_1) is the intensity of the irradiating r.f. field (expressed in angular units) and $\omega_1-\omega_I$, is the frequency of B_1. ω_I is the nuclear Larmor frequency also in angular units. The shape of the resonance thus depends on (γB_1). At low values of (γB_1) such that $(\gamma B_1)^2 T_1 T_2 \ll 1$, L is a Lorentzian function and T_2 may be evaluated from the line-width, $\Delta\omega$

$$\Delta\omega = 2/T_2 \quad \text{or} \quad \Delta\nu(\text{Hz}) = \frac{1}{\pi T_2}. \tag{14.2}$$

The peak amplitude is also a function of (γB_1), i.e.

$$\text{peak amplitude} = \frac{M_0(\gamma B_1)T_2}{1+(\gamma B_1)^2 T_1 T_2}. \tag{14.3}$$

This equation has a maximum value when $(\gamma B_1)^2_{\text{max}} = 1/T_1 T_2$. Knowing T_2, T_1 may be evaluated if $(\gamma B_1)^2_{\text{max}}$ is known. One way of determining this is to use a sample of known T_1 and T_2.

Finally we note that an alternative form of eqn (14.1) is

$$\frac{(\gamma B_1)}{\text{peak height}} = a T_1 (\gamma B_1)^2 + a/T_2. \tag{14.4}$$

where a is a constant depending on the instrumental constants and the concentration of the sample. A plot of $(\gamma B_1)/$peak height versus $(\gamma B_1)^2$ gives a straight line of slope $a T_1$ and intercepts a/T_2 on the ordinate and $1/T_1 T_2$ on the abscissa. This is probably a better method than the above for it gives a value of T_2 and can be used over a more limited range of (γB_1), which again must be known.

14.5.2. *Pulse techniques*

Instead of applying the irradiating r.f. field B_1 continuously it may be applied in pulses. In biochemical systems pulse techniques have been most used to measure the relaxation rates of the solvent water hydrogens. The concentration of water is so much larger than that of any component in the system that effectively a single resonance is observed. However with the use of Fourier transform techniques pulse methods can be used to determine relaxation rates of individual resonances of more complex molecules. A brief description of the basic principles is given to provide an introduction to the more advanced texts.

We shall assume that the reader is familiar with the use of a rotating frame of co-ordinates to describe n.m.r. experiments as discussed in Chapter 2. We noted in Chapter 2 that a burst (or pulse) of radiation applied for a time $t_{\pi/2} = \pi/2\gamma B_1$ along the y'-axis resulted in the nuclear magnetization being turned through $90°$ into the x'-direction. The n.m.r. system detects the magnetization in the $x'y'$-plane. We also considered (in Chapter 2) the sample as being composed of a large number of infinitesimal volume elements. Over each element the field B_0 is perfectly homogeneous but because of inhomogeneities in the field B_0 may vary from element to element. This means that each volume element will precess at a slightly different angular velocity ($\omega = \gamma B$). As seen from the rotating frame the magnetization vectors of the individual volume elements will start to fan out and the resulting loss of phase produces a decrease in the voltage observed in the detecting coil (Fig. 14.1). The decay of the n.m.r. signal in the $x'y'$-plane is referred to as a *free induction*

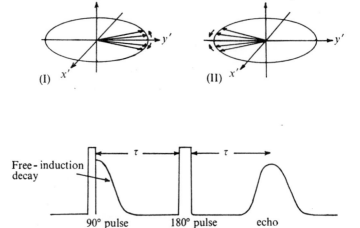

FIG. 14.1. Dephasing of nuclear spins (I) followed after time τ by application of 180° pulse leading to rephasing (II) and consequent echo.

decay. This is in fact the Fourier Transform of the absorption line of the sample (see below).

The free induction decay eventually reaches zero, but if during the 'fanning out' process a 180° pulse is applied at a time τ, the loss of phase due to inhomogeneities in B_0 is reversed and at time, 2τ, an echo is produced whose shape is simply that of two free induction decays placed back to back. The echo is referred to as a spin echo.

The above behaviour applies only if T_1 and T_2 are infinitely long. For example, in addition to the loss of phase caused by magnetic field inhomogeneities (denoted T_2^*) there is also that due to the natural T_2 processes. Clearly the relative magnitudes of two effects are important in trying to measure T_2. For example, the magnetic field inhomogeneities may be so large and the dephasing due to this so fast that it is almost impossible to apply the 180° pulse fast enough to overcome this effect.

The usual way of measuring T_2 in spin-echo experiments is due to Carr and Purcell (1954). It consists of a 90° pulse followed by a 180° pulse at time τ and then further 180° pulses at time intervals 2τ. By this method only the decay due to the natural T_2 processes is monitored while that due to the magnetic field inhomogeneities is largely eliminated. The method however suffers from the disadvantage that if the 180° pulses are not precise, errors on the nuclear magnetization can be cumulative (i.e. the magnetization in the $x'y'$-plane immediately before and after a 180° pulse will then be different). The free induction signal then decays to zero too rapidly. This problem can be largely overcome by using a modification due to Meiboom and Gill (1958) which introduces a 90° phase shift for B_1, between the 90° and 180° pulses.

To measure T_1, a 180° pulse is first applied; the nuclear magnetization is thus along the $-z$ direction. The magnetization now decays towards its

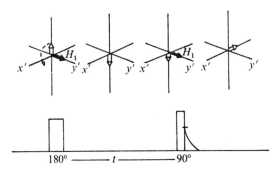

FIG. 14.2. 180° pulse followed by decay of magnetization from inverted value. 90° pulse allows value after time t to be measured.

initial value with a time constant T_1. After an arbitrary time τ, a 90° pulse is applied which turns the magnetization into the $x'y'$-plane, so that a free induction signal is observed (Fig. 14.2). The observed height is a measure of the resultant magnetization (i.e. that in $+z$ and $-z$ direction). The system is then allowed to come to thermal equilibrium by waiting about $5T_1$ and the experiment repeated but with a different value of τ.

In this way the recovery of the nuclear magnetization after the 180° pulse can be plotted. Since the initial height of the free-induction decay is proportional to the magnetization then

$$\ln \frac{(ht)_e}{(ht)_r} = e^{\tau/T_1} \quad \text{or} \quad \ln[(ht)_e - (ht)_r] = \tau/T_1 ; \tag{14.5}$$

$(ht)_e$ and $(ht)_r$ are the initial heights at thermal equilibrium and at time τ. Figure 14.3 shows an illustration of a T_1 determination. At the null point in Fig. 14.3, $(ht)_{\text{null}} = \frac{1}{2}(ht)_e$ and thus

$$\frac{1}{T_1} = \frac{\ln 2}{\tau_{\text{null}}}. \tag{14.6}$$

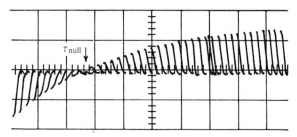

FIG. 14.3. Measurement of T_1 by a combination of 180°/90° pulse pairs.

In practice the null method is often used. The determination of the null point is often made easier if a 180° pulse is applied a time τ' after the 90° pulse. The disappearance or null of the echo may be easier to monitor.

14.5.3. *Measurement of* T_1 *in the rotating frame (called* $T_{1\rho}$)

Here the method is to align the magnetization initially along B_0 along the rotating field B_1. One method of doing this is as follows (see Fig. 14.4). (a) B_1 is applied along the y'-axis for a length of time equivalent to a 90° pulse. (b) B_1 is then *not* turned off but shifted in phase by 90°, (c), so that it too lies along the y'-axis. (d) The magnetization is then said to be locked along B_1 and relaxation occurs along B_1. B_1 is then turned off after a time τ. A free induction decay is observed in the $x'y'$-plane, the height of which measures the remaining magnetization after the time τ. By varying τ, the time constant for the decay of magnetization along B_1 may be measured. Since $B_1 \ll B_0$, this is equivalent to measuring T_1 in a very low magnetic field.

14.6. Fourier Transform (FT) n.m.r.

For many years high-resolution n.m.r. spectra were recorded by applying B_1 continuously and scanning the excitation r.f. frequency through the nuclear precession frequencies and measuring the r.f. absorption at each point. In FT n.m.r. the entire spectrum of frequencies is excited simultaneously. This may be achieved by applying B_1 as a strong r.f. pulse which

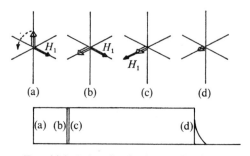

FIG. 14.4. Relaxation in the rotating frame.

excites a transient n.m.r. response that contains *all* the frequency components of the spectrum at the same time. This response is, as we have seen above, the free induction signal. The important point to emphasize is that all the information normally available in the high-resolution spectrum is present

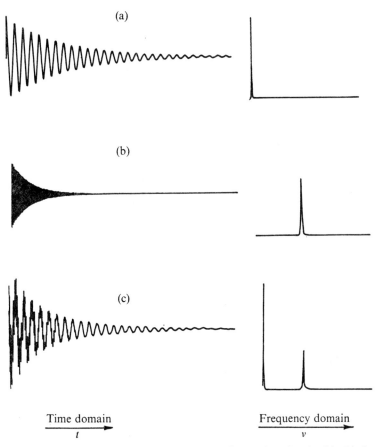

(a)

(b)

(c)

Time domain
t

Frequency domain
v

FIG. 14.5. Free-induction decay and Fourier transform of a signal with (a) long T_2 (b) short T_2 and (c) a combination of (a) and (b).

in the free-induction signal. The analysis of this response in terms of chemical shifts etc. is not really possible without resort to a mathematical device known as a Fourier transformation. This converts a response from the time domain to the frequency domain (see Fig. 14.5). The transformation is usually carried out by a computer. The use of a computer (CAT) also enables a large number of free-induction signals to be added to improve the signal-to-noise ratio before Fourier transformation.

Generally the advantage of a Fourier transform experiment is the efficiency of collecting the information. For illustrative purposes let us consider 10^3 s as being typical of the time taken to record a conventional spectrum using continuous wave irradiation. If the FT experiment takes 4 s, then *during the same time* 249 more scans can be accumulated leading to an increase in signal-to-noise ratio of $\sqrt{249}$ under ideal conditions.

There are other advantages of having a transient response, in that it is less susceptible to distortions due to artificially introduced time constants and too high a level of B_1. Also, the 'ringing' (or wiggles) normally observed after passing through a resonance using continuous-wave methods is absent. Additionally, of course, the relaxation rates of each line in the spectrum can be measured. For example, the recovery of all the resonances after being inverted by a 180° pulse may be monitored for the complete spectrum in a manner analogous to the determination of T_1 described above. There seems little doubt that in the future this technique will probably make obsolete the continuous-wave methods at present used for determining relaxation rates. Clearly the important applications of FT n.m.r. are for samples where the concentrations are low and particularly for ^{13}C n.m.r. where the low natural abundance (1%) has prevented proper exploitation of this nucleus.

14.6.1. *Convolution and deconvolution and Fourier Transformation*

One of the great advantages of the Fourier transform technique is that it is possible to alter the line shape in the frequency domain spectrum by manipulations in the time domain. (These manipulations are carried out by a computer.) Alterations can, of course, be carried out in the frequency domain but it is often difficult to allow for effects of time constants, wiggles etc. leading to some unavoidable distortion.

Multiplication of a free-induction decay by an exponentially *decaying* function results in line broadening, but increase the signal-to-noise ratio markedly. This procedure is called *convolution*. Most of the noise contribution appears in the 'tail' of the free-induction decay where the signal has more or less died out. If this tail is forced towards zero (by multiplication by a small number) the amount of noise in the transformed spectrum will be less, but the lines somewhat broadened (Fig. 14.6). The free induction decay has a time constant T_2', where

$$\frac{1}{T_2'} = \frac{1}{T_2^*} + \frac{1}{T_{2,c}}$$ (14.7)

where T_2^* is the natural time constant and $T_{2,c}$ that of the convolution function.

Deconvolution is simply the opposite of this, i.e. multiplication of a free induction decay by an exponentially *increasing* signal (Fig. 14.6). Thus

$$\frac{1}{T_2''} = \frac{1}{T_2^*} - \frac{1}{T_{2,c}}$$ (14.8)

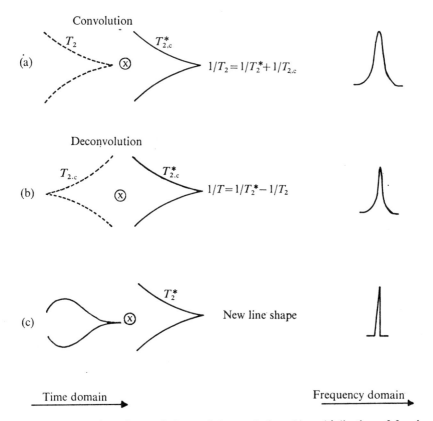

FIG. 14.6. Illustration of convolution and deconvolution. (a) multiplication of free induction decay by (a) an increasing exponential; (b) a decreasing exponential, and (c) a combination of (a) and (b).

Campbell, Dobson, Williams, and Xavier (1972) have used these techniques to sharpen up the line-widths in native lysozyme (see p. 101). The theory can be explained as follows: Consider two separate experiments, involving two different convolution functions C_1 and C_2: Then we may write

$$\frac{1}{T_2'} = \frac{1}{T_2^*} + \frac{1}{T_{2,c_1}} \quad \text{and} \quad \frac{1}{T_2''} = \frac{1}{T_2^*} + \frac{1}{T_{2,c_2}} \tag{14.9}$$

which give Lorentzian line-widths in the frequency domain which are proportional to

$$\frac{T_2'}{1 + \Delta\omega^2 T_2'^2} \tag{14.10a}$$

and

$$\frac{T_2''}{1 + \Delta\omega^2 T_2''^2}. \tag{14.10b}$$

By subtracting (in a computer) $K \times$ eqn (14.10b) from eqn (14.10a) and then using the F.T. technique we obtain a line shape which is considerably different from Lorentzian. The approximate line-shape is proportional to

$$\frac{T_2' + \Delta\omega^2 T_2' T_2''^2 - KT_2'' - KT_2'' T_2'^2 \, \Delta\omega^2}{(1 + \Delta\omega^2 T_2')^2 (1 + \Delta\omega T_2'')^2} \tag{14.11}$$

By choosing K such that $K = T_2''/T_2'$, the line shape becomes proportional to

$$\frac{T_2' - KT_2''}{1 + \Delta\omega^2 (T_2' + T_2'')^2 + \Delta\omega^4 T_2'^2 T_2''^2} . \tag{14.12}$$

From this expression we may note the following: (1) the signal intensity at resonance is proportional to $T_2' - KT_2''$ i.e. there will be a loss in signal-to-noise; (2) the half-width is narrower and is approximately $(T_2' + T_2'')$; and (3) the term in $\Delta\omega^4$ in the denominator means that the intensity drops very sharply to zero after resonance. The line-shape is almost triangular.

The importance of this technique is that it allows spin–spin couplings which may be partially resolved or present as shoulders to become more resolved, thus providing an additional aid to assignment.

References

CARR, H. Y. and PURCELL, E. H. (1954). *Phys. Rev.* **94**, 630.

CAMPBELL, I. D, DOBSON, C. M., WILLIAMS, R. J. P., and XAVIER, A. V. (1973). *J. mag. reson.* in press and in: International Conference on Electron Spin Resonance and Nuclear Resonance in Biology and Medicine and Fifth International Conference in Magnetic Resonance in Biological Systems, *Ann. N.Y. Acad. Sci.*, in press.

MEIBOOM, S. and GILL, D. (1958). *Rev. sci. Instrum.* **29**, 688.

Some further useful references

Spin decoupling

WALTER, R., GLICKSON, J. D., SCHWARTZ, I. L., HAVRAN, R. J., MEIENHOFFER, J., and URRY, W. D. (1972). *Proc. natn. Acad. Sci.*, **69**, 1920, and references therein.

Nuclear Overhauser effect

NOGGLE, J. H. and SCHIRMER, R. E. (1971). *The nuclear Overhauser effect*, Academic Press, New York.

BALARAM, P., BOTHNER-BY, A. A., and DADOK, J. (1972). *J. Am. chem. Soc.*, **94**, 4015.

——, —— and BRESLOW, E. (1972). *J. Am. chem. Soc.*, **94**, 4017.

INDOR

GIBBONS, W. A., ALMS, H., SOGN, J., and WYSSBROD, H. R. (1972). *Proc. natn. Acad. Sci.*, **69**, 1261.

Pulse techniques

FARRAR, T. C. and BECKER, E. D. (1971). *Pulse and Fourier Transform n.m.r.* Academic Press, New York and London.

SHORT BIBLIOGRAPHY

Below we list some useful review articles dealing specifically with the applications of magnetic resonance spectroscopy to biochemistry.

COHEN, J. S. (1972). Nuclear magnetic resonance investigation of the interactions of biomolecules. In *Experimental Methods of Biophysical Chemistry*, Vol. 6, Chapter 12, 521.

DWEK, R. A., WILLIAMS, R. J. P., and XAVIER, A. V. (1972). The applications of paramagnetic ions in biological system. In *Metal ions in biological systems* Vol. 6 (ed. H. Siegel) Dekker, New York in press.

JOST, P. and GRIFFITH, O. H. (1972). Electron spin resonance and the spin labelling method. In *Methods in Pharmacology*. Vol. 2 Chapter 7. (ed. V. Chignell) Appleton-Century-Crofts, New York.

MCCONNELL, H. M. and MCFARLAND, B. G. (1970). Physics and chemistry of spin labels. *Quart. Rev. Biophys.* **3**, 91.

MCDONALD, C. C. and PHILLIPS, W. D. Proton magnetic resonance spectroscopy of proteins. In *Fine Structure of Proteins and Nucleic Acids*, Vol. 4. (ed. G. D. Fasman and N. Timasheff) Dekker, New York (1970) page 1.

MILDVAN, A. S. and COHN, M. (1970). Aspects of enzyme mechanism studied by nuclear spin relaxation induced by paramagnetic probes. *Advances in Enzymology*, **33**, 1 (ed. F. F. Nord) Interscience, New York.

ROBERTS, G. C. K. and JARDETZKY, O. (1970). Nuclear magnetic resonance spectroscopy of amino acids, peptides and proteins. In *Advances in Protein Chemistry*, **24**, 448.

SHEARD, B. and BRADBURY, E. M. (1970). Nuclear magnetic resonance in the study of biopolymers and their interactions with ions and small molecules. In *Progress in Biophysics and Molecular Biology*, **20**, 187.

SYKES, B. D. and SCOTT, M. (1972). NMR studies of the dynamic aspects of molecular structure and interaction in biological systems. *Ann. Rev. biophys and Bioeng.* **1**, 27.

WÜTHRICH, K. (1970). Structural studies of hemes and hemoproteins. In *Structure and Bonding*, **8**, 53.

AUTHOR INDEX

SUBJECT INDEX

SUMMARY OF SYMBOLS

THE nomenclature in n.m.r. and e.s.r. has not, as yet, been systematically organized. Frequently the same symbol is used in many different contexts. I have tried to keep as far as possible to those symbols in currrent usage—but inevitably to avoid confusion some have had to be changed. Where the same symbol is used in different contexts, the meaning is usually obvious and I have not thought it necessary to introduce unfamiliar symbols, but have given a page number so that the reader can become aware of this difficulty.

Not every symbol in the book is summarized here—only those that are frequently used or those for which there could be some confusion. Throughout the book the subscripts A and M denote *bulk* and *bound* sites respectively. *The use of an asterisk* (*) denotes parameters relating to the presence of a macromolecule, but this convention is only used when it would appear that some ambiguity could arise.

Greek symbols

β	Bohr magneton, 0.92731×10^{-20} erg G^{-1}
β_N	nuclear magneton, 0.50504×10^{-23} erg G^{-1}
γ_I	gyromagnetic ratio of nucleus of spin I
γ_H	gyromagnetic ratio of proton, 2.6735×10^4 rad s^{-1} G^{-1}
γ_S	gyromagnetic ratio of electron, 1.7598×10^7 rad s^{-1} G^{-1}
$\Delta\gamma$	measure of the anisotropy of the electron g value, s^{-1} G^{-1}
δ	chemical shift—p.p.m.
Δ	angle between rotation axis and the distance vector of nuclei in question (see page 187)
ε	angle between electric field gradient and the magnetic field (see page 331)
ε	nuclear Overhauser enhancement (Chapter 14)
ε^*	proton relaxation enhancement of solvent water molecules in the presence of a macromolecule
ε_b	characteristic enhancement of metal–enzyme binary complex
ε_t	characteristic enhancement of ternary metal–enzyme–substrate complex
ε_q	characteristic enhancement of quaternary metal–enzyme–substrate–substrate—complex
ε_s	characteristic enhancement of metal–substrate binary complex
ζ	angle through which magnetization is rotated by a pulse of rf (see page 28)
η	viscosity of solvent, poises (or g cm^{-1} s^{-1})
$\tilde{\eta}$	asymmetry parameter (see page 328)
θ_i	angle between principle axis and the distance vector r (see pages 59 and 100)
θ'_R	angle between applied magnetic field and vector joining two unlike spins (Chapter 12)

$\bar{\mu}$	average number of metal ions bound per molecule of enzyme
μ_{eff}	effective magnetic moment of metal ion (see page 177)
μ_N	nuclear magnetic moment
$\mu_m \ (\equiv \mu)$	maximum component of μ_N along magnetic field
ν_a	frequency (in Hz) of central resonance component of quadrupolar nucleus in an asymmetric environment
ν_m	frequency shift (in Hz) between outer and central resonance component
ν_I	nuclear Larmor frequency (in Hz)
$\Delta\nu$	line width (in Hz) (see also page 46)
$\Delta\nu_E$	line width contribution due to presence of enzyme
$\Delta\nu_P$	line width contribution due to presence of paramagnetic ion
ρ_c^π	electron density expressed as the fraction of one unpaired electron on ring carbon atom adjacent to the proton for which hyperfine interaction is observed
σ	screening constant
$\sigma'_{\text{ani}}; \sigma''_{\text{ani}}$	anisotropic contributions to shielding constant of a neighbouring cylindrically unsaturated bond
σ_D	diamagnetic screening constant
σ_P	paramagnetic screening constant
τ	spin label correlation time (see page 291)
τ_c	dipolar correlation time
τ_D	diffusional correlation time
τ_e	hyperfine (or scalar) correlation time
τ_J	correlation time for changes in J quantum number
τ_M	chemical exchange lifetime
τ_R	rotational correlation time
τ_S	electron spin relaxation time
$\tau_{1,S}$	electron spin lattice relaxation time
$\tau_{2,S}$	electron spin spin relaxation time
τ_v	correlation time for modulation of zero field splitting
τ_{rot}	correlation time for rotation of a methyl group about its axis (see page 141)

$$\frac{1}{\tau_{c,i}} = \frac{1}{\tau_R} + \frac{1}{\tau_{e,i}}; \qquad \frac{1}{\tau_{e,i}} = \frac{1}{\tau_{i,S}} + \frac{1}{\tau_M}$$

where $i = 1$ or 2 (see page 184)

$\tau_{c,1}; \tau_{c,2}; \tau_{c,3}$	correlation times along different axes (see pages 141 and 187)
ϕ	angle between x-axis and the projection of the radius vector r in the xy plane (see pages 59 and 100)
ϕ	dihedral angle (see pages 34ff and 131)
χ	angle between main symmetry axis and the electron and nuclear dipole distance vector (see page 188)
χ_i	anisotropic susceptibilities in different directions
ω_I	nuclear Larmor frequency (rad s^{-1})
ω_S	electron Larmor frequency (rad s^{-1})

$\Delta\omega$	chemical shift of nucleus A, in bulk environment relative to its shift in a pure solution of A
$\Delta\omega_A$	chemical shift of nucleus in *bulk* (A) site
$\Delta\omega_M$	chemical shift of nucleus in *bound* (M) site

Roman symbols

A	mass number of a nucleus
$A/\hbar$	hyperfine coupling constant (rad. s^{-1})
A_{xx}, A_{yy}, A_{zz}	hyperfine coupling constant along different directions
$A = \frac{1}{3}(A_{xx} + A_{yy} + A_{zz})$	
a	effective radius of complex
a_1	radius of ligand molecule
a_2	radius of paramagnetic ion complex
B_0	main magnetic field
B_1	secondary (weak) magnetic field perpendicular to B_0
ΔB_{obs}	line width (in gauss) of nucleus in presence of protein (see page 242)
ΔB_0	line width (in gauss) of nucleus in absence of protein (see page 242)
B	constant in Bloembergen-Morgan equation containing the value of the resultant electronic spin and the zero field splitting parameters (see page 178). See also B^0 and B$'$ on pages 181–182 where B^0 and B$'$ represent different zero field splittings
B_{local}	local magnetic field at nucleus corrected for screening or shielding by the electrons around it
B_μ	local magnetic field at a nucleus caused by the presence of other nuclei of magnetic moment μ
δB	e.s.r. line width of observed spin in the presence of a second paramagnetic centre (see page 315)
δB_0	e.s.r. line width is the absence of a second paramagnetic centre (see page 315)
C	dipolar interaction coefficient between two electron spins (see page 315) $= \dfrac{g\beta^3\mu^2}{\hbar r^6}\tau$
C	contact interaction in Solomon-Bloembergen equations (e.g. see page 220)
d	distance of closest approach between a ligand molecule in the bulk solutions and the paramagnetic ion (or complex)
D	dipolar interaction constant in the Solomon-Bloembergen equations (e.g. see page 220)
$\mathscr{D}$	$= \frac{1}{2}(\mathscr{D}_I + \mathscr{D}_s)$
$\mathscr{D}_I$	diffusion coefficient of the ligand molecule
$\mathscr{D}_s$	diffusion coefficients of paramagnetic ion (or complex)
D_x, D_y, D_z	components of the ligand field
E	nuclear Zeeman energy (page 11)

E_M	activation energy for chemical exchange
E_R	activation energy for rotation
E_s	activation energy for electron relaxation
E_V	activation energy for motion characterized by
eq	electric field gradient
eQ	electric quadrupole moment
g	Lande g factor—(often used in e.s.r. to define line position)
g_{xx}, g_{yy}, g_{zz}	components of g
$g_\parallel = g_{zz}$	component of g along z axis, if g is axially symmetric
$g_\perp = g_{xx}, g_{yy}$	component of g along x or y axis, if g is axially symmetric
g_N	nuclear g factor
h	Planck's constant, $6 \cdot 625 \times 10^{-27}$ erg. s
$\Delta H^\ddagger$	enthalpy of activation
I	nuclear spin (spin angular momentum)
I	moment of inertia (page 27)
J	rotational quantum number (pages 27 and 65)
J	spin spin coupling constant (Hz)
J_{AX}	spin spin coupling constant between nuclei A and X
$J(\phi)$	value of coupling constant for a given dihedral angle ϕ (see page 34)
$J(O)$	spectral density (intensity of fluctuations) at zero frequency
$J(\omega)$	spectral density at frequency ω
K_I	inhibitor–enzyme dissociation constant
K_D	metal–enzyme dissociation constant
K_1, K_2, K_3, K_A, K_S	see page 260
K_A'	see page 271
$k_{on}(k_1)$	associative rate constant
$k_{off}(k_{-1})$	dissociative rate constant
k	Boltzmann's constant, $1 \cdot 3804 \times 10^{-16}$ erg. deg^{-1}
m	magnetic quantum number
M_0	equilibrium magnetization
N_s	number of spins per ml of solution
P_M	mole fraction of ligand nuclei bound
q	number of ligand nuclei bound
Q	constant relating spin density and hyperfine interaction (see page 56)
r	distance of nucleus from second nucleus or from a perturbing (e.g. paramagnetic) centre
R	gas constant per mole
R	reduction in dipolar coupling from anisotropic rotation (see page 187)
S	total electron spin
$\Delta S^\ddagger$	entropy of activation
T	absolute temperature (K)
T_i	relaxation time
$i = 1$ or 2,	$i = 1$ spin lattice relaxation time
	$i = 2$ spin spin relaxation time

$1/T_i$	relaxation rate
$1/T_{i,A}$	relaxation rate of *bulk* (A) site
$1/T_{i,A}$	outer sphere relaxation rate
$1/T_{i,E}$	enzyme contribution to relaxation rate
$1/T_{i,M}$	relaxation rate of *bound* (M) site
$1/T'_{i,M}$; $1/T''_{i,M}$	relaxation rate of *bound* site (see pages 344 and 345)
$1/T_{i,obs}$	observed relaxation rate
$1/T_{i,p}$	paramagnetic contribution to the relaxation rate
$1/T_{i,scalar}$	scalar relaxation rate
$1/T_i$	spin lattice relaxation rate in the rotating frame
$T_2(m)^{-1}$	spin spin or transverse relaxation rate of each line of a spin label spectrum—characterized by the nuclear spin state m.
$\bar{V}$	partial molar volume
Z	charge number of a nucleus